The Latin American Studies Book Series

The Latin American Studies Book Series promotes quality scientific research focusing on Latin American countries. The series accepts disciplinary and interdisciplinary titles related to geographical, environmental, cultural, economic, political, urban and health-related research dedicated to Latin America. The series publishes comprehensive monographs, edited volumes and textbooks refereed by a region or country expert specialized in Latin American studies.

The series aims to raise the profile of Latin American studies, showcasing important works developed focusing on the region. It is aimed at researchers, students, and everyone interested in Latin American topics.

Submit a proposal: Proposals for the series will be considered by the Series Advisory Board. A book proposal form can be obtained from the Publisher, Andrea Brody-Barre (andrea.brody-barre@springer.com).

Roberta Marx Delson

Scientists or Spies?

The American Museum of Natural History in World War I Latin America

Springer

Roberta Marx Delson
Kings Point, NY, USA

ISSN 2366-3421 ISSN 2366-343X (electronic)
The Latin American Studies Book Series
ISBN 978-3-032-18116-9 ISBN 978-3-032-18117-6 (eBook)
https://doi.org/10.1007/978-3-032-18117-6

This Springer imprint is published by the registered company Springer Nature Switzerland AG
The registered company address is: Gewerbestrasse 11, 6330 Cham, Switzerland

PBS interviewer to physicist Frederick Seitz, April 3, 2006:

"Is there any science that's not wrapped in politics?"

Seitz: "Oh, there are some things. The disappearance of the frog…I don't think that had any political repercussions."

"The task of intelligence—trying to find out secrets the enemy is giving away without realising it—turned out to be what these professors had been training at all their lives. The French playwright Molière created a character who is astonished to discover that he has been speaking prose all his life without knowing it. So it was with the professors of the OSS. They'd been spies, detectives, heroes all their lives without knowing it. And when they answered the call of a global crisis, they discovered new aspects of their own potential."

Elyse Graham, Book and Dagger: How Scholars and Librarians Became the Unlikely Heroes of World War II (New York, Ecco Press, 2024), p. 298.

For the memory of AMNH Anthropology Curator Craig Morris (1939–2006), who truly understood the importance of Latin America to the American Museum of Natural History.

And for the memory of my mother, Ruth Lynn Sperber Marx (1917–2011), botanist and biologist, from whom I first learned about trees and science.

Foreword

A modern reader can be forgiven for assuming that Latin America played no role in the Great War that erupted in Europe in 1914. The horrific bloodshed of the land, air, and sea battles that raged across Europe, Africa, and parts of Asia was absent—though Mexico engaged in a bloody civil war between 1910 and 1917, US troops invaded several Latin American countries, and German U-boats harassed shipping in the Atlantic and Caribbean. Nonetheless, compared to the trauma that engulfed Europe, Latin America seems tranquil.

The English language scholarship on the war has tended to follow that assumption. Most of the research on US foreign policy, for example, tends to emphasize the post-World War II era.[1] Latin American historians do examine US military and economic intervention, but the literature on the economic and political role Latin American countries played during the war is often muted.[2] In this cacophony of silence, perhaps the least studied aspect of the war in Latin America is the role of espionage and sabotage.

Two seminal works include Barbara Tuchman's *The Zimmerman Telegram* (1958) and Friedrich Katz's book, *The Secret War in Mexico* (1981). Jamie Bisher's *The Intelligence War in Latin America, 1914–1922* (2016) remains the most thorough study yet, with a few more books and a smattering of articles trickling in.[3] Despite the quality of these works, they do not represent anything like the quantity of studies produced for World War II and the Cold War. This silence often reflects the paucity of sources. Though some post-war memoirs exist, much of the archival material remained classified until relatively recently, and much that once existed in Latin American nations has disappeared or been destroyed. This is one of the things that makes *Scientists or Spies* such a contribution to the existing literature. Professor Delson has uncovered hitherto unknown records associated with the American Museum of Natural History that reveal how the confluence of science and politics allowed scholars at one of the United States' most prestigious museums to play important roles in the covert war the US waged with Germany and eventually Japan, in Latin America between 1914 and 1918.

In the early months of that war, Germany held the upper hand. Germany was far better prepared than the Allies or the United States for covert operations. They had already constructed logistical and intelligence networks that often relied on their extensive expatriate communities scattered throughout Latin America. Prior to the war, Britain and Germany were both connected to their empires and their embassies across the Atlantic through a network of underwater cables. But Germany had also created the most advanced wireless system in the world with wireless stations in Africa and Latin America. They equipped their naval vessels with wireless as well, which meant they could communicate over 1000 miles without the British being able to intercept the transmissions.[4]

In 1914, with the commencement of hostilities in Europe, German agents fanned out around the globe, often operating out of existing embassies and consulates. They focused their efforts on disrupting Allied supply lines, sabotaging defense industries and shipping, securing the neutrality of Latin American nations, and fomenting war between the United States and Mexico to divert US resources away from Europe. These agents targeted Allied shipping, sabotaged Allied vessels, fostered relationships with German expatriate communities, established wireless transmitters, and distributed propaganda intended to encourage Latin American leaders to adopt pro-German policies—or at least to remain neutral.

There were good reasons for Germany and the United States to believe Latin American countries might support the German cause. Several Latin American countries, especially Argentina, Brazil, Chile, Paraguay, and Uruguay, had large Italian and German populations with a smattering of Japanese communities scattered throughout the region. The belligerents in the war actively sought to recruit expatriates living in Latin America to the war effort and were quite successful.[5]

All Latin American countries whose export economies depended on trade with Europe faced the potential for serious economic setbacks from being cut off from the trans-Atlantic trade because of the British blockade and German U-boat activity. Many Latin American political and military figures also admired German military discipline and identified with German culture and politics. The 1910–1917 Revolution in Mexico, just on the US border, seemed to create a real possibility of drawing the US into an international war. These conditions created the undertones for the covert war in Latin America.

Equally important was the ongoing US military and economic intervention in Latin America. The United States had long treated the nations of Latin America like huge warehouses from which they could extract raw materials, enrich US businessmen, and dump cheap US goods. Throughout the late nineteenth and early twentieth centuries, the US imposed itself on Latin America with military interventions to support US interests. Between 1900 and 1914, the US intervened militarily more than twenty times in Latin America. By 1914, the US and Latin American countries were entwined in a network of market exchanges that benefited the elite in both regions. US businessmen launched an all-out economic invasion of Latin America at the same time the US military was imposing itself as the guardian of US entrepreneurs. The US also spent the early twentieth century establishing constabularies in Haiti, the Dominican Republic, Panama, and Nicaragua, ostensibly to provide stability, when the real motive was to keep the flow of raw materials and agricultural products to the United States open and the prices low.[6]

The US domination of Latin American markets through military interventions, constabulary forces, and aggressive banking and business practices prior to the war created a situation in which US financiers and politicians had vested interests in Latin America that they would seek to protect. The war also created new opportunities. The US became one of the main suppliers of the Allied Forces with weapons, ammunition, food, and medical supplies. The source for much of these materials was Latin America.[7]

Though most Latin American nations could not afford to ignore US demands for neutrality, and after 1917, for open support of the Allies, countries like Mexico charted their own course. Still engulfed in a cataclysmic civil war, it was not obvious to most Mexicans that alignment with the US was in their interest. After all, it had barely been fifty years since the US had forcefully seized half of Mexico's national territory in the Mexican-American War of 1848. Germany had never done such a thing, although German agents had been active in Mexico since at least 1910, and they had even tried to smuggle the exiled president José Victoriano Huerta Márquez (President from February, 1913, to July, 1914) back into Mexico, as they would do in 1917 with Vladimir Lenin into Russia. However, US agents intercepted Huerta at the Mexico-Texas border, and he died in a US prison in 1916.[8]

José Venustiano Carranza (President from 1917–1920) had distinct pro-German tendencies and accepted German advisors in 1916. When the German State Secretary for Foreign Affairs, Arthur Zimmerman, presented Pres. Carranza with a proposition to ally himself with Germany on the promise of the return of the territory Mexico lost to the US, it was a tempting, if unrealistic, idea. In the end, Carranza refused, but the letter and its blatant attempt to turn one of the United States' closest neighbors into invaders catapulted the US into the war. The US already had good reason to fear the instability of the Mexican Civil War when the anonymous 1915 Plan of San Diego surfaced, encouraging Latinos, African Americans, and Asians in the southwest to rise up and kill all Anglo-Americans over sixteen years of age.[9] These fears seemed to be realized when Pancho Villa's forces crossed the border near Nogales and killed American citizens on American soil in 1916.[10]

After their wars of independence in the early nineteenth century, Latin American states found themselves deeply indebted to European states, such as Great Britain. Consequently, investment capital simply did not exist to upgrade and modernize their export economies. With so few options, they tended to base those economies on the extraction of raw materials and the exploitation of the rural labor force. Since Latin America had long been the source of such raw materials for European colonizers, European states and the United States automatically turned to it for war materials.[11]

In February of 1917, a US agent photographed a Japanese ship offloading machinery and ammunition at Manzanillo for the construction of a munitions plant in Mexico City, which made the Zimmerman Telegram, revealed a few weeks later, even more inflammatory.[12] All the belligerents in World War I needed nitrates, saltpeter, and copper for their munitions, oil and rubber for their vehicles, lumber, sisal for ropes, grease, leather, food (especially meat, grains, sugar, and coffee), and medical supplies, such as quinine. Latin American nations had made themselves indispensable as suppliers of these goods to Europe and the United States. There was much at stake. The war machines of the Allies and the Central Powers could not function without these goods. Those who gained the most ready access to them held the advantage. It was this dynamic that drove the Atlantic war and the war of espionage and subterfuge all throughout Latin America. And the United States had placed itself perfectly to extract these goods and ship them to the Allies in Europe. Great fortunes were to be made as the United States became a leader in global finance and commerce.

The British blockade cut Germany off from trade with the US and Latin America, which encouraged Germany to complain about the US breach of neutrality for its trade with the Allies. But they also engaged in espionage and sabotage against US industries and transportation, including their U-boat campaign. German agents, coordinated out of the German embassy, successfully caused a dock explosion in July of 1916 in Jersey City, New Jersey.[13] German saboteurs probably sank the collier USS *Cyclops* with its 306-man crew.[14] In 1917, the US Office of Naval Intelligence became concerned about the possibility of German U-boat stations in Central America and reports of suspicious naval activity off Argentina and Uruguay.[15] The Caribbean also proved a vital and strategic transit point for shipping through the newly opened Panama Canal, and therefore, a vital region for German U-boat activity.

The United States responded with covert operatives of its own who spied on suspected German agents, created relationships with high-ranking officials, encouraged Latin American states to remain neutral, and after 1917, to join the Allies. Because the US had not developed an extensive network of spies in Latin America, it had to rely on individuals with knowledge of the local languages, cultures, and geographies. Since 1919, when the renowned German-American anthropologist Franz Boas publicly exposed four US-based archaeologists for spying on behalf of the US,[16] we have known that academics with expertise in Latin America were recruited by various intelligence agencies to engage in espionage, using their legitimate academic pursuits as cover.[17]

Scientists or Spies sheds new light on how the nascent spy infrastructure in the United States deployed academics with special linguistic skills and perfect covers because they were not fake, but legitimate scientific expeditions. Archaeologists, anthropologists, botanists, ornithologists, and paleontologists undertook undercover work while pursuing their own research agendas.

Professor Delson demonstrates that far from being an afterthought for US intelligence agencies, Latin America was on the front lines. Indeed, a secret, covert war was being waged for the allegiance of Latin American elites and politicians and for the vital supplies necessary to prosecute the war in Europe.

Despite Franz Boas's public exposure of the role some academics played in spy work for the United States in Latin America, a veil of obscurity has descended over these spies and their activities. As we have already noted, a few works have appeared, but World War I espionage has not garnered the same attention from scholars of US foreign policy as World War II and Cold War espionage have. Part of the reason, as Prof. Delson shows, is the difficulty in following the crumbs of this history, which left a thin paper trail, barely visible a hundred years later, that was never meant to be seen by outsiders. Through fortuitous discovery and careful investigation, Prof. Delson has brought to light an important aspect of the covert war in Latin America that is little understood or appreciated. She has contextualized these spies and their activities, highlighting how their academic pursuits dovetailed with their spy activities. The result is a fascinating window into the early years of US international covert operations that is rarely seen and poorly understood.

It is difficult to say who won the covert war in Latin America. Three nations—Brazil, Cuba, and Panama—declared war on Germany in April of 1917. Five other states joined in 1918 shortly before the war ended. Only Brazil sent any military forces, and none of them saw combat. Seven nations maintained their neutrality, including Mexico, Chile, Argentina, and Colombia.[18] The United States emerged from the war as the big economic winner in Latin America. European investment had dried up, and European shipping was in shambles. The US had crafted closer economic, political, and military bonds with Latin American states, and US businessmen were well-positioned to profit from the war's ending. Many of the academic spies Prof. Delson has studied also profited from the war in that they were able to continue their academic pursuits despite restrictions on such expeditions to Latin America, while the government learned the incredible value of deploying the country's brightest minds in espionage work. This relationship between scientists and the US government often proved to be mutually beneficial, although Prof. Delson writes that the benefits to the American Museum of Natural History were ambiguous.

After 1929, the US ended most of its intelligence gathering until the onset of World War II. When the US entered the war, the country turned to its academics once again and sent them against the Nazi regime.[19] And so, academics- and scientists-turned-spies returned to the field in the service of their country. What Elyse Graham has said about the academic spies of World War II applies also to those from World War I:

> …they were experts…in unsuspected disciplines that turned out to be supremely useful for intelligence work. Disciplines that taught them how to work with messy data. How to sift through paper evidence like newspapers, leaflets, and novels. How to gather clues from unlikely sources like advertisements and society columns. How to read a lot in just a little time. How to look at a pile of cracked and curling pages and see a treasure hunt. How to evaluate claims. How to tell stories. How to come up with audacious methods of solving problems using unlikely data…. How to understand another country—because the past, too, is another country—on its own terms. If you're going to fight a war, you can't just put boots on the ground; You also have to put ground under the boots, in the form of maps, blueprints, local knowledge.[20]

The scientists Prof. Delson examines did just that. Scientists or Spies sheds new light on this obscure history and shows how these often mutually beneficial relationships developed in the early twentieth century and how they set the stage for the next, much expanded effort at turning academics into spies.

Professor Delson's book joins the growing body of work on the covert war in Latin America and makes a lasting contribution to the broader field of Latin American history.

James E. Wadsworth
Professor of History
Stonehill College
Easton, USA

Notes

1. For one study that does focus on the pre-World War II era, see Jeffrey M. Dorwart, Office of Naval Intelligence, The Birth of America's First Intelligence Agency, 1865–1918 (Annapolis, Naval Institute Press, 1979).
2. Some works that address the First World War in Latin American include Alan McPherson and Yannick Wehrli, eds., Beyond Geopolitics: New Histories of Latin America at the League of Nations (Albuquerque, University of New Mexico Press, 2015); Jane M. Rausch, Colombia and World War I: The Experience of a Neutral Latin American Nation during the Great War and Its Aftermath, 1914–1921 (Lanham, Lexington Books, 2014); Albert Bill, South America and the First World War: The Impact of the War on Brazil, Argentina, Peru and Chile (Cambridge, Cambridge University Press, 1988); Stefan Rink, Latin America and the First World War, trans. Christopher W. Reid (New York, Cambridge University Press, 2017); Thomas Baecker, "The Arms of Ypiranga: The German Side," The Americas, vol. 30, no. 1 (1973), pp. 1–17; Phillip Dehne, "How Important was Latin America to the First World War?," Iberoamericana, vol. 14 (2014), pp.151–164; Sandra Kuntz Ficker, "El impacto de la Primera Guerra Mundial sobre el comercio exterior de México," Iberoamericana, vol. 14 (2014), pp. 117–137; Frederick C. Luebke, Germans in Brazil: A Comparative History of Cultural Conflict during World War I (Baton Rouge, Louisiana State University Press, 1987); Francisco Luiz Teixeira Vinhosa, O Brasil e Primeira Guerra Mundial: A Diplomacia Brasileira e as Grandes Potências (Rio de Janeiro, Instituto Histórico e Geográfico Brasileiro, 1990).
3. Three of the earliest works include Barbara W. Tuchman, The Zimmerman Telegram (New York, Ballantine, 1958); Friedrich Katz, The Secret War in Mexico: Europe, the United States and the Mexican Revolution (Chicago, The University of Chicago Press, 1981) and Charles H. Harris III and Louis R. Sadler, The Border and the Revolution: Clandestine Activities of the Mexican Revolution, 1910–1920 (Silver City, High-Lonesome Books, 1988). Probably the most thorough work is Jamie Bisher, The Intelligence War in Latin America, 1914–1922 (Jefferson, NC, McFarland and Company, 2016). See also Bill Mills, Treacherous Passage: Germany's Secret Plot against the United States in Mexico

during World War I (Lincoln, University of Nebraska Press, 2017); Charles H. Harris III and Louis R. Sadler, The Archaeologist was a Spy: Sylvanus G. Morley and the Office of Naval Intelligence (Albuquerque, University of New Mexico Press, 2003); Friedrich Schuler, Secret Wars and Secret Policies in the Americas, 1842–1929 (Albuquerque, University of New Mexico Press, 2010).
4. John Keegan, Intelligence in War: Knowledge of the Enemy from Napoleon to Al-Qaeda (New York, Alfred A. Knopf, 2003), 105.
5. Rink, Latin America…, pp. 43–47.
6. Martha K. Huggins, Political Policing: The United States and Latin America (Durham, Duke University Press, 1998), 26–40.
7. Thomas L. Pearcy, The History of Central America (New York, Palgrave Macmillan, 2006), 81-86; and Rink, Latin America…, pp. 25–32.
8. Rink, Latin America…, p. 55.
9. *Ibid.*, p. 56.
10. Michael C. Meyer, et al., The Course of Mexican History (New York: Oxford University Press, 1999), p. 527.
11. Rink, Latin America…, pp. 15–21.
12. Bisher, The Intelligence War…, p. 100.
13. Michael J. Lyons, World War I: A Short History (Upper Saddle River, New Jersey, Prentice Hall, Inc., 2000), p. 225.
14. Bisher, The Intelligence War…, pp. 238–239.
15. *Ibid.*, p. 106.
16. Franz Boas, "Scientists as Spies," The Nation, 109 (December 20, 1919), p. 797.
17. E.g., Harris and Sadler, The Archaeologist…
18. Rink, Latin America…, p. 123, Table 3.1 and pp. 137–138.
19. Elyse Graham, Book and Dagger: How Scholars and Librarians Became the Unlikely Spies of World War II (New York, HarperCollins, 2024), 2.
20. *Ibid.*, p. 297.

Preface

What is the influence of politics on science, or science on politics? Are the boundaries between the two fields of knowledge absolute or are they inextricably, and inevitably, interwoven? In the late 1950s the eminent scholar and scientist C. P. Snow famously postulated that at least with regard to undergraduate curricula, a wide gap existed between humanistic/political studies and the sciences.[1] Scientists, according to Snow, might have experienced some exposure to a broader range of subjects in order to qualify for their university degrees, but those who studied the humanities remained almost universally ignorant of the sciences. He suggested that this situation had resulted in the development of two distinct bodies of learning or, as he preferred to call them, "two cultures." This binary division, criticized at the time for being far too simplistic, nonetheless gained wide credence amongst a range of mid-century scholars. By extension, Snow seemed to be suggesting that true science could exist independently of politics, as well as the obverse. Despite or perhaps because of the obvious roles that physics and chemistry had played in the political outcomes of the two World Wars, Snow's notion of a dichotomy stuck, appealing to a peace-starved, mid-century intelligentsia tired of wartime.

Much of his thinking permeated the 1960s. For those of us who were students of "science" during these halcyon days (in my case, anthropology, and for my husband, paleontology/geology) the divide which supposedly existed between science and politics seemed logical, and perhaps even inviolable. For example, as a beginning graduate student in Anthropology, I was awarded a "National Defense" Foreign Language Fellowship, for which I agreed to study the Portuguese language, then considered "exotic." Despite its categorization as a "National Defense" grant, this funding did not impose any requirements on my work during my studies at Columbia University, nor while I engaged in dissertation research in Portugal and Brazil. In truth, nothing in my work contributed in any way to the security of the United States, nor was I required to accomplish anything which could be construed as contributing to the "national defense" other than to pass my course work and learn Portuguese. My husband's graduate work in the natural sciences was similarly financed by the National Science Foundation, again without any overlapping political guidelines. Thus, the two of us proceeded through years of graduate school, including during the defense of our dissertations, continuing to believe in the existence of "neutral science" and its distance from politics. This was clearly a sheltered bubble.

Following the completion of his degree in paleontology, my husband was invited to participate in a six week trip in 1975 to China as part of a group of scientists from the United States (a result of President Nixon's rapprochement with that nation). Upon returning home, he received what we both considered to be a shocking phone call from an unknown caller identifying himself as a representative from the CIA. Would my husband be willing to discuss what he had seen in China and provide photos and/or insights into his visit, the caller asked him? My husband answered with a firm "no" and that, seemingly, was the end of the matter. But what about the other colleagues with whom he had traveled to China? Had they cooperated with the government? Had they expected to do so in advance of their trip?

Unbeknownst to either of us, there had already been a long standing collaboration between academe and the CIA prior to 1975,[2] so perhaps others had not been as naïve as we had. In our case, however, the privileged sanctuary of our graduate academic world had been forever breached by this unexpected and truly unwelcome phone call. I had almost forgotten about my husband's unnerving experience with the government's clumsy attempt to ferret information from him when I stumbled across an unusual cache of photographs filed in the iconographic archives of the American Museum of Natural History (AMNH). As a Research Associate in the Division of Anthropology, over the course of many years I had spent considerable time in this archive locating images to illustrate my research. But on this occasion, while I searched for photographs of Caribbean sugar fields, I unexpectedly encountered several images of United States Marines taken during the initial stages of their occupation of the Dominican Republic (1916–1917). These stunning photographs, filed with others of a more standard ethnographic content, depicted Marines drilling, disembarking from ships in the harbor, playing baseball and lounging in front of their barracks. Some of them were shown walking about with their arms wrapped around local women (Fig. 1, left; see also Figs. 10.8–10.10).

I had previously thought that I was thoroughly familiar with the broad spectrum of anthropological and biological images which the AMNH possesses. But these views of U.S. Marines, both candid and posed, were seemingly totally out of context. I was even more intrigued to learn that these images had been taken by a bird collector, one Rollo H. Beck (Fig. 1, right), a freelancer from California on assignment in the West Indies for the American Museum of Natural History, rather than by a military photographer/liaison embedded with these troops. Mr. Beck figures prominently in this book: his assignment, which lasted from 1912–1917, took many turns and twists, including a previously unscheduled move from South America to the Caribbean, where he had clearly used his camera to great advantage. Initially engaged to collect sea birds for the Museum in the Southern Hemisphere, in mid-1916 Beck received instructions from the Museum administration to reverse course and begin tracking land birds in the very sections of the Antilles which coincidentally were in the crosshairs of the United States government. As I would later learn (see Chap. 10) from reading Beck's personal correspondence and the incomplete field notes he penned

Fig. 1 Left, U. S. Marine walking in the street with two local women, his arm around one of them, La Vega area, Dominican Republic, December, 1916; photo by R. H. Beck. This and similar photos set off the study leading to this book. AMNH Asset ID 236935. Right, Rollo H. Beck, October, 1917. AMNH Asset ID 36747.

while pursuing birds (both land and pelagic), the collector was not only keenly aware of the U.S. military presence in (and the strategic importance of) the Caribbean, but he also communicated with, and was advised on security matters during his stay by, the highest ranking of U.S. Marine Corps officers. Moreover, the U.S.M.C. extended this courtesy to him beyond the confines of the Dominican Republic, giving him access to their officers stationed in Haiti and Cuba as well.

When I discussed Beck's photographs with my colleague Prof. Mario Bick of Bard College, he shook his head knowingly. Was I aware, he wondered out loud, that Franz Boas, the distinguished Columbia University anthropology professor, and a former member of the Anthropology Department at the AMNH, had agonized over the number of colleagues in his field who had, according to him, been recruited for intelligence work by the United States government during World War I? I was not, but I was instantly struck by the eerie similarity between this early instance of twentieth century cooperation between the U.S. government and scientists, and the China quandary my husband had faced in 1975. Moreover, as these extracurricular activities to which Boas had referred occurred during the same interval in which Beck had been assigned to South America and the Caribbean (i.e., during World War I), Mario speculated that the Californian bird collector might have been involved in some form of intelligence gathering as well.

My colleague's prescience would eventually lead me far beyond Beck's experiences with the Marine Corps to a much broader quest to identify those scientists from the American Museum of Natural History who had indeed worked for the U.S. government in Latin America as intelligence assets during the time frame of World War I. As it happened, there were quite a few Museum men who provided data to government agencies on those countries in the Western Hemisphere which were familiar to them from their previous research. Technically still on the institution's payroll, but with salaries supplemented with government funds, these researchers were essentially "loaned" to the United States Government by the Museum and assigned to surveillance activities in Latin America. Their wartime activities as informants and the consequences of their actions are the subject of this book. Boas, I would learn, was not only correct when he exposed the non-scientific ventures of his academic colleagues, but he had sounded a pioneering warning of the dangers to intellectual integrity when science and politics crossed paths.

As I continued to ruminate over the roles of science and politics in Latin American history, I realized that one could argue, as I do in Chap. 1, that connections between science and politics are not new and can be traced back to antiquity.[3] Even in Latin America, cooperation between scientists and government agencies was present during the entire time of the European conquest, when metropolitan governments eagerly subsidized the colonial missions of naturalists in an ill-concealed effort to inventory local resources. Later, Latin American scientists would play a pivotal role in the drive for independence from colonial powers, a result of their having spread nativist political ideas among the intelligentsia. Part of the background I provide in this book, therefore, identifies several historical periods of overlap between science and politics in Latin America in the time frame prior to World War I. Hopefully, my narrative underscores how parallel interpretations of Latin American history may be obtained by examining the work of scientists in the hemisphere, a perspective not often explored in traditional historiography. One thing is starkly evident from my findings: those incidences of the melding of politics with scientific personnel which occurred during the World War I period far exceeded the collaborations which had preceded them. The undeniable truth is that the U.S. government became dependent on their "scientist-spies" to provide intelligence on Latin America which was otherwise unobtainable; quite a few of the agents who participated in this intelligence gathering were drawn from the ranks of the American Museum of Natural History.

I am equally concerned here with the emergence of what I call "scientific Pan Americanism", a sense of common purpose among scientists in the hemisphere who sought to prevent the further deterioration of resources in their native countries. Ironically, this movement emerged during the very time that political skepticism of heavy-handed United States policies in Latin America was at its height, i.e., at the turn of the twentieth century. Regardless of other Pan American initiatives encouraged by Washington officials in such subjects as commerce or law, Latin American scientists would not be pushed into superficially accepting broad Pan American cooperation with their United States colleagues without first defining their own objectives.

On a personal level, I have been fascinated by the interplay of familiar themes of Latin American history and the activities of the American Museum of Natural History in that region; the conjunction of science with political events in the Americas is a subject which historians need to explore further. For the contents and interpretation of the data herein I, alone, am responsible.

Roberta Marx Delson
Kings Point, New York, USA

Notes

1. C.P. Snow delivered the Rede lecture on "The Two Cultures" on May 7, 1959. His ideas were reprinted in The Two Cultures and the Scientific Revolution (London, Cambridge University Press) in 1959. Recently, the emphasis on today's STEM education is being challenged by a movement toward reintegration of humanities subjects in the curricula of all students. Cf. Nicole LaPorte, "Da Vinci Was a Double Major," Town and Country Magazine, May 2023, p. 93.
2. I am grateful to an anonymous reviewer for the suggestion to look at Karen M. Paget's Patriotic Betrayal: The Inside Story of the CIA's Secret Campaign to Enroll American Students in the Crusade Against Communism (New Haven, Yale University Press, 2015). Paget's (and her husband's) experiences predate my husband's trip to China by several years but verify the undeniable existence of cooperation between academe and the CIA.
3. For example, the classical mathematician Archimedes helped his fellow Syracusans in their struggle against the Romans and lost his life as a result.

Acknowledgments

In my efforts to illuminate this history I am indebted to many scholars and friends. Above all, I am grateful to Mario Bick, of the Department of Anthropology at Bard College, whose insights helped frame my project. Additionally, I am beholden to the numerous colleagues at the AMNH who were willing to help a fellow staff member. I would single out Barbara Mathe of the main Library (since retired) for facilitating my access to the Museum's extensive written and photographic archives and for providing the necessary expertise to locate relevant materials; and Senior Research Services Librarian Mai Reitmeier, who worked to provide image files for many of the figures included in the book. Jannatul Ahmed and Matthew Heenan of AMNH Global Business Development and Roberto Lebron of AMNH Communications facilitated the use of these images and graciously waived the usual fees. Additionally, I thank Tom Baione, Harold Boeschenstein Director of the Museum Library, as well as the rest of his most accommodating staff; Kirsten Mable of the Anthropology Department Archives; Mary LeCroy of the Ornithology Library; Pat Brunauer of the Mammalogy Department Archives; and Judith Galkin, Susan Bell and Ivy Rutsky for allowing me entrée into the Vertebrate Paleontology archival collection. I am also grateful to the successive Chairs of the AMNH Division of Anthropology for supporting me as a colleague, especially Drs. Enid Schildkrout, Laurel Kendall and, currently, Ashley Hammond. In point of fact, the Museum community has shown enormous interest in my project as well as outstanding cooperation; for this I am most appreciative.

Outside the AMNH, I would single out Dr. Constance Areson Clark of The Worcester Polytechnic Institute, Massachusetts, for sharing with me her fascination with Henry Fairfield Osborn, President of the AMNH before and during World War I; and Dr. James Ginther, Chief Archivist of the United States Marine Corps in Quantico, Virginia, for discussing bibliographic material on the period of USMC occupation of Haiti and the Dominican Republic. Additionally, the staffs at the New York Historical Society; the Columbia University Library Section of Rare Books and Manuscripts (especially Celeste Brewer, Tara Craig, David Olson and Jocelyn Wilk); the National Archives (NARA) offices in Washington D.C. (which declassified the ONI records concerning AMNH staff scientists in the 1990s), College Park and New York; as well as Peter Lewis at the American Geographical

Society were also enormously helpful. I can further single out Melanie Rosenberger, who undertook research on my behalf at NARA in College Park. Priscilla Mollard and Heather Yager of the California Academy of Sciences in San Francisco were most gracious in sending me copies of the correspondence of Rollo Beck and his wife Ida, while Seth Cotterell of the library and archives provided several images from the Beck collection and waived the fees for including them in this book. The staffs of the Carnegie Institution (Washington, D.C.) and the American Philosophical Society (Philadelphia, PA) were more than accommodating in securing documents during my limited stay at their organizations. To my former colleagues in the History Department of Drew University (where I was teaching Latin American History when the idea of this book was first conceived) as well as the University Seminar on Brazil at Columbia University, I express my gratitude for their pertinent suggestions and encouragement. I also acknowledge the invaluable commentaries and suggestions from three anonymous reviewers of my preliminary manuscript and especially Prof. James E. Wadsworth, who read both the early version and the entire final manuscript and made a number of powerful suggestions, as well as writing the Foreword to this book. By incorporating their ideas into the book, I believe the end product is greatly enhanced. I further thank the editors of the Springer series in Latin American Studies for accepting my work in a science-focused assemblage. I owe a real debt to Aaron Schiller, Springer Editor for Earth Sciences, Geography and Sustainability, who shepherded the book from its preliminary stages through review and into production. I am grateful to Tina Silva, Gowtham Chakravarthy Vijayakumar R, Sonam Yadav and their colleagues in Springer book production for the well-designed end product you are reading. To all of these individuals, I offer my thanks.

This book has taken many years to research and complete amidst a series of personal crises; I can only express my eternal gratitude to my husband, Dr. Eric Delson, former Chair of the Anthropology Department of Lehman College and fellow Research Associate at the American Museum of Natural History, for his patience and understanding. He has listened with rapt attention to my discoveries as I became increasingly familiar with some of the outstanding historical figures of the AMNH. His enthusiasm was and is infectious. Other members of my family, including my late mother Ruth Lynn Sperber Marx (who fostered my interest in science), and my ever-patient son William Delson and his wife Lisa Levy, have all played their part in this saga.

Roberta Marx Delson passed away suddenly and unexpectedly in February, 2025, soon after completing the text of this book. With the aid of our son William, I have edited the manuscript, updated the acknowledgments and incorporated some of the images she had selected for the volume. I hope we have done her justice.

Eric Delson

Introduction

Scientists as Spies and Their Interaction with Politics

In an October 16, 1919 letter to the editor of The Nation,[1] the distinguished German-American anthropologist Franz Boas (Fig. 2) railed against the activities of scientist/scholars who, he insisted, had acted as agents of the United States government during World War I. His seminal essay, "Scientists as Spies", famously alleged that "…[A] number of men who follow science as their profession, men whom I refuse to designate any longer as scientists, have prostituted science by using it as a cover for their activities as spies."[2]

The principal target of Boas' contempt was the noted archaeologist Sylvanus Morley, although Thomas Gann, John Held Jr., Samuel Lothrop, Arthur Carpenter and Herbert J. ("Joe") Spinden (of the American Museum of Natural History) were similarly excoriated for working with the U.S. Office of Naval Intelligence (ONI) in Mexico and Central America, as Harris and Sadler have recently discussed.[3] Boas, of course, was correct in his assertion that such scientists/scholars had worked for the government; Naval Intelligence had, in fact, specifically recruited these men for their potential as "undetectable" intelligence assets. In Morley's case his "cloak" was his position at the Carnegie Institution in Washington, D.C., a credible cover which allowed him to engage in espionage for the United States government in the Yucatan while appearing to continue his investigations of Mexican archaeology.[4] The Carnegie was fully aware of this arrangement and agreed to pay Morley the difference between his former full-time salary and what Naval Intelligence was prepared to offer for his services.[5] Indeed, it was the authenticity of Morley's cover as "Carnegie archaeologist" that made him invaluable to the Navy, especially when the transparency of such commonly-used double identities for spies in the field as "cattle-man", "mining-man" and "timber-man" began to raise suspicion.[6] As an archaeologist excavating in the field, Morley was able to routinely report back to Naval Intelligence on such matters as transportation, communication, and the "disposition of the indigenous population of the Yucatan," as well as singling out those who might be spying for the Germans

Fig. 2 Franz Boas, ca. 1915; downloaded from Wikimedia Commons (public domain)

or whose sympathies might lie in that direction.[7] Yet, for all intents and purposes, Morley could continue to truthfully represent himself as a bona fide archaeologist on assignment in the Yucatan for the Carnegie Institution.

Morley's actions in Mexico, however, were just the tip of the iceberg. One can only imagine the barbs Boas would have hurled at the host of individuals other than those mentioned in his essay who had similarly concealed their assignments as government agents in Latin America during World War I while they ostensibly pursued academic research. Like Morley, these recruits (a cadre which included natural scientists, geographers, economists and regional specialists) were temporarily "loaned" to the U.S. government and dispatched on secret intelligence reconnaissance missions.

Morley's adventures in sleuthing were, in retrospect, hardly unique. The only real differences between him and other scientists working for the government were the various agencies to which they were assigned: while some agents worked directly for the Navy (ONI) or the Department of State, others volunteered for service with the War Trade Board (created in 1917 by presidential decree to impose an embargo on German shipping), the American Red Cross (associated with the War Department), the Council of National Defense (a government think tank), the Bureau of Foreign and Domestic Commerce and the rather mysterious "Inquiry", headquartered at the American Geographical Society in New York. More than a few of the scientists who took part in these Latin American fact-finding missions were directly affiliated with the American Museum of Natural History, including the above mentioned Mayan specialist, archaeologist Herbert Joseph Spinden.

When Boas "outed" his colleagues in his post-war diatribe, he was promptly censured by the American Anthropological Association for his supposed lack of patriotism. But the reality was that Boas was no turncoat; he had patiently waited until the conclusion of the war before venting his frustration with those scientists whom he deemed hopelessly corrupted. Even with his earlier suspicions of his colleagues' dual identities, an opinion which he privately voiced to several members of the Anthropology department at the AMNH, Boas had chosen to control his outrage, revealing his concerns publicly only when wartime hostilities ended.[8]

But should Boas have been astounded that his fellow scientists were engaged in intelligence gathering in the first place? After all, the very men he had accused of scientific "whoredom" arguably represented the *crème de la crème* of the United States intellectual community, including the very small number of citizens sufficiently credentialed to work in those areas of Latin America where Washington so urgently required intelligence. If it is easy to understand the Government's desire to recruit such superb assets to aid the war effort, the more pressing issue becomes what motivated these scientists to volunteer for government duty and their host institutions to allow them to do so?

Whatever one's opinion of these "spying" activities may be, there is no denying the invaluable service that Boas' revelation has continued to provide the scientific community by unequivocally identifying the minefields which result when science and politics overlap. As relevant today as they were over a hundred years ago, the ethical questions posed by Boas keep haunting the anthropological community, as witnessed by the American Anthropological Association's decision in 2005 to create an Ad Hoc Commission on the Engagement of Anthropology with the U.S. Security and Intelligence Communities.[9] Other academic associations have struggled with the same issues; the National Academy of Sciences and the American Psychological Association have each conducted similar self-examinations of their own involvement in government-related research. Moreover, while the geographical focus of today's concerns with intelligence-gathering masquerading as academic research is more likely than not to lean toward the Middle East (including Afghanistan, Iraq, Iran, Israel, Syria, Egypt and Pakistan), or Eastern Europe, there is an unmistakable sense of déjà vu when we recall Boas' commiserations over the scientists who operated inter alia as agents in World War I Latin America. In truth, concerns over intellectual compromise as a result of scientists (or other academics) engaging in spying are as relevant and compelling today as they were a century ago: *Plus ça change, plus c'est le même chose.*

Were we, however, to extend Boas' critique beyond the ethical questions which he raised to a consideration of the broadest possible implications of his letter to The Nation, we would find ourselves confronting an even more troubling consideration—can science ever be neutral and devoid of political content, or, is science inevitably political?

History provides us with an answer. The simple truth is that scientists/explorers have long bowed to political imperatives; they have universally offered (or been coerced to provide) their services to their home governments, especially in a colonialist context (see Chap. 1 on Latin America). Similarly, when Napoleon ordered 151 scientists and artists into Egypt to chart the course of his empire and to preserve evidence of his success by taking possession of objets d'art, he indelibly welded science to his expansionist scheme. In truth, Napoleon's use of scientists and other non-political actors to serve the demands of state are hardly unique in historical context; such collaborations have existed since antiquity.[10]

Despite this evidence, some historians continue to suggest that prior to World War I the importance of science to national interests was underappreciated and that, even with the outbreak of global hostilities in 1914, policymakers remained "skeptical… especially in the Anglo-American world, of the value of science for technology."[11] Even if that had been the case (and it is frankly doubtful), within a very few months following the start of the War, scientists were enthusiastically collaborating with politicians on efforts which reached far beyond technology. In retrospect, it was only logical that the confluence of political exigencies and the great pool of talent represented by the intellect of scientists would expand during wartime, and that such heretofore "isolated" scholars (including those in the natural sciences) would be enticed to play a critical role in wartime planning. There was no uncertainty in anyone's mind that World War I was going to be a technical war; some later scholars have even described it as the "chemists' war" (especially as regards the use of poisonous gas during the conflict) as opposed to the dominance of the "physicists" who developed the bomb and thus shaped the outcome of World War II.

But the historical long view suggests that scientists, including chemists, engineers and technicians, have contributed to the "science of war" from at least as far back as the Renaissance (especially on questions of fortification), or even to antiquity. What distinguished the cooperation of scientists and politicians in World War I from earlier collaborations was the broad realization of just how much individuals trained in scientific methodology might contribute to winning the War. Scholars whose work focused on careful, scientific observations, evaluation of data and reaching evidence-based conclusions (even if the evidence was scant) were now suddenly attractive to a broad spectrum of politicians who actively recruited them to function as valuable advisors to and sometimes collaborators with government agencies. As Shaul Katzir had recently noted, during World War I scientists began "providing technical solutions, or at least partial remedies, to serious military and civilian problems [there was an] encounter and collaboration of scientists, often from academia, with military officers, industrial researchers, engineers, managers and governmental officials on war-related projects."[12]

Parenthetically, the United States was not the only country which enlisted the services of "men of science" for World War I service. It was the British who led this effort when they created a government advisory council of scientists in 1915, a year after the War had begun. Later, this preliminary council was reconstituted as the Department of Scientific and Industrial Research; separate advisory scientific teams worked alongside, and specifically for, the Admiralty and the Ministry of Munitions.

The French government went further than the British, linking universities, with their intellectual manpower, to defense circles, while in the enemy nation of Germany, the Kaiser Wilhelm Foundation for the Science of War Technology brought together the many scientists who had already been working for the government and placed them under one convenient umbrella.[13]

In short, during World War I hundreds of scientists (and other learned individuals) were recruited to work for government agencies in Europe and the United States. In the latter, President Wilson created the Council of National Defense in October of 1916, with its headquarters in the nation's capital. Wilson's Council functioned as a government think tank throughout the period of World War I, soliciting input from the country's most talented scholars and businessmen (banker Bernard Baruch and Samuel Gompers of the American Federation of Labor, for example) on broad issues of government concern even before the United States entered the fray. There was widespread support for this effort: the New York Times enthusiastically endorsed its mission, opining that "The organization of the Council likewise opens up a new and direct channel of communication and cooperation between business and scientific men and all departments of the Government."[14] Later, President Wilson would create a smaller, less visible and ultra-secret think tank in New York City, known as the "Inquiry," which tasked eminent scholars with preparing an outline for the post-World War I world.[15]

These authorized collaborations of scientists and politicians in the United States were allowed to partly lapse during the period between the two World Wars, although the Council of National Defense was briefly revived in the 1940s. On the other hand, informal cooperation between scientists and politicians, which had been highly heralded during wartime, never actually ceased throughout the critical inter-war period.

During World War II, the formal bonds between science and the U.S. government were re-established, although such connections were not always transparent. When the War concluded in 1945, partnerships between scientific academics and government officials continued, but without a specific umbrella organization (like the Council of National Defense) to foster that intimacy. Some observers remained leery of these connections: Vannevar Bush (Fig. 3), who was known for his association with the Atomic Project, submitted a carefully worded polemic entitled "Science—the Endless Frontier," to President Truman in 1945, for the express purpose of convincing him to "keep[ing] politics out of science."

Notwithstanding Bush's plea, exactly the opposite took place. Recently H. Holden Thorp, Editor–in-Chief of Science, has argued that the hope for scientific neutrality went nowhere because, in Thorp's words, "[S]cience has always been political."[16] The post War world witnessed scientists in the United States, seemingly oblivious to potential conflicts of interests, accepting monies from a mix of governmental and political sources, including industry and business. This funding has come from the National Science Foundation, the National Institutes of Health and increasingly from private foundations and business corporations; each organization had its own agenda. And yet, it has become glaringly obvious that when science and politics overlap, neutrality in such relationships is easily

Fig. 3 Vannevar Bush, 1940–1944. Source U. S. Office for Emergency Management; downloaded from Wikimedia Commons (public domain)

compromised. We must ask ourselves, is autonomy in research even possible under such conditions, or is scholarly independence just illusory at best?

Despite the lack of consensus on such critically important questions, recent speculation over the consequences of politics driving science has generated an impressive literature. This writing has focused on the implications (good and bad) of involving chemists, physicists, mathematicians and social scientists in the creation of global *realpolitik*, with the goal now being "to equip decision-makers in Washington, D.C., with an evolved idea of what science and technology can be and [to create] policy recommendations to push American science and technology to new heights."[17] While differing on what the impact of World War II ultimately meant for ethical scientific research, most observers accept the view that the blurring of the boundary between science and politics has not only increased exponentially in the recent past, but has become normative in the twenty-first century. The recent Covid pandemic ironically confirmed that conclusion: perhaps it was inevitable that the scientists involved in stemming the spread of this disease in the United States would be repeatedly criticized for being "too political."

Science and Politics in World War I Latin America and the AMNH

My focus in this book, however, is not on the post-World War II era, nor is it on the continuing collaboration between government and science spawned during that conflict. Similarly, although I reference previous contact between scientists and politicians which occurred throughout the colonial period of Latin America, and beyond, in this manuscript I am focused on the early modern elision of science and politics which took place during World War I. Even before the United States Declaration of War in 1917, Washington recognized the potential of scientists, including those who ostensibly operate in an intellectual sphere far detached from political concerns (ornithologists or mammalogists from the American Museum of Natural History, for example), to provide vital assistance to politicians. Throughout the course of this book, I explore the development of this convergence by viewing the AMNH, and those of its scientists who acted as agents for Washington agencies, as a microcosm of this phenomenon, especially as it played out in Latin America during World War I.

By focusing on the conjunction of scientific impartiality and political imperative within the context of the American Museum of Natural History, I examine what this "cooperative experiment" portended for the institution, its ongoing agenda and future plans, as well as for its scientists who were recruited by the government. Of course, it is a truism that beginning in the nineteenth century "Great natural history museums [have been] among the world's premier institutions of scientific research, training and education. The research produced by these museums, based on their collections of biological, geological, and anthropological specimens, has been of incalculable importance in formulating and testing the most fundamental theories and principles of these and related disciplines."[18] No self-respecting natural history museum would ever have consented to deviate from this model of research, and I am not suggesting that the activities of the AMNH compromised this premise during World War I. In point of fact, those scientists from the Museum who accepted covert government assignments in Latin America never fundamentally altered their personal research interests, nor were they asked to do so. But each man formulated a research project geared for the countries to which the government had assigned them, which, in turn, functioned as a convenient cover for them to collect the politically sensitive data so desperately sought by Washington politicians.

In seeking intelligence on Latin America, the Washington bureaucracy scarcely imagined it would be necessary to recruit scientists who conducted the type of research which is the purview of natural history museums, e.g., studies of South American oceanic birds or Caribbean coral reefs or fauna of the Brazilian *pantanal* wetlands. In fact, most of the scientific research projects conducted by the AMNH staff were sufficiently esoteric as to be far beyond the grasp of the Washington agencies for which these scientists would work.

AMNH scientists had already been conducting fieldwork in such strategically important regions of the globe since the turn of the century, which made pressing them into double duty as spies an irresistible proposition. The story that unfolds here is not intended as an exposé on sleuthing, nor is it an attempt to pass judgment on the dual loyalties of those scientists who simultaneously worked for the U.S. government and the American Museum of Natural History. Instead, my purpose in this study is to understand why and under what circumstances these interactions took place. What were the ramifications of this cooperation; were research goals and/or ethics compromised in the process? Did the AMNH administration which, at its highest levels, gave tacit approval to the intelligence gathering activities of its staff, appreciate the challenges that such arrangements represented to its researchers, or were such issues conveniently overlooked in the name of patriotism or some other goals? Was the Museum in some manner the unintended beneficiary of these collaborations? What, if any, were the implications for the future of scientific cooperation in the Americas?

As I see it, the Museum provides a perfect locus for examining the growing cooperation between science and politics in the years surrounding World War I. To a large degree, the staff members who participated in government intelligence efforts considered themselves not just scientists, but veritable guardians of society, typifying a special breed of men recently identified by Richard Slotkin as the "figurative sons of Roosevelt, well-born, Ivy-League-educated, virile, vigorous sportsmen and hunters, physically fit, and used to the outdoors."[19] The book is construed as a snapshot in time of these extraordinary individuals as well as their broad range of involvement from 1914 to 1919 in Latin America. The fact that their Museum-sponsored expeditions to Latin America occurred in conjunction with the outbreak of World War I in Europe begs the question of how these three separate tropes, i.e., the "War", "Science" and "scientific Pan Americanism," became intertwined.

While AMNH scientists pursued other global investigations prior to World War I (for example, the Congo Expedition of Lang and Chapin, 1909–1915; or the Jessup North Pacific Expedition, 1897–1902), my focus in this book is on the Latin American "theatre" for two principal reasons: the first is that the Museum was deeply committed to natural history research in the Western Hemisphere from the 1890s onward and had already made its mark in Latin America in several fields (including ornithology) before the onset of the War. The second reason is that Latin America, and especially those nations deemed essential to the war effort because of their strategic location, or professed neutrality, or the primary goods which they produced, or their affinity for German causes (or all of the above) was where the Museum's research activities, coincidentally, were concentrated. It was in these countries that science and politics became symbiotic. Regardless of their individual areas of expertise or their rank within the museum hierarchy, a select group of scientists from the AMNH, encompassing the gamut from paleontologists and scholars of modern-day animals to anthropologists and archaeologists, all reported with alacrity from their locations in Latin America to government agencies (or, sometimes, to contacts within the Museum) on what they observed in situ. In their "cloaked" capacities these AMNH natural scientists supplied United States government agencies with much invaluable on-the-spot knowledge about our Latin American neighbors, especially with regard to those which had chosen to remain neutral during the War, rather than support the Allied cause. And while they were gathering intelligence, these scientists continued to collect on the Museum's behalf, not infrequently becoming the recipients of scathing criticism from the Museum Administration if their efforts in this regard seemed to be faltering.

This book does not purport to be an institutional history of the American Museum of Natural History, nor does it present an in depth analysis of the careers of its scientists/agents. In recent years many excellent studies have done just that, including Constance Areson Clark's book on Henry Fairfield Osborn's role in the debate on evolution, and Brian Regal's and Ronald Rainger's examinations of Osborn's long, distinguished scientific career. Monographs about other AMNH scientists, such as Lowell Dingus and Mark Norell's book on Barnum Brown, have increased our understanding of the Museum's sense of purpose and its relationships with its staff. Yet another approach has been to examine the histories of individual departments of the Museum (for example, Stanley A. Freed's comprehensive two volume history of the Department of Anthropology).[20]

Beyond the AMNH, the roles of several anthropologists from other institutions in spying in World War I have been explored in the carefully documented study by Harris and Sadler cited above. Quetzil E. Castañeda has considered the political mission of the Carnegie Institution in Washington, D.C. during World War I. Finally, Jamie Bisher has recently published an exhaustive examination of German activities in Latin America during World War I and the recruitment of agents (from the United States and elsewhere) as a response to those threats.[21] Where the present discussion differs from these studies is in its focus on the relationship of politics to science, and the specific objectives realized by the American Museum of Natural History through its cooperation with Washington during the World War I time frame. It aims to encourage a conversation about the implications of these actions while also calling attention to the role of the Museum in fostering cooperation between scientific communities in the Americas.

Structure of This Book

This book is divided into three parts. Part I covers the background to what would become the broad cooperation of AMNH scientists with the United States government. Chapter 1 opens with a wide overview of intellectual networking in the scientific communities of the Americas, beginning with the colonial period, through the nineteenth century, and ending with the years preceding World War I. The stated objectives of exchanges in the colonial era may have been the collection of new data and the sharing of ideas, but contacts between scientific colleagues in the Americas (North and South) simultaneously provided unique conduits through which to negotiate politics. This was especially the case during the throes of independence struggles at the end of the eighteenth and into the nineteenth century.

The brilliant rapport between like-minded scientists created during this time would, unfortunately, deteriorate rapidly in the second half of the nineteenth century, and break down entirely in the period following the Spanish-American War (1898), when negative feelings toward the United States reverberated throughout the hemisphere. Leery of United States investors in their region, Latin American scientists were the first to sound the alarm about the destruction of local environments and resources. Ultimately, these same scientists would purposefully exclude their North American colleagues from several important scientific conferences which were held at the turn of the century in Buenos Aires, Montevideo and Rio de Janeiro.

But by 1908, as seen in Chap. 2, the pendulum slowly swung back, with scientists in the United States and Latin America now championing a "scientific" version of Pan Americanism, a goal explicitly fostered by Leo S. Rowe, an eminent political scientist from the University of Pennsylvania. As a result of his influence, the Second Pan American Scientific Congress was convened in Washington. D.C. in late 1915 and brought together scientists from all over Latin America and the United States. If its primary function was to foster intellectual exchange and camaraderie amongst scientists in the hemisphere, it was also the first venue from which the State Department sought to recruit scientific men to work for government service in Latin America.

Clearly, some of the North American scientists who did their research in Latin America (and who were not all inevitably spies) had always functioned as intellectual "ambassadors" to the various nations in which they unearthed or observed their specimens, exchanged collections or did their research. They were the friendlier faces of the United States, and the cooperation of scientists with their counterparts in Latin American research institutions did much to offset virulent anti-U.S. attitudes which had festered since the mid-nineteenth century. Such scientists also conveniently fostered cooperation between the nations, a goal vital to the success of the Allied cause during World War I.

Chapter 3 examines the specific interests of the American Museum of Natural History in Latin America prior to World War I and during the years dominated by the global conflict. Scientists from the Museum explored, excavated and did their research in almost every corner of the hemisphere, strongly encouraged by the very determined (and socially connected) president of the institution, Henry Fairfield Osborn (often known as HFO). As a scientist, Osborn was interested in pursuing Latin American fieldwork from an evolutionary perspective; his obsession with the subject was such that he would bypass the Museum's self-imposed moratorium on overseas research during wartime to allow his staff to work secretly as agents for the U.S. government in Latin America, provided they continued their research in the region for the Museum. It was thus with Osborn's express approval that AMNH personnel became deeply involved in covert intelligence gathering in Latin America during World War I. In his view, this wartime work would also set the stage for advancing important Pan American contacts among his scientists, a move which he imagined would augur well for his post-war research plans for the region.

Chapter 4 discusses the concerns of the United States government with German activity in Latin America as well as the role of the Office of Naval Intelligence (ONI) and the State Department in recruiting capable agents (including from the Museum) to ferret out intelligence. Newly headquartered in an office in downtown New York City, the local activities of the ONI ramped up at roughly the same time as Osborn's views on German hostilities were evolving. The latter would move from doubter of the merits of the European War to *Überpatriot*; President Osborn subsequently involved his entire institution in numerous wartime activities, including spying. Chapter 5 explores the specific response of the scientific departments at the AMNH to the war as "Preparedness" became the order of the day.

Part II of this book (Chaps. 6–10) presents micro-histories of Museum personnel who had been involved in Latin American research before WWI and who were dispersed throughout the region during the War, where they simultaneously served as government intelligence assets. My specific interest here is in how they attempted (sometimes unsuccessfully) to fulfill their research obligations to the Museum while carrying out their government assignments. The list of such individuals is notable not only for its breadth but because it reads like a veritable "who's who" of natural history scientists of the period. Included are Barnum Brown of dinosaur fame, who reported on petroleum reserves in Cuba; Frank Chapman, the distinguished ornithologist who served as an American Red Cross Commissioner in South America; George Cherrie, an ONI agent who followed his participation in Theodore Roosevelt's AMNH-sponsored expedition to Brazil with his own trip into the Brazilian *pantanales* from where he reported back on what he deemed suspicious German activity; Herbert "Joe" Spinden, the anthropologist who worked with Sylvanus Morley in gathering data on Central America for the ONI; Charles R. Eastman, an ichthyologist and ONI agent whose work in Brazil supplied Washington with vital information on the shipment of manganese out of that country to Germany; and Henry Crampton, who worked at the Council of National Defense and linked the AMNH to Washington. All of these individuals (and others, including Rollo H. Beck) would engage in activities quite distinct from either their official museum duties or areas of expertise, and they did so with the express approval of President Osborn. We can thus interpret Osborn's entry in the Trustees Report of Spring, 1917 which casually mentions that "several members of the scientific staff had been called upon for important and hazardous service"[22] as a masterful stroke of understatement, designed to camouflage the actions of an administrator who should have known better than to authorize their service.

Part III (Chap. 11) concludes with an assessment of this important period in the Museum's history. It considers the implications of the Museum's scientific involvement in Latin America and the reasons why the AMNH modified its research agenda after 1918 to now promote explorations in Asia. Although Henry Fairfield Osborn cogently observed "our future scientific relations with South America are of great political importance,"[23] the postwar period saw the AMNH redirecting its efforts toward collecting in the Far East rather than concentrating on the Western Hemisphere.

Throughout this text I have relied on data from the Central Archives of the American Museum of Natural History, as well as the Museum's iconographic holdings. Various satellite departmental archives, including Anthropology, Mammalogy, Ornithology and Vertebrate Paleontology yielded valuable letters and diaries. Outside the AMNH, I have consulted collections and public records from the National Archives (Washington, Maryland, and New York sites), which declassified its ONI records from World War I in the 1990s, as well as the Rare Book collection at Columbia University, New York Historical Society, California Academy of Sciences, American Geographical Society and Carnegie Institution in Washington, D.C. among other sources. Because American Museum of Natural History scientists traveled widely, their impact, as well as their written record, is broadly scattered.

Perhaps a word of caution is in order: this account may be interpreted in several ways. Some may view this discussion of the elision of science with politics as a portrait of misguided patriotism on the part of the American Museum of Natural History, from participants on its scientific staff through the upper echelons of its administration. Others may consider the evidence presented here as quintessential confirmation of President Henry Fairfield Osborn's enormous ambition for himself, as well as for the institution he oversaw. More than a few Museum personnel have continued up to the present time to deny "rumors" that have long circulated at the institution regarding the AMNH scientists who acted as intelligence assets for the U.S. government during World War I. Such ignorance may be explained both as the result of the Museum's determination to keep these activities secret, as well as the federal government's continuation of the classified status of such agent participation until the 1990s.

My major purpose in writing this account, however, is to open a conversation about an extraordinary period in the chronicle of the American Museum of Natural History, a time in which the liminal boundaries between the scientific establishment and the government unraveled and dissolved, perhaps forever. It is written neither as an indictment of, nor as a polemic justifying, either Boas' or Osborn's positions. Rather, I hope this monograph will help broaden our understanding of Osborn's actions in the context of the Museum's history, as well as the reasons why those of the institution's resident scientists who agreed to be intelligence assets believed their government work truly mattered. Until the global conflict was resolved, these men were convinced that the smooth course of Latin American/United States relations, if not the actual outcome of the War and the future of AMNH scientific endeavors in the region, hung in the balance.

Notes

1. Franz Boas' essay, "Scientists as Spies", was published in The Nation, 109 (December 20, 1919), p. 797.
2. *Ibid..*, as cited in George W. Stocking, Jr., ed., The Shaping of American Anthropology, 1883–1911: A Franz Boas Reader (New York, Basic Books Inc., 1974), p. 336.
3. Paul Sullivan, Unfinished Conversations: Mayas and Foreigners Between Two Wars (New York, Alfred A. Knopf, 1989), p. 132. See also Charles H. Harris III and Louis R. Sadler, The Archaeologist was a Spy: Sylvanus G. Morley and the Office of Naval Intelligence. (Albuquerque, University of New Mexico Press,

2003), and David L. Browman, "Spying by American Archaeologists in World War I (with a minor linkage to the development of the Society for American Archaeology)," Bulletin of the History of Archaeology, vol. 21, no. 2, 2011, pp. 10–17.

4. Harris and Sadler, The Archaeologist…, p. 134.
5. *Ibid.*, p. 132.
6. *Ibid.*, p. 134.
7. *Ibid.*, p. 133.
8. Boas outlined his concerns to Pliny Goddard of the AMNH Anthropology Department (August 7, 1917, Division of Anthropology Archives, hereafter DAA) and colleagues at other institutions, according to Harris and Sadler, The Archaeologist…, pp. 284–288. In the same manner, although Boas was against U.S. involvement in the War initially, once the nation declared war against Germany, he stopped his public anti-war speeches. See Herbert S. Lewis, "The Passion of Franz Boas," American Anthropologist, June, 2001, Vol. 103 (2), p. 457. Boas was a member of the AMNH Anthropology Department from 1896 to 1905, leaving after disagreements with museum administrators. At the same time, he had joined Columbia University, where he later consolidated all anthropology faculty in an Anthropology Department and remained there until retirement.
9. Annual Report 2006, the American Anthropological Association, pp. 10–13. David Price, "Anthropologists as Spies," The Nation, November 20, 2000., pp. 24–27 pointed out that as of 2000 the American Anthropological Association still had no proscription against espionage despite heated debates during the Vietnam War into the 1990s.
10. See Nina Burleigh, Mirage: Napoleon's Scientists and the Unveiling of Egypt, (New York, Harper Collins, 2007). Earlier, the contribution of Archimedes toward defending Syracuse against attacks by the Romans was noted. One might also cite the defense work of Leonardo da Vinci on fortifications.
11. Shaul Katzir, "In War and Peace: The technological promise of science following the First World War," in Jaume Navarro, ed., special issue "The Promises of Science. Historical Perspectives," Centaurus, 59 (2017), p. 224.
12. *Ibid.*, p. 225. See also Peter J. Bowler and Iwan Rhys Morus, Making Modern Science: A Historical Survey (Chicago, University of Chicago Press, 2008), p. 465.
13. *Ibid.*, pp. 466–7.
14. The Council of National Defense was created by order of President Wilson on October 11, 1916. Cf. New York Times, October 12, 1916, p. 10. See also William J. Breem, Uncle Sam at Home: Civilian Mobilization, Wartime Federalism, and the Council of National Defense, 1917–1919, (Westport, Conn., Greenwood Press, 1984).
15. See Lawrence E. Gelfand, The Inquiry: American Preparations for Peace, 1917–1918, (New Haven, Yale University Press, 1963). After the War concluded, the International Union of Pure and Applied Physics (IUPAP) attempted to promote science as a way of sponsoring universalism and peace.

Although today the union is over one hundred years old, its task has been anything but easy. Cf. Roberto Lalli and Juame Navarro, "How physicists strove for peace,". Nature, Vol. 633, 19 September 2024, pp. 515–517. Many of the historians, librarians and scientists recruited to work for the U.S. government during World War II through the OSS remained on after the War at the newly created CIA. Cf. Elyse Graham's Book and Dagger: How Scholars and Librarians Became the Unlikely Heroes of World War II (New York, Ecco Press, 2024).

16. H. Holden Thorp, "Science has always been political," in Science, 17 July 2020, Vol. 369, Issue 6501, p. 227.
17. A new task force formed from almost 70 leaders of science, academia, industry, etc. has been recently (as of 2024) formed to explore ways in which the United States can maintain a strategy for the growth of science and technology. It is known as VAST, "Vision for American Science and Technology." Cf. Sudip S. Parikh, Marcia K. McNutt and Darío Gil, "A new vision for American science," Science, 29 November, 2024, Vol. 386, p. 947. See also David Nugent's fascinating discussion of social sciences in the United States and their relationship to imperialism and corporations or foundations (such as the Rockefeller or Carnegie) funding in the post World War I period. David Nugent, "Knowledge and Empire: The Social Sciences and the United States Imperial Expansion", in Identities: Global Studies in Culture and Power, Vol. 17, (2010), pp. 8–9. Another relevant paper on the crossover between science and politics is Clark A. Miller, "An Effective Instrument of Peace: Scientific Cooperation as an Instrument of U.S. Foreign Policy, 1928–1950," OSIRIS 2006, 21, pp. 133–160. In The New Yorker, 5 January 2022, Rivka Galchen explores the reasons why scientists have shared their knowledge with unfriendly nations. Cf. "Why Scientists Become Spies: Access to information only goes so far to explain the curious link between secrets and those who tell them."
18. Gregory C. Mayer, Jerry A. Coyne, Jonathan B. Losos, Johannes Foufopoulos, Neil Shubin, Douglas J. Futuyma, Benjamin C. Campbell, and Scott V. Edwards, "Museums' Role: Increasing Knowledge," in Science, 8 March 2013, Vol. 339, pp. 1148–1149.
19. Richard Slotkin, Lost Battalions: The Great War and the Crisis of American Nationality, (New York, Holt, 2005), p. 76.
20. See Constance Areson Clark, God—or Gorilla: Images of Evolution in the Jazz Age, (Baltimore, Johns Hopkins University Press, 2008); Brian Regal, Henry Fairfield Osborn: Race, and the Search for the Origins of Man, (Aldershot, England, Ashgate (Routledge), 2002); Ronald Rainger, An Agenda for Antiquity: Henry Fairfield Osborn and Vertebrate Paleontology at the American Museum of Natural History, 1890–1935, (Tuscaloosa, University of Alabama Press, 1991); Lowell Dingus and Mark Norell, Barnum Brown: The Man Who Discovered *Tyrannosaurus rex*, (Berkeley, University of California Press, 2010); Stanley A. Freed, Anthropology Unmasked:

Museums, Science, and Politics in New York City, 2 vols. (Wilmington, Ohio, Orange Frazer Press, 2012).

21. Harris and Sadler, The Archaeologist…; Quetzil E. Castañeda, "The Carnegie Mission and the Vision of Science: Institutional Contexts of Maya Archaeology and Espionage," pp. 27–60 in Regna Darnell and Frederic W. Gleach, Histories of Anthropology Annual, Vol. 1, (Lincoln, Nebraska, 2005); Jamie Bisher, The Intelligence War in Latin America, 1914–1922, (Jefferson, North Carolina, McFarland and Company, 2016).
22. HFO report to Trustees of AMNH, May 7, 1917.
23. Osborn to Frank Chapman, New York, April 18, 1918. Central Archives (hereafter CA), AMNH, file 1069.

Contents

Abbreviations

AMNH	American Museum of Natural History
Beck CAS	Files of Rollo H. Beck, California Academy of Sciences
BJ	AMNH Beck Journal
BPC	Beck photographic collection, AMNH special collections
CA	AMNH Central Archives
CAS	California Academy of Sciences
CUSMC	Columbia University Special Manuscript Collections
DAA	AMNH Division of Anthropology Archives
DMA	AMNH Department of Mammalogy Archives
DOA	AMNH Department of Ornithology Archives
DVPA	AMNH Department of Vertebrate Paleontology Archives
EC AMNH	Executive Committee of the American Museum of Natural History
NA, RG	National Archives, Record Group
NYHS/HFO	New York Historical Society, Diaries of Henry Fairfield Osborn

List of Figures

Part I
On the Eve of War

Chapter 1
The Interaction of Latin American and United States Scientists on the Eve of WWI

During a June 15, 1908 meeting conducted in the formal diplomatic rooms of the Department of State, Secretary of State Elihu Root addressed ten of the nation's leading scientists. These individuals had been chosen to represent the United States at the First Pan American Scientific Congress, an event which would be held later that year in Santiago de Chile. A leading proponent of Pan Americanism in its broadest sense, Secretary Root tried to reassure these men that their "errand was scientific and not diplomatic." But, at the same time, he didn't hesitate to remind them that not only were their trips being financed by Congress, but that many of the issues to be discussed in Chile would lay the groundwork for sessions of a more overtly political nature which would be held in 1910 during a Pan American conference in Buenos Aires. He admonished the assembled group that the mere fact the United States had been invited to send a delegation to the Chilean meeting of Latin American scientists was a stunning achievement.

He need not have done so; all of the men in the room were acutely aware that the convening boards of the three previously held Latin American scientific congresses had drawn an exclusive perimeter around their minions, not even bothering to invite U.S. representatives. "For the first time," Root reminded the gathered scientists, "under the influence of the new Pan-American entente, an all-American congress is to be held, and it is very desirable that we should contribute our fair share and that the work which you do and the associations which you form should contribute toward the establishment of permanently good relations and continue the work of doing away with the misapprehensions, jealousies, and suspicions which have so largely influenced the views entertained for this country in some of the Latin-American countries."[1] Among the distinguished nominees chosen to attend the conference were representatives from the Smithsonian Institution in Washington, D.C., Prof. Hiram Bingham of Yale University (and, later, of Machu Picchu fame) and Dr. Bernard Moses from the University of California. The group included archaeologists, political scientists and specialists in the natural sciences, all of whom were chosen for their expertise in Latin America.

R. M. Delson, *Scientists or Spies?*, The Latin American Studies Book Series,
https://doi.org/10.1007/978-3-032-18117-6_1

Surprisingly, none of the nominees at Root's meeting represented the American Museum of Natural History, notwithstanding the considerable fieldwork which was concurrently being conducted by the AMNH in Latin America. The State Department omission must have been unsettling to the AMNH administration; in any event, this would be the last time that American Museum of Natural History personnel would not be included in any short list of scientific candidates with relevant Latin American *bona fides*. This was especially so after Henry Fairfield Osborn, whose interest in Latin America stemmed largely from an evolutionary perspective,[2] became President of the Museum in 1908. Under his guidance the AMNH would become a dominant, if not the pre-eminent, U.S. natural history museum conducting scientific research in the region.

The scientists who met with Secretary Root in 1908 understood the goals he had outlined in his address. Their presence in Washington was by itself confirmation (even without the participation of the AMNH) of the nation's growing fascination with Latin America. Simply put, interest in our hemispheric neighbors had been flourishing in lockstep with the increasing political and economic importance of the region to the United States government and business interests. The Spanish-American War of 1898, which brought U.S. troops first into Cuba and then to Puerto Rico, paved the way for later U.S. involvement in the West Indies and Central America. These later actions, initiated under military control, were taken in supposed fulfillment of the 1904 Roosevelt Corollary to the Monroe Doctrine, which condoned intervention in the Americas in the event of "chronic wrongdoing."[3] But, at the same time, such involvement also resulted in unprecedented economic opportunities for United States investors in what emerged as transformed political landscapes. Contact with these newly "opened" regions furthered the desire for more concrete scientific knowledge of these countries. The late geographer Neil Smith wryly observed that "[I]n the early years of the twentieth century, it [Latin America] was the leading destination for U.S. direct investment and a playground for U.S. explorers and archaeologists."[4]

Coincidentally, while Latin Americans remained wary of growing U.S. interests in the region, it was painfully evident that the number of scholars and scientists in the United States who were truly knowledgeable about Latin America was actually quite limited. Not only did Root have a small pool of candidates from which to choose his delegates for the Chilean meeting, but universities were simultaneously scrambling to expand their ranks with qualified scholars who were familiar with the region. Much of what was known about Latin America was, according to geographer Isaiah Bowman, quite frankly "rubbish." He expressed that sentiment in a 1915 exchange with his colleague John Clark from the Carnegie Institution, suggesting to him that "[c]ollege instruction in the geography and history of South America has grown fully as rapidly as instruction in Spanish. The data obtainable are either of the gushy variety supplied by John Barrett [of the Pan-American Union] or the popular variety supplied by various serial books of a superficial order… What we need for thoughtful people and for serious college instruction is a series of real interpretations based on thorough work…".[5]

Faulty knowledge about Latin America, however, was not from lack of interest; on the contrary, the U.S. public eagerly sought information on their southern neighbors. That interest had first been sated with glimpses of Latin American exotica[6] made possible with the advent of still photography in the late nineteenth century, but rapidly transpiring events at the turn of the century demanded more precise and accurate factual data. As every schoolchild knew, it was newspaper coverage (sensationalist, to be sure) of the protracted Cuban struggle for independence from Spain, fought between 1895 and 1898, that created a groundswell of public support for U.S. intervention in that island. The dramatic explosion of the USS *Maine* on February 15, 1898 was merely a pretext for the U.S. to enter Cuba's ongoing war, as it conveniently reinforced the already-held convictions about Spanish brutality relentlessly depicted in North American dailies. In the years following the conclusion of the Spanish-American War, academics (including scientists) would conduct research in areas of ever-increasing political and economic importance to the United States. Under the watchful leadership of its President, Henry Fairfield Osborn, the AMNH would take its place at the front of the queue.

In general, historians of Latin America have focused on the political and pecuniary dimensions of turn of the century United States expansion into Latin America, rather than researching the scientific implications of such advances. Historians point out that in the usual order of affairs, military invasions by the United States were closely followed by large infusions of private U.S. capital into infrastructure and land holdings. This resulted in fortunes being made for businessmen savvy enough to invest in newly acquired colonies (for example, Puerto Rico) or in those countries (Nicaragua, Haiti, Cuba, and the Dominican Republic) in which the prolonged presence of the U.S. Marine Corps would stabilize the political and economic climate sufficiently to insure profitable returns. Such investments are fascinating from a researcher's point of view, and their histories are broadly represented by some excellent studies on the politics of invasion and capital expansion into sugar plantations, railroads and other forms of infrastructure.[7] By contrast, the corresponding scientific aspects of this turn of the century U.S. expansion into Latin America are relatively unknown, although they are no less deserving of attention; these, too, were equally influenced by political considerations.

It would be wrong, however, to assume that U.S. scientists became involved with their counterparts in Latin America only after the 1898 Spanish-American War had concluded. In point of fact, interactions between North and South American scientists had begun as early as the eighteenth century and continued during their nations' respective struggles for independence, from 1776 to 1824; by that time not only had the United States broken away from England, but most of the former Latin American colonies had declared their freedom from Spain and Portugal. What began as wholesome cooperation between scientists in the United States and elsewhere in the Americas, however, slowly deteriorated, collapsing altogether by the mid-nineteenth century, as referenced by the Secretary of State in his 1908 briefing. Thus, the sea change in attitude toward U.S. scientists, which the invitation to participate in the First Pan American Scientific Congress in Chile represented, should not be underestimated. To fully understand the implications of the so-called "Pan American entente"[8] of the first decade of the twentieth century, a brief examination of its very complicated "prehistory" is warranted.

Science in the New World

From the 1500s to the 1700s, the concept of "scientific research" in the Western Hemisphere more often than not served as a blind for advancing the interests of metropolitan countries in their respective colonies. Antonio Barrera-Osorio has recently argued that, long before most Latin American colonies achieved their independence in the early nineteenth century, science in the Americas had been purposely harnessed to political aims.[9] During the colonial period, state-sponsored scientific expeditions ostensibly embraced the cause of epistemology by collecting and researching the unknown in overseas territories; in truth they masked a more compelling goal of cataloguing valuable colonial resources upon which the metropolitan economies were becoming increasingly dependent. It is not surprising to learn, therefore, that evidence of politically motivated reconnaissance forays is available almost from the time of Columbus; such ventures lasted through more than three centuries of colonial Iberian rule in the Americas. Those intrepid naturalists willing (or stalwart enough) to venture into remote, often dangerous, regions of the colonies were eagerly supported by their home governments, resulting in mutually satisfying arrangements which lasted well into the politically-charged atmosphere of the late eighteenth century.

But that was not the only connection between science and politics. As the eighteenth century unfolded, scientific activities in the Americas were increasingly influenced by the overtly liberal politics of the Enlightenment. Leading scientists of the time would meet (often secretly) within the chambers of learned societies, or academies, to discuss new and inherently subversive, nationalist ideas, including those which recognized the natural laws of mankind along with their corollary arguments for self-government. In these settings, as Juan José Saldaña has observed, "science took on a leading role in social transformation of the region, and it became one of the cultural and material agents of change."[10]

It was also within the confines of these academies that contacts between the scientific communities of the Spanish and Portuguese colonies and like-minded brethren in North America were initiated and nurtured. In retrospect, the inter-hemispheric exchanges between such institutions could be interpreted as the tentative first steps toward a "scientific Pan-Americanism." The academies in which the scientists met were not, of course, schools in the strict sense of conferring degrees on young students but rather, as exemplified by the well known American Philosophical Society in Philadelphia, gathering places where learned individuals could meet and hold discussions. In that role, they did much to foster communication among some of the best scientific minds of the hemisphere.

For those who have an interest in historiography, it is worthwhile noting that the study of intellectual exchanges between such academies was one of the earliest themes of the emerging discipline of Latin American history as it took shape in the United States in the first half of the twentieth century. In his pioneering essay published in 1942, the distinguished historian Harry Bernstein observed that "[A] new intellectual link, forged from the effects of the Enlightenment, united scientists in America."[11] Recognition of their mutual struggles against metropolitan constraints encouraged scientists in North America to support the Spanish American bid for independence. Bernstein singled out Dr. Samuel Latham Mitchill of Columbia University, who was later named the director of the Lyceum of Natural History in New York (the forerunner to the New York Academy of Sciences), as "the first North American to reconcile scientific enlightenment and democracy by endorsing the Latin American struggle for liberation."[12]

Among the distinguished North Americans whose work figured broadly in enlightened discussions in Latin America was Benjamin Franklin, who corresponded with many members of scientific societies within the hemisphere, and whose works on "optics, waves and heat rays" were translated into Spanish. Alexander Garden, a South Carolina botanist, and Dr. Benjamin Smith Barton, a Philadelphia physician interested in early indigenous cultures, were among a host of other North American scientists in direct contact with their Latin American counterparts in the late 1700s. And the exchange worked in reverse: the Guatemalan scientist and physician, Dr. José Felipe Flores, for example, traveled north from the city of Antigua, Guatemala to meet in person with his scientific colleagues in Philadelphia.[13]

Through the early years of the nineteenth century, the newly independent nations of the Americas continued to endorse contacts between their respective scientific communities. A United States diplomatic mission of 1817, for example, was instructed to encourage direct contact between the New York Lyceum of Natural History and noted Latin American naturalists, including Bartolemé Muñoz, director of the Cabinet of Natural History in Buenos Aires.[14] Such cooperation, forged in an atmosphere of mutual interests, boded well for a peaceful Pan-American melding of scientific and political objectives, and gave rise to some promising early hemispheric exchanges.

But sustaining such idealized cooperation between scholars proved elusive after political relations between the United States and Latin America soured midway through the nineteenth century. Perhaps rupture was inevitable given the unilateral nature of the 1823 Monroe Doctrine, in which the United States single-handedly assumed the role of protector of the hemisphere from European interlopers. A few decades later, maintaining mutual trust would be even harder to sustain when, in the post-1848 era, the United States proclaimed victory in the Mexican-American war, absorbing over half of Mexico's national territory in the process.

By the mid-nineteenth century, this catastrophic shift in the political landscape, as well as in the balance of power, impinged on all aspects of hemispheric relations, including the once-heralded partnership between the North American and Latin American scientific communities. The increasing role of U.S. investors in exploiting Latin American natural resources, and the repeated but unsuccessful early attempts of the United States government to take over the Spanish colonies of

Cuba and the Dominican Republic, as well as the Danish Virgin Islands, resulted in a re-evaluation of the United States and its scientists by Latin American intellectuals. The once trusted northern neighbor was now viewed as a threat, earning the famous moniker of the "colossus of the North."[15]

The stinging vitriol which Latin Americans leveled against the United States for its political actions also reverberated within scientific settings. From this point on, the exchange of knowledge was no longer confined to the venues of enlightened academies but, instead, replaced by a highly resented, one-way flow of ideas and technology which moved hegemonically either north to south (i.e., within the hemisphere) or east to west from Europe. Although technically no longer European colonies, Latin American nations emerged from the 1820s even more dependent on export-led economies than during the colonial period, even if trade was no longer limited to the metropolitan countries. "Informal empires" (sometimes referred to as "neo-colonialist") increasingly characterized Latin America, with the British emerging as the single most important source of capital. This new phase of the Latin American economy was reliant not only on foreign investment for growth, but on foreign "scientific know-how;" the curriculum in local technical schools as well as the technological equipment imported from abroad rapidly replaced historic patterns.[16] The use of imported technology was experienced in every context, becoming as characteristic of coffee production as it was of sugar agriculture, animal husbandry or mining. By the second half of the nineteenth century, this phenomenon had eclipsed the preeminence of the natural sciences as an avenue of exchange between nations: it had been superseded by an emphasis on the more technological sciences (including engineering, agriculture and mining). Moreover, this shifting paradigm clearly paralleled, while it simultaneously advanced, the interests of investors from the more "progressive" nations, including those from the United States as well as Europe.

Cynicism flourished in this new atmosphere. For thoughtful observers of Latin American landscapes it was obvious that the imported, steam-spewing, behemoth mechanical engines which crushed cane more efficiently than older methods, or spun sisal into rope more effectively than ever before, bore responsibility for the destruction of local natural resources. No wonder that Latin American scientists, who had unhesitatingly shown their nationalistic tendencies in the late eighteenth century, now called attention to what they, themselves, had witnessed in the course of their careers, i.e., the despoiling of forests, the fouling of rivers, the replacement of small farms by large consolidated estates (often foreign-owned) and the inevitable destruction of the native landscape, phenomena which they attributed to foreign interests. These issues, raised subtly at first by scientists in the post-1850 era, became their all-consuming rallying cry by the turn of the century, coincident with the 1898 war between the United States and Spain over Cuba. At this critical juncture Latin American scientists recognized that they would be compelled to either take the lead in the conservation of their nations' resources, or forever be held responsible for the dire consequences which would befall their respective countries.

Surprisingly, one of the most glaring irritants to the Latin American scientific community was the rapidly expanding international traffic in local bird plumage; it was widely understood that the slaughter of the Latin American bird population was conducted for the sole, frivolous reason of allowing fashionable women world-wide to adorn their hats with exotic feathers. Beginning in the early nineteenth century, and still the mode in the first years of the twentieth, the seemingly never-ending quest for rare feathers had resulted in an ever widening demand for more unusual plumage (even the tanned skin of exotic birds was said to be universally coveted for gloves).[17] This issue became a cause célèbre among those who wished to conserve their nations' resources and, as a result, a developing Latin American ornithological consciousness became intrinsically linked to nationalism. So when the widely-heralded publication of Aves do Brasil (Birds of Brazil) by Emílio Goeldi appeared in 1894, the study not only "revolutionized" ornithological knowledge in Brazil, it sounded the alarm about the likely depletion of that country's unique avian patrimony.[18] It didn't take much to triangulate from the dangers present in the ornithological example to a realization that other valuable resources were similarly facing rapid extinction. In that manner, the recognition of the inherent dangers of the plumage trade became the tipping point in a high-minded nationalist rhetoric decrying the ever-accelerating loss of Latin American natural resources not just of birds, but of petroleum, manganese, minerals, and, in its broadest extension, even cultural artifacts.

Perhaps it was in this last category that the growing ill-will between scientific communities reached the breaking point. As Robert Aguirre points out in his study of the importance of Latin America to Victorian culture, the raiding of cultural patrimony (especially, but not entirely, by the British) was accompanied by cultural denigration of the peoples of Latin America by Europeans and North Americans alike. On the one hand, museum exhibits based on Latin America mounted in "advanced" nations created "visually seductive panoramic displays that portrayed foreign lands as rich, available and conquerable...," but, on the other, they "represent[ed] indigenous Central Americans and Mexicans as hopelessly backward and unknowable, [and] Creole elites and *mestizos* as untrustworthy and both as incapable of appreciating or protecting the cultural riches they had inherited from their ancestors."[19]

Infuriated with this new phase of harnessing "science" (in this case, archaeology) to political/cultural agendas, a move which chided Latin Americans for not recognizing the need to protect their national treasures, a small cadre of scientists once again positioned themselves at the forefront of what became distinctly nationalist campaigns. Parallels between their activities and those of their forebears during the period of the Enlightenment and Independence were obvious; the difference was that now, post-1850, Latin American scientists stood as a bulwark not against colonial masters but in opposition to wider foreign intrusion. Vowing to protect their own patrimonies against destructive interests, their collective disgust would eventually lead to the convening of regular scientific meetings in Latin America. Such convocations purported to be "Pan American," but very conspicuously excluded scientists from the one nation of the Americas perceived to be ruinous to Latin American interests, i.e., the United States.

Given the complex interaction of science with politics in the Americas, a phenomenon in evidence from at least the 1700s onward, it is not surprising that Latin American scientists were attuned to the rapidly widening abyss which had emerged between the scientific communities of the United States and Latin America. It was against this background of diminished camaraderie with their North American colleagues that three important scientific meetings were convened: the first gathering was held in Buenos Aires, Argentina, in 1898, followed soon after by another in Montevideo, Uruguay, in 1901, and culminating in a third meeting in Rio de Janeiro, Brazil, in 1905. Scientists from the United States were purposely excluded from all three of these convocations.

This was no mere omission; it was clear from the programs of these meetings that the exclusion of scientists from the United States was intended to be obvious. The excuse offered by organizers of these Latin American scientific meetings for not including their U.S. colleagues was predicated on what they termed the "preliminary" nature of their meetings. From the brief, published remarks on the situation, the organizers of these meetings appeared to suggest that scientists from "more advanced" nations, such as the United States, should not be asked to convene with their Latin American counterparts until it was determined that such "experiments," had been successful.[20] But as a rationale, the "fear of looking foolish" was probably far less important in the exclusion of scientists from the United States than the perceived *realpolitik*: organizers suspected that North American colleagues would seek to determine the parameters of such scientific exchanges, if not set their very agendas.

It would take a full decade after their initial meeting in Buenos Aires for Latin American scientists to confront their suspicions and agree to relax previous strictures on inviting North American colleagues to their gatherings. The 1908 meeting, slated to be held in Chile, the same meeting for which Secretary Root had so carefully screened candidates, would be the first to include invited scientists from the United States. The impact of this First Pan American Scientific Congress cannot be underestimated; indeed, future relations among scientific interests in the Americas would be dependent on the success of the venture.

Notes

1. These remarks are contained in Report of the Delegates of the United States to the Pan-American Scientific Congress held at Santiago, Chile, December 25, 1908, to January 4, 1909, (Washington:1909), pp. 7–8. Secretary of State Root was a leading proponent of the Pan American Union. See Castañeda, "The Carnegie…," p. 41.
2. Osborn, by training a paleontologist, had convinced the AMNH of the necessity of an expedition to Patagonia to either prove or disprove Argentine theories that the southernmost part of that country was the "center of vertebrate evolution and distribution." See Rainger, An Agenda…, pp. 92–93.
3. The term "chronic wrongdoing" was used by President Theodore Roosevelt in an address to Congress on December 6, 1904. This has come to be known as the Roosevelt Corollary to the Monroe Doctrine and it served as the justification for intervention in Haiti, the Dominican Republic and Nicaragua. The Latin American response was to label such intervention as "dollar" or "gunboat" diplomacy.

4. Neil Smith, American Empire: Roosevelt's Geographer and the Prelude to Globalization. (Berkeley, University of California Press, 2003), p. 54.
5. I. Bowman to John Clark, Division of Economics and History, The Carnegie Institution, American Geographical Society Archives, June 2, 1915. To remedy this lacuna, Bowman published a geography textbook in 1915 entitled South America: A Geography Reader. The same year, the sociologist Edward A Ross produced South of Panama which considered the issues of race and labor, among other subjects. These two contributions are analyzed in Ricardo D. Salvatore, Disciplinary Conquest: U.S. Scholars in South America, 1900–1945. (Durham, Duke University Press, 2016, chapters 7 and 8).
6. See Robert M. Levine, Images of History: Nineteenth and Early Twentieth Century Latin American Photographs as Documents. (Durham, Duke University Press, 1989) or Wendy Watriss and Lois Parkinson Zamora, Images and Memory, Photography from Latin America. (Austin, University of Texas Press, 1998).
7. Two recent examples include Thomas O'Brien, The Century of U.S. Capitalism in Latin America (Albuquerque, University of New Mexico Press, 1999) and César J. Ayala, American Sugar Kingdom: The Plantation Economy of the Spanish Caribbean, 1898–1934 (Chapel Hill, University of North Carolina Press, 1999).
8. See Report of the Delegates of the United States to the Pan-American Scientific Congress, Santiago, Chile, December 25, 1908, to January 4, 1909, (Washington:1909), 8.
9. This is one of the major themes of Antonio Barrero-Osorio, Experiencing Nature: The Spanish American Empire and the Early Scientific Revolution. (Austin, University of Texas Press, 2006). Recently Hal Langfur, Adrift on an Inland Sea: Misinformation and the Limits of Empire in the Brazilian Backlands (Stanford University Press, 2023) has suggested that even scientific expeditions could be manipulated, especially where royal patronage was involved. Cf. p. 298.
10. Juan José Saldaña, "Science and Public Happiness during the Latin American Enlightenment", p. 87, in Juan José Saldaña, ed. Science in Latin America: A History (Austin, University of Texas Press, 2006).
11. See Harry Bernstein, "Some Interamerican Aspects of the Enlightenment" in Arthur P. Whitaker, ed., Latin America and the Enlightenment, 2nd ed. (Ithaca, Cornell University Press, New York, 1961), p. 57. Recently some historians have begun to question the direct connection between writings of eighteenth century Latin American enlightenment figures and the events of the Independence period, yet the influence of such thinkers on the political realm has not been completely dismissed. See Sophie Brockmann, The Science of Useful Nature in Central America: Landscapes, Networks and Practical Enlightenment, 1784–1838 (Cambridge, Cambridge University Press, 2020).
12. Bernstein, "Some Interamerican…", p. 59.
13. *Ibid.* pp. 57–58.
14. *Ibid.* p. 65.
15. In the words first used by the Nicaraguan poet Rubén Darío in his poem "Á Roosevelt" ("Ode to Roosevelt"), published in 1904, the United States had become the "Colossus of the North".
16. Such welding together of agribusiness and science was also evident in global perspective, leading one researcher to coin the phrase "plantation science." See Prakash Kumar, "Plantation science: improving natural indigo in colonial India, 1860–1913," British Journal of the History of Science, Vol. 40 (December 2007), pp. 537–565.
17. On this point see Regina Horta Duarte, "Pássaros e Cientistas no Brasil: Em busca de proteção, 1894–1938" Latin American Research Review, Vol. 41, No. 1, February 2006, p. 7. According to her, in Rio de Janeiro, a single ranch exported annually 20,000 hummingbird (*beija-flor*) skins to France. See also her "Birds and Scientists in Brazil: In Search of Protection, 1894–1938," in Martha Few and Zeb Tortorici, Centering Animals in Latin American History (Durham, Duke University Press, 2013), pp. 270–301 in which she expands the theme of nationalism and the study of birds. Another study which considers the feather trade is Scott Weidensaul, Of a Feather: A Brief History of American Birding, (New York, 2007). On another curious note, the self-proclaimed "feather king," Isidor Cohnfeld,

located his feather factory in New York City. According to the New York Herald, it was "the finest feather factory in the world." See Christopher Gray, "The 'Feather King': A NoHo Prequel," New York Times, Sunday, May 27, 2012.

18. Duarte, "Pássaros…", p. 4. A society for the protection of Animals and Plants had been formed in Cuba in 1882, while in Argentina an Ornithological Association of La Plata was begun in 1916. Cf. Duarte, "Pássaros…,", p. 8.
19. See Robert D. Aguirre, Informal Empire: Mexico and Central America in Victorian Culture, (Minneapolis, 2005), p. xvi. Anne Secord's "Corresponding Interests: Artisans and Gentlemen in Nineteenth Century Natural History," The British Journal for the History of Science, Vol. 27, No. 4, Dec., 1994, pp. 383–408, points out how many nineteenth century collectors in Britain bought their specimens from natural history agents in London and elsewhere.
20. James Brown Scott, Second Pan American Scientific Congress: Final Act and Interpretive Commentary Thereon. (Washington, 1916), p. 43.

Chapter 2
Pan Americanism, Leo S. Rowe and the Second Pan American Scientific Congress

The use of the term "Pan American" as a unifying construct for the broad diversity which is Latin America was hardly original when it was appropriated in 1908 by the First Pan American Scientific Congress. On the contrary, the historic roots of the concept date as far back as the 1820s, when the term "Pan American" referred to an ephemeral, but nonetheless hoped-for, political consensus among American nations. While it is true that scientists in the eighteenth century had manifested a precocious unity in their support for enlightened political ideas, by the 1820s (i.e., after most of the Latin American nations had declared their independence from Spain and Portugal), these scholars had been marginalized in the political discourse. As a result, most historians credit Simón Bolívar, renowned as the "Liberator of South America," with developing the broad outline of Pan Americanism as a political concept, rather than the scientific community.

Buoyed by the personal enthusiasm he had generated from assisting the independence movement in Latin America (only Cuba, Puerto Rico, and for a brief period, the Dominican Republic, remained part of the Spanish Empire after 1824), Bolívar called for the convening of an inter-hemispheric congress in 1826, during which representatives from across the Americas would meet as one body, to determine their future course of action. His intention was to host representatives from the newly independent American nations (including the United States) at a mid-way point in the hemisphere; he chose for this venue the province of Panama, then part of the nation of Colombia. Once convened in this locale, he anticipated that the attendees would develop a unified, Pan American political stance, which could subsequently be presented as a *fait accompli* to the outside world.

Bolívar's dream of an inclusive congress, however, remained just that; only a handful of nations, including Mexico, the Central America Confederation, Colombia and Peru, actually sent delegates to the meeting in Panama. Travel difficulties, mutual suspicions and the financial uncertainties experienced by the newly independent nations affected overall attendance. Although the United States agreed to participate in the meeting, its delegates were unfortunately delayed and

R. M. Delson, *Scientists or Spies?*, The Latin American Studies Book Series,
https://doi.org/10.1007/978-3-032-18117-6_2

they missed the event altogether. For the few attendees who actually arrived in Panama, discussions about future self-assistance failed to result in a mutual understanding of how such policy might be implemented. Notwithstanding these disappointing results, the most enduring aspect of this first meeting was that the concept of "Pan Americanism" had been coined and formally accepted as a viable goal by all the attendees.[1]

Over the course of the nineteenth century, three other (non-scientific) Pan American congresses were convened (in Lima, Peru in 1847; Santiago, Chile in 1856; and again in Lima in 1864). The objective of each conference was to acknowledge the mutual political goals of the attending nations, and to advance a unified position against European invasion; yet, in each instance, the creation of a broad, well-defined Pan American movement faltered. This failure was at least partly attributable to the United States, which declined to attend meetings held after 1826 primarily because it had already unilaterally declared itself the sole guarantor of hemispheric protection against European aggression with President Monroe's December 2, 1823 presentation of the Monroe Doctrine to the United States Congress.

The conferences which followed the initial "Pan American" meetings were primarily geared toward codifying international law as it applied to the Americas, but collectively did little to advance hemispheric consensus. Held in Lima (1877) and in Montevideo (1888 and 1889),[2] these meetings were avoided by the United States, which continued to reference the Monroe Doctrine as the unique legal instrument via which Washington preserved for itself the unilateral right to protect the Americas from outside interference.

It must have come as a huge surprise to the Latin American political community, therefore, when the United States unexpectedly announced plans to host its own "Pan American" congress at the end of the nineteenth century. Instead of choosing to meet in a Latin American country, the venue for this meeting would be Washington, D.C.; its organizers proposed an emphasis on commercial ties which bound the Americas together, rather than political (or scientific) objectives. Although remaining suspicious of the motives of the United States, seventeen nations did agree to send delegates to the October, 1889 convocation, where they were lavishly entertained at the expense of U.S. taxpayers. Before the meeting began, these delegates had been treated to a lengthy, six week train journey through the United States which was meant to impress upon them the breadth of the prospering U.S. economy. When the attendees finally got down to business in Washington, some actual grievances were discussed, including the ongoing Tacna-Arica land ownership dispute between Chile and Peru. But the overarching theme of the congress, to foster some vague notion of inter-American cooperation, really applied only to commercial interactions. This meeting was colloquially referred to as the "Pan American Conference."[3]

While short on recognizable achievements, the Conference could, however, claim to have laid the "groundwork for future agreement on copyrights and trademarks, improved transportation in the Americas" and the creation of a Pan American commercial bureau which would be under the direct supervision of the United States State Department during its initial ten years.[4] Initially ignored by both Latin American nations and the United States, the Bureau somehow managed to survive its precarious beginning, eventually emerging into what became known as the Pan American Union; its successor is today's Organization of American States.[5]

Five subsequent Pan American meetings were convened, beginning with a meeting in Mexico City in 1901, and ending with a gathering in Havana[6] in 1928. The emphasis in each of these meetings continued to be on trade; little was actually achieved in the quest for Pan American political equality, a difficult proposition in light of the quasi-imperialistic course of action in the region pursued by the United States after the 1898 Spanish American War. The takeover of Puerto Rico as a colony in that year plus repeated interventions in Haiti, Nicaragua, Mexico, the Dominican Republic and Cuba before the Havana meeting did little to persuade wary Latin American observers that a "Pan Americanism" of political equality did or even could exist.

The situation would remain essentially unaltered until Franklin Delano Roosevelt became President of the United States. Breaking with previous rhetoric, Roosevelt announced a "Good Neighbor" policy during an April 12, 1933 speech which he delivered at the newly constructed and elegant Pan American Union building in Washington, D.C. He defined this new initiative by suggesting that "[T]he essential qualities of a true Panamericanism must be the same as those which constitute a good neighbor, namely, mutual understanding and, through such understanding, a sympathetic appreciation of the other's point of view." The President then called for an end to armed intervention by the United States in Latin America, and the beginning of a new era of mutual cooperation.[7]

But Roosevelt's vision was far in the future... Simply put, in the period prior to World War I, notwithstanding the Washington meeting and the creation of the Pan American Union, strong suspicion of the United States continued unabated among the Latin American nations.

The Role of Leo Rowe

Attempts by Washington politicians and commercial interests to allay the misgivings of Latin Americans, as well as to acquaint U.S. citizens with their southern neighbors, centered on the staging of what were referred to as "Pan American" themed exhibitions, huge events held on fairgrounds in the United States. These large, showy fairs were in essence commercial expositions, the purpose of which was to provide an introduction to the diversity of the Western Hemisphere to a largely uninformed U.S. audience; once again, the emphasis was on business connections rather than political equality, or scientific exchange. The first of these fairs, the World's Industrial and Cotton Centennial Exposition, opened in New Orleans in 1884, enchanting visitors with its recreations of Mexican and Central American style gardens. Nonetheless, the real focus of the effort was to highlight the role of New Orleans in transporting goods within the hemisphere.

Succeeding this exposition were the North, Central, and South American Exposition of 1885–1886, located at the same New Orleans Fairgrounds, and the World's Columbian Exposition of 1893 held in Chicago. Finally, the northern city of Buffalo, New York, an industrial powerhouse due to the enormous amount of electrical energy generated by the Niagara Falls, orchestrated a well-attended spectacle. This 1901 exposition, however, had an equally minimal impact on improving North–South relations in the hemisphere.[8]

For the entirety of the nineteenth century therefore, and into the first decades of the twentieth, the notion of a workable Pan Americanism, one in which all the nations were deemed equal partners, remained a mutable concept. Two scholars from the United States attempted to rectify this situation by believing in, and promoting a vision of, what inter-hemispheric cooperation might look like. One of these individuals was John Barrett, an early head of the Pan American Union, which was headquartered in Washington. Barrett argued for what he termed "Practical Pan Americanism," a vaguely-defined term by which he meant that there should be numerous exchanges on a variety of topics within the Americas.[9] His efforts resulted in, among other things, the publication of the Pan American Bulletin, a journal which featured a compilation of articles on diverse subjects running the gamut from economics to archaeology. To its (and Barrett's) credit the Bulletin was published in Portuguese and Spanish as well as English. But other than this publication, Barrett's impact on interhemispheric relations was limited and, as noted in the criticism of his work in Chap. 1, he was not widely regarded as a major player.

The other U.S. scholar whose work was closely associated with the notion of Pan Americanism approached the concept from an academic perspective, envisioning relations between nations as an exercise largely rooted in intellectual exchanges between scholars. Well known from the joint positions he held as Professor of Political Science at the University of Pennsylvania and President of the American Academy of Political and Social Science (1902–1930), Leo Stanton Rowe (Fig. 2.1) commanded great respect in academe. Although technically not a student of Latin America, his expertise on governmental development led President McKinley to appoint him to the commission charged with revising the laws of the newly acquired Puerto Rican colony in 1900. Rowe viewed this assignment through the lens of what he, himself, would later acknowledge was that of a "progressive imperialist."[10] Convinced on one level of the need to understand local cultures, Rowe was nonetheless unequivocal in his desire to "Americanize" the Puerto Rican people.[11]

Yet, just a few years later, in 1906, during his first visit to South America, Rowe's thinking underwent a powerful transformation. As he traveled through Argentina, Brazil, Bolivia, and Chile, Rowe observed what he considered "progress" in urban projects, education and local governments.[12] Such progressive style governments, he opined, would inevitably lead to the flourishing of democracy.[13] Given his distinct approval of their nations (one of the few North Americans to voice such opinions), Rowe was widely celebrated during his visit. More importantly, after the decision had been reached to invite North Americans to the 1908 "First Pan American Scientific

Fig. 2.1 Left, Leo S. Rowe, 1917, Bain News Service; right, Robert Lansing, 1917, Harris & Ewing, photographers; both downloaded from Wikimedia Commons (public domain)

Congress" in Santiago, Chile, Rowe was an obvious first choice to attend the meeting as a formal United States delegate.

Rowe's ideas not only swayed Latin American scientists, they provided a blueprint for how all Americans might work together. He called upon Washington to recognize his unique approach to Pan Americanism, which emphasized "moral and intellectual ties,"[14] over commercial ones. Writing in the influential journal The North American Review in April, 1909, Rowe convincingly argued that "[C]lose and organic relations must be established between the universities of North and South America, between scientific organizations…with each recurring conference, unity of action is extended over a larger number of subjects, and over subjects of increasing importance."[15] His warning over the ever-increasing number of scientists Germany was supplying to South America proved a prescient harbinger of the coming decades, when the need for American scientists from the North and the South to cooperate with each other as a counter to German influence was suddenly, and unavoidably, obvious.[16]

Was Rowe single-handedly responsible for the change in attitude of Latin American scientists toward their North American brethren? Probably not. But against a backdrop of international meetings which had purposely excluded the participation of their United States colleagues, Rowe's vision of a broad Pan American brotherhood of intellectuals was not just widely embraced by Latin American scientists, but adopted as the *leitmotif* of their future meetings. The tentative "thaw" in relations, which began in the context of the 1905 Rio gathering where the decision to invite U.S. participants to attend future scientific meetings was first announced, gained speed at the 1908 gathering in Chile. Despite no official explanation being offered for what essentially constituted a 180° reversal in attitude, Secretary Root's small band of scientists from the United States, which included Leo Stanton Rowe, was warmly welcomed.

Against a backdrop of the political antagonism which Latin Americans bore toward the United States in the first decade of the twentieth century, the Chilean scientific meeting was a truly remarkable accomplishment; the presence of the universally admired Leo Rowe at the congress likely contributed to its success. Not only did U.S. delegates attend in person, but several papers by U.S. colleagues presented *in absentia* were equally well received. This meeting would henceforth be referred to as the "First Pan American Scientific Congress", a linguistic *tour de force* which conveniently avoided having to explain the purposeful absence of United States scientists at earlier convocations.

As groundbreaking as these changes in attitude were, the Santiago meeting further departed from the past when its membership unexpectedly voted to hold their next international gathering in Washington, D.C. rather than in Lima, Peru, the venue which had initially been chosen. Considering the decades-long suspicions of the United States held among scientists, as well as the relative failures of numerous parallel, contemporary "Pan American initiatives" (such as the commercial fairs) to create a true Pan American spirit, this decision represented an extraordinary departure from the past. It was likely that Rowe also played a major role in this outcome, although the published notes of this meeting do not provide any explanations for the substitution of Washington over Lima as the meeting site. Fittingly, while he was still in Santiago, Rowe was chosen to become a member of the Executive Committee of the group and thus given a significant role in organizing the forthcoming Washington meeting.[17]

The outbreak of World War I unavoidably delayed the timing of the Second Pan American Scientific Congress, which was not held until just after Christmas, 1915. Nonetheless, the meeting proceeded with panache and was universally acknowledged a success. Delegates who attended the Washington event could allocate their time between attending a broad program of academic sessions during the daytime and social events in the evening; these offerings lasted from post-Christmas through the first week of January, 1916. Most significantly, the need to foster Pan American scientific exchange became the underlying theme of the Congress which, in turn, encouraged a palpable spirit of camaraderie among the scientist-attendees. This gathering, therefore, merits closer scrutiny for what it reveals about Pan Americanism in the scientific and intellectual sense, as well as its emphasis on the changing nature of United States-Latin American scientific relations. Coincidentally, the Washington meeting also marked the "official" debut of the AMNH into the Latin American scientific world, notwithstanding the substantial fieldwork the Museum had already conducted in the region for more than two decades.

The Second Pan American Scientific Congress and the AMNH

When the Second Pan American Scientific Congress (SPASC) finally convened in Washington, D.C., the event was far more elaborate than any of the previous scientific meetings held in Latin America. The Washington venue attracted hundreds of delegates, many of whom arrived with their families in tow, both from Latin America and across the United States. With many social occasions and dinners scheduled each evening, attendees were kept busy after the day's formal sessions had concluded. The purpose of these soirees was to encourage the intermingling of representatives, a goal which apparently was achieved with considerable enthusiasm.

The Second Pan American Scientific Congress was held concurrently with the 19th Congress of Americanists[18] (although the latter met separately at the Smithsonian Institution, also in the nation's capital). Meetings at both locations provided opportunities for leading U.S. scientific institutions to highlight and share the results of their ongoing Latin American investigations with their Latin American counterparts. It was anticipated that most of the major universities in the United States would send representatives to the SPASC, but it was the prominent scientific institutions, including the Carnegie Institution, the Smithsonian, and the AMNH, which dominated the science congress by the sheer numbers of their attendees. A look at the list of distinguished attendees from the American Museum of Natural History underscores the importance attached to this meeting by its president, Henry Fairfield Osborn; among the eight scientists chosen to represent the institution were Clark Wissler, the Curator of Anthropology, and Herbert J. Spinden, from the same department; Henry E. Crampton, Curator of Invertebrate Zoology at the AMNH and Professor of Zoology at Columbia University; and Frank M. Chapman, the Curator of Ornithology. The AMNH sent (and paid the expenses of)[19] its top men, clearly thrilled at the opportunity to stake out a real presence at this gathering and further its agenda for Latin America.

From the outset, it was obvious that the SPASC meeting would be as political as it was scientific. Even before the meeting details had been finalized, Secretary of State Robert Lansing (Fig. 2.1) shared his assessment with Osborn that "[T]his great Congress, although Scientific in name, comprehends many of the principal branches of human activity, including such interesting topics as commerce, finance, transportation, public health and sanitation, mining and metallurgy, international law, engineering, education, conservation, etc.; and indications now point to a greater attendance of representative Latin-Americans than have ever before participated in a Pan-American meeting."[20] Lansing urged Osborn to host some of the more important Congress delegates at the New York museum at the meeting's conclusion, a request with which the Museum president eagerly complied. On his own, Osborn had already decided to invite the most prominent Latin American scientists to his city to not only view the Museum's extensive collections of artifacts and specimens from the region, but to coordinate with them afterwards on future research projects in their respective countries. The invitations sent to these attendees would serve a two-fold purpose: they would reduce the fear that Osborn had of other institutions (such as the Smithsonian) dominating future collaborations with the Latin American scientific community by making a priori arrangements with visiting representatives while they were in New York; and they served as proof of the Museum's eagerness to broaden an already important relationship which it had been nurturing with the State Department. In responding to Lansing's office on Osborn's behalf, George H. Sherwood, the Museum Secretary, succinctly laid out the importance the Museum attached to hosting the congress members. Setting the date of January 13, 1916, for hosting the New York event, Sherwood thought that such invitees would be "personally interested in the work of the American Museum of Natural History…since the Museum for several years has been carrying out important scientific explorations in South America."[21]

Just as Lansing had predicted to Osborn, the papers presented at the Washington meeting embraced a wide spectrum of topics ranging from natural history to finance. There was much fanfare for the renewed spirit of scientific cooperation (in the mold of Pan Americanism) which informed the Congress, as well as an unspoken agreement by Latin American scientists to accept their United States counterparts as partners. Particularly noteworthy were the sessions held in conjunction with Section III of the meeting ("Conservation of Natural Resources"), in which many of the grievances expressed by Latin American scientists in the late nineteenth century were openly addressed. This section stressed the need for preservation of Latin American natural resources, including petroleum; viewed against the reality of ongoing United States-Latin American relations, the papers presented in this section were extraordinarily high-minded, idealistic and perhaps, in retrospect, a bit naïve. In a proposal intended to bolster local economies, for example, one delegate boldly (and precociously) advocated for a transformation of Latin American land distribution, a goal which he proclaimed achievable by breaking up plantation holdings which would subsequently be reorganized into small-scale agricultural production units.[22] His proposal not only ignored the blatant reality that much of North America's investment capital in Latin America at this time was tied up in large-scale agribusinesses, but the painful truth that almost everywhere in Latin America, beginning in the nineteenth century, there had been a tendency toward land consolidation under single proprietors, rather than land reform through the creation of a viable peasantry.

But this was not the only progressive goal proposed at the Congress. The meeting's "Final Act" included a position paper advocating curricular reform in the United States which, had it been implemented, would have resulted in students having a requirement to study Spanish across the country. Conversely, it was suggested that Latin American educators be obliged to instruct their students about the history and customs of the United States. Resolutions concerning the environment and government protection of patrimony (the subject of considerable contention earlier) were also touted as a lofty ideal to which all could aspire.[23]

The State Department's interest in such high-minded objectives was limited. While the goal of the Congress was to create a transcendent "scientific Pan Americanism," the U.S. government viewed the Congress as an ideal opportunity to advance some very specific political, rather than scientific, objectives. Ostensibly,

Lansing's aim was to erase the mutual suspicions which had long existed between the Latin American and the U.S. scientific communities; that much was evident in his appeal to Osborn to host delegates after the meeting had concluded. But, beyond this, the State Department also hoped to parlay its involvement in the meeting into a means of confronting the adverse affects the ongoing European War was having on Latin American-United States relations.

The Second Pan American Scientific Congress and the War

By the time the Second Pan American Congress met in December, 1915, the on-going fighting in Europe had already seriously impacted shipping across the Atlantic. The fact that the United States was not yet an active participant in the European conflict, as well as the disruption of traditional trade routes across the Atlantic, represented an opportunity for U.S. investors to make substantial inroads into previously closed Latin American markets at the expense of German and even British interests.[24] Wartime threats to existing oceanic shipping, the result of sudden, and unprovoked, German U-boat attacks, gave southern bound trade routes, which sailed safely within the Americas, a long sought-after but heretofore unattainable advantage.

At the same time Washington was growing increasingly alarmed at the encroachment of German businesses into the Latin American region. Commercial outposts owned by German companies were viewed as proof positive of the Reich's intent to prevail in the Americas, a position optimistically predicated upon Germany's "eventual" wartime victory. It was becoming clear that both the United States and Germany had similar plans to dominate trade in the Americas once the War ended. If the commercial ambitions of the United States were to predominate, it would be necessary to first convince most of the Latin American nations to drop their positions of neutrality vis à vis the European War and embrace the Allied side. Accomplishing this goal would require reliable data on German activity in the hemisphere, information which up to this point was distressingly hard to obtain.

It was with this broader aim of acquiring intelligence that the State Department viewed the meeting of the Second Pan American Scientific Congress as a venue for ferreting out and recruiting promising U.S. academics, individuals who did their research in Latin America, and who might be persuaded to pass on what they observed about pro-German sentiments in Latin America to Washington. Scientists (including representatives from the AMNH) who had previously spent substantial time in the Latin American "field," and thus were intimately familiar with the territory, language and customs of the region, were considered prime reconnaissance sources. Many of these scientists were also fluent in German, the scientific language of the day, and/or they had degrees from German universities, which made observing pro-German activities even easier. Additionally, many of the natural scientists who collected specimens of living species in Latin America were also good shots with a rifle and conveniently traveled with guns for their own protection as well as to bring down those specimens; it was thought that they would know how to defend themselves. Holding these quintessential qualifications, attendees at the Second Pan American Scientific Congress, who were there strictly for academic reasons, were now reimagined as a potential pool of candidates from whom Washington politicians might recruit knowledgeable and willing informants.

The State Department objectives at the Congress therefore purposely melded academic/scientific interests with politics. In this context it is instructive to parse the real meaning of the address delivered by F. W. Hodge (the delegate from the Smithsonian Institution's Bureau of American Ethnology) when he spoke at the SPASC on "What the United States Government has done for the Science of Anthropology."[25] The ethnologist began his remarks by lauding the Smithsonian for its interest in ethnographic and statistical records and noting that a substantial portion of current anthropological field research had already been supported by several allied governmental departments, including the U.S. Geological and Geographical Survey of the Territories, the Bureau of the Census, and the Army Medical Museum. He seems to have been subtly suggesting that the *quid pro quo* which had been ongoing at that august government organization might serve as a role model for future government collaborations with other institutions.

In truth, Hodge was simply acknowledging the obvious: by late 1915, coeval with the Congress, the government was poised to become even further embedded into the specific scientific community which earned its living by focusing on research in Latin America. The Second Pan American Scientific Congress thus provided the State Department with a hunting ground for enticing such talent to work in tandem with the government.

Was Germany a real enough threat to the Americas to warrant this recruitment? While some recent scholars have cast doubts on the success of (or even desire for) the expansion of German interests in Latin America during World War I,[26] Jamie Bisher's 2016 study of German intelligence activities in the region renders such conclusions moot. Bisher confirms that the German mission in Latin America had three, distinct components: to advance pro-German propaganda, to conduct "sabotage against war material manufacturers and shippers," and to acquire as many neutral passports as they could through their consuls or agents.[27] In short, Washington's fears of German activities in the hemisphere were not unfounded; the growing attention which Germany paid to the protracted Mexican Revolution only served to increase that apprehension. Additionally, areas of Latin America which had large concentrations of German immigrants (for example, Chile and Brazil) had begun to voice decidedly worrisome pro-German sentiments as the war was being prosecuted in Europe.

The State Department's assessment that the Second Pan American Scientific Congress might prove valuable for locating intelligence agents was prescient. One need look no further for confirmation of this appraisal than to the recruitment of

Richard T. Ely, professor of Political Economy at the University of Wisconsin. At the SPASC, Ely delivered the keynote address for Section III on the Conservation of Natural Resources[28]; within a few months of the conclusion of the Congress, he was appointed Assistant Director of the Intelligence Division of the War Trade Board, a body which oversaw commerce and the activity of foreign firms globally, but which emphasized trade in Latin America. Nor would it be a coincidence that a number of other attendees at the Washington meeting (including Spinden, Crampton and Chapman of the AMNH) would be approached by U.S. agencies in the near future to gather intelligence and report back on the very nations in which they conducted their scientific research.

For the United States Government, and the State Department in particular, the meeting of some of the best scientific minds in the hemisphere which took place at the Second Pan American Scientific Congress was critical for both explicit and hidden reasons[29]: for public consumption, the meeting aimed to reverse years of suspicion toward scientists in the United States, heralding a return to the heady days of the post-independence period, with its vibrant North/South scientific cooperation. In this regard, it is no surprise that the optimistic conclusions of the "Final Act" of the Congress read like a gushy paean to Leo S. Rowe's published work on "intellectual Pan Americanism."

But, in service to a more hidden agenda, the State Department also envisioned the Congress as a fertile nursery in which to sow the seeds of a future wartime collaboration between the United States government and scientific academe, especially with regard to Latin America. This carefully cultivated intimacy between scientists and politicians, with its subtle but obvious call to patriotic duty, would affect many academic and science institutions in the United States, beginning in 1916, and especially after Washington's declaration of war against Germany, in April, 1917. But not all institutions would respond to the U.S. government's request with the alacrity and enthusiasm of the American Museum of Natural History in New York. Just why and how the Museum (which began ramping up its research agenda in Latin America in the 1890s) responded this way, and the specific role of AMNH President Henry Fairfield Osborn in recognizing such political imperatives, are the subjects of the remainder of this book.

Notes

1. See Frederico G. Gil, Latin American-United States Relations (New York, Harcourt, Brace, Jovanovich, 1971), pp. 145–150, *et passim.*
2. *Ibid.*, p. 146.
3. Joseph Smith, "The First Conference of American States (1889–1890) and the Early Pan American Policy of the United States," in David Seining, ed., Beyond the Ideal: Pan Americanism in Inter-American Affairs (Westport, Conn., Praeger, 2000), pp. 24–26.
4. *Ibid.*, p. 26, *et passim.*
5. Gil, Latin American…, p. 147.
6. *Ibid.*, pp. 148–152, *et passim.*
7. *Ibid.*, p. 156.

8. See Robert Alexander Gonzales, Designing Pan-America: U.S. Architectural Visions for the Western Hemisphere (Austin, University of Texas Press, 2011), pp. 19–65 *et passim* for discussions and plans of these fairs.
9. John Barrett's ideas centered on the United States and its opportunities in Latin America. While he was interested in education, he was not focused on intellectual exchange. See John Barrett, "Practical Pan-Americanism," North American Review, September, 1915, pp. 1–11.
10. Ricardo D. Salvatore, Disciplinary Conquest: U.S. Scholars in South America, 1900–1945, (Durham, Duke University Press, 2016), p. 137.
11. *Ibid.*, p. 141.
12. *Ibid.*, p. 142.
13. *Ibid.*, p. 143.
14. *Ibid.*, p. 144.
15. Leo Stanton Rowe, "Our Interest in a United America," North American Review, April, 1909, p. 593.
16. *Ibid.*, p. 592.
17. His appointment is commented upon in James Brown Scott, Second Pan American Scientific Congress: The Final Act and Interpretative Commentary Thereon, (Washington: 1916), p. 45, a volume which assesses the history and scope of the entire 1915 Second Pan American Scientific Congress in Washington, D.C.
18. The XIX International Congress of Americanists met from December 27–31, 1915 at the Smithsonian. Unlike the Second Pan American Scientific Congress, and the ones which had preceded it, the "Americanists" was an international association which had begun in Europe in 1875. Their discussions included history, economy and anthropology and were far broader than strictly scientific interests. Nevertheless, the American Museum of Natural History, deeply involved in Latin American research, hosted the first meeting of this group in the United States, in 1902. Cf. Chap. 3.
19. Both Chapman and Spinden were fully funded. Director Lucas to Herbert Spinden December 23, 1915 and Director to Chapman, December 23, 1915. CA, File 632, AMNH.
20. Robert Lansing to H. Fairfield Osborn, Washington, November 23, 1915. CA, File 632, AMNH.
21. As early as July 1915, Osborn was intent on inviting delegates to New York so that "we should establish most friendly relations with the South American Republics." Osborn to Director Lucas, July 7, 1915. CA File 632, AMNH. The reception for the delegates was held on January 13, 1916. Secretary Sherwood to William Phillips, Third Assistant Secretary of State, confirmed the date, December 31, 1915. CA File 632, AMNH. See also account in *Notes* in the American Museum Journal, Vol. XVI, No. 1, (January, 1916), p. 76.
22. Proceedings of the Second Pan American Scientific Congress, Washington, U.S.A., Section III, Conservation of Natural Resources, (Washington, 1917).
23. Scott, Second Pan American...
24. David M. Kennedy, Over Here: The First World War and American Society (Oxford, 1980), p. 302.
25. F. W. Hodge, "What the United States Government has done for the Science of Anthropology" in Proceedings of the Second Pan American Scientific Congress, Washington D.C. Dec. 27, 1915 to Jan. 8, 1915, Section I Anthropology, (Washington, 1917), pp. 168–174. For a recent view of how a handful of early anthropologists (the "action anthropologists"), including Sol Tax, became involved in helping indigenous communities see Ira Bashkow, "There's more to anthropology's past than most of us know," American Anthropologist, Vol. 125, pp. 177–180, 2023.
26. For example, the work of Nancy Mitchell, The Danger of Dreams: German and American Imperialism in Latin America (Chapel Hill, University of North Carolina Press, 1999). See Chap. 3 for further discussion of German interests in Latin America.

27. Bisher, The Intelligence War…, p. 31.
28. Richard T. Ely, "Conservation and Economic Theory," in Proceedings of the Second Pan American Scientific Congress…, Section III, pp. 10–21.
29. Recent historiography has criticized this congress for not being sufficiently "scientific." See Rodrigo Fernos, Science Still Born: The Rise and Impact of the Pan American Scientific Congresses, 1898–1916 (New York, iUniverse, 2003), p. 23. I would argue that this was precisely the goal of U.S. participation in the Congress, i.e., to develop political ties with Latin America, in conjunction with scientific ones.

Chapter 3
Under New Leadership: Henry Fairfield Osborn and the AMNH in Latin America

Secretary of State Elihu Root had omitted AMNH scientists from the list of candidates for the U.S. delegation to the 1908 First Pan American Scientific Congress. Was this a reflection of his lack of knowledge regarding the Museum's previous research activities in Latin America? Or had it occurred because the institution was far better known for its fieldwork in the western regions of North America and the Congo, among other places? Regardless of his reasoning, the truth was that the Museum had been funding serious collecting expeditions to Latin America from as early as the 1890s, more than a decade before Secretary Root would convene his chosen representatives.

Interest in Latin America had spread rapidly across the entire American Museum of Natural History community at the turn of the century. In the twenty-year interval between the conclusion of the Spanish-American War in 1898 and the 1919 Versailles peace settlement following World War I, with the blessing of senior administrators including President Henry Fairfield Osborn, ever increasing numbers of AMNH scientists fanned out to all corners of Latin America, a region valued for its unusual and (at least to U.S. audiences) relatively unknown flora and fauna. The AMNH researchers who flocked to the region represented the broad spectrum of natural history disciplines including paleontology, geology, ichthyology, mammalogy and ornithology as well as archaeology and anthropology. In the course of their research, these scientists traversed the length and breadth of Latin America, from Mexico to Chile and through the West Indies.

The impressive number of AMNH-sponsored ornithological expeditions to Latin America in this time frame attests to the Museum's powerful commitment to fieldwork in this region. Between 1901 and 1915, twenty-one separate avian-collecting parties were dispatched by the Museum to Panama, Colombia, Ecuador, British Guiana, Venezuela, Bolivia, Argentina, Mexico and Brazil, averaging out to more than one major collection trip per year![1] Additionally, in 1912, the noted Californian bird collector Rollo H. Beck left the States for South America where he assumed the leadership of the "Brewster-Stanford" expedition, a

R. M. Delson, *Scientists or Spies?*, The Latin American Studies Book Series,
https://doi.org/10.1007/978-3-032-18117-6_3

multi-year, privately-sponsored project to collect sea birds from Peru, Chile, the Falkland Islands, Argentina and Brazil. At first only nominally associated with the Museum, within several months of Beck's arrival in South America, the parameters of his assignment had expanded to include amassing the definitive collection of South American pelagic (oceanic) birds in the United States for the New York institution.[2] The twin objectives of the AMNH were to display representative specimens of these heretofore unknown and/or unusual species to the public in new exhibitions, while staff scientists simultaneously undertook studies of the larger group of specimens which the expedition had collected.

Although Museum scientists did not abandon fieldwork in other parts of the globe, the early twentieth century emerged as a particularly opportune moment for North Americans to pursue research within Latin America. Given the post-1898 *realpolitik,* sending scientists to conduct research in the region was proving far easier than it had been in the past; e.g., several territories were now under direct control of the United States (Puerto Rico, and the Virgin Islands after 1917) or had U.S. troops stationed in them at various times (Cuba, Haiti, Mexico, the Dominican Republic and Nicaragua) from 1898 to 1918. Accessibility, as well as the powerful lure of the unknown (and potentially exhibitable), drew AMNH scientists to the Caribbean and parts of Central America, where the off and on presence of U.S. Marines sometimes guaranteed safety. They also traveled to South America for collections, where U.S. diplomatic relations were considerably more tenuous.

From the large number of expeditions authorized by the Museum administration, it was obvious that the AMNH meant to distinguish itself from the other North American natural history institutions which were similarly engaged in Latin American fieldwork. Domination of research in this region would be earned by a two-pronged approach: first, the Museum was obliged to hire outstanding scientists whose interests lay chiefly in Latin America, and secondly, the institution had to commit to presenting original collections of heretofore unknown Latin American materials to the public. In fulfilling these objectives, the AMNH would be forced to vie with its sister scientific institutions, including the Smithsonian and Carnegie Institution in Washington, D.C., the Peabody Museum at Harvard, the Field Museum in Chicago and the New York Academy of Sciences, among others, both in the hiring of staff and the advancement of collection opportunities.[3] Like the AMNH, these institutions also saw the great promise represented by territories which had been newly "opened" to the United States after the Spanish-American War. The aim of the AMNH was to accelerate the scope of the institution's Latin American expeditions, which would be geared to either matching the collecting efforts of their competitors or, hopefully, exceeding them.

The AMNH administration met this challenge head on by the turn of the century. Archaeologists whose work was centered on the complex civilizations of pre-Columbian Latin America (Marshall Saville and Herbert Spinden) were invited to join the Museum staff and were subsequently put in charge of new excavations arranged with the Mexican government. Another significant hire was Frederick Ward Putnam, a distinguished anthropologist who brought years of prior experience studying Latin American cultures with him to the Museum. It was

Putnam who organized both the Anthropology Building at the Chicago World's fair of 1893, where exhibits had been staged on Central and South American archaeology and ethnology, and the anthropological component of the new Field Museum in Chicago.[4] The AMNH was successful in persuading Putnam to leave his then current post at the Peabody Museum at Harvard and take a position in New York, where he was appointed Curator of Anthropology (1894–1903; he was allowed to maintain his affiliation with the Cambridge, Massachusetts institution). Arriving in New York, Putnam at once involved the AMNH in several major Latin American expeditions, including missions to Peru, Bolivia and Mexico. In conjunction with the Peabody Museum, Putnam sent AMNH archaeologists to join the Harvard staff at their ongoing excavation at Copán, in Honduras.[5]

It was not all smooth sailing. Looming large as an impediment to any North American natural history museum's plans for collecting in Latin America was Colombia's (and, by extension, the rest of Latin America's) resentment over President Theodore Roosevelt's heavy-handed approach to the region. When Roosevelt encouraged the 1903 breakaway of Colombia's province of Panama, a move which not only resulted in the independence of the region, but also to the building of the Panama Canal under United States control, the former President incurred the enduring wrath of the Colombian government (and others). In attempting to counter Colombian resentment, the AMNH and other North American scientific institutions would surely have to tread cautiously; here the Museum had a distinct advantage in the careful, behind the scenes diplomacy which AMNH ornithologist Frank M. Chapman had perfected. While he collected bird specimens in Colombia (from 1911 onward), Chapman's sympathetic courtship of the Colombians did much to overcome reluctance in Bogotá, and to convince local officials to permit him and his AMNH colleagues to continue the pursuit of bird specimens in their country.

Where careful calculations failed, and in cases where other Latin American nations continued to exhibit leeriness of the United States, the Museum increasingly relied on the State Department to facilitate entrée. With such carefully executed plans, by the turn of the century the AMNH had managed to explore vast expanses of Latin America and would continue their expeditions through the following two decades, including during the critical war years of 1914–1918. Additionally, by cultivating contacts with relevant scientists in host countries, the AMNH carved out a pivotal role for itself in sponsoring ongoing Pan American scientific cooperation, expanding the concept of "scientific Pan Americanism" in ways that Secretary Root could only have dreamed of.

With equally timely astuteness, as early as the 1890s the AMNH administration recognized that the New York public wanted greater and more accurate knowledge of the cultures and biology south of the border. In recognition of how this potentially expanded audience might impact the AMNH, the Museum stretched its resources to cover the cost of multiple lengthy explorations to Latin America. Destinations for research trips now included the Bahamas, where coral reefs were studied; the West Indies, where the focus was on volcanoes; and several nations of South America where invertebrates were collected. The noted Swiss archaeologist

Adolph Bandelier who on his own had previously, but sporadically, sent enticing material from Peru and Bolivia to New York was now, at Putnum's suggestion, directly supported by the Museum. Similarly, the Norwegian anthropologist, Carl Lumholtz, whose ethnographic work on the Huichol Indians of northern Mexico was highly regarded, received funding from the AMNH for the expeditions he conducted during 1894–1897.[6]

On the other hand, funding archaeological and anthropological expeditions to Latin America outright did not mean that the AMNH would reject future donations offered by well-meaning individuals; Museum resources were still far too limited to allow for such prohibitions. Accordingly, the Administration graciously accepted an archaeological collection from the so-called Duke of Loubat who, despite his title, was actually a philanthropist from the United States. A noted collector of ancient Mexican and Central American artistic treasures, Loubat's generous donation to the Museum provided the core of the material displayed in a new Hall of Mexico, a project officially opened to the public in the first decade of the new century.[7]

By the end of the 1890s, however, it became increasingly apparent that if the AMNH was to achieve pre-eminence in its Latin American archaeological collections over other institutions it would need to focus on and display artifacts which had been systematically obtained through the careful excavations of its own staff.[8] In one of its first ventures of this kind, the AMNH signed a cooperative agreement with strongman President Porfírio Díaz of Mexico allowing Museum personnel to excavate at Palenque and Mitla under the direction of staff archaeologist Marshall Saville. Per the terms of this accord, the actual artifacts unearthed at the site would remain in Mexico, while detailed casts would be forwarded to the AMNH to be displayed in the new Hall of Mexico. Saville's finds (albeit copies), along with Loubat's donated collections of authentic artifacts, would now be organized into what quickly emerged as a blockbuster New York exhibition.[9]

From this point on, the pace of Museum involvement in research in the Americas never slackened. The newly defined Latin American orientation of the Museum's agenda was reflected in the extraordinary decision of the International Congress of the Americanists to convene their first ever meeting in the United States at the AMNH in 1902. The choice of the Museum for the meeting's venue over other contenders (like the Smithsonian, for example), was ample recognition of the esteem in which the AMNH was already held for its burgeoning Latin American scientific collections. Building on its growing reputation as a serious center for Latin American natural history studies, Museum staff began to offer free, in-house instruction on Brazil, Chile and Peru (places with which staff scientists were personally familiar) to New York City public school teachers. These staff members also participated in lecture circuits outside the institution itself and were considered an important presence at the Pan American Exposition of 1901, in Buffalo, New York. Evidence from the Museum's archives suggests that all of these events were well attended.[10]

Finally, a truly unique component of the AMNH commitment to introducing all things "Latin American" to the New York public was the 1908 opening of a Mexican-themed restaurant in the Museum's basement. This singular eatery replicated, to scale, the inner courts of the Mitla complex recently excavated by Marshall Saville.

Its decoration included reproductions of furniture and friezes from the Mitla site executed in colors which accurately reflected the site, allowing the restaurant to provide an incomparable educational as well as gustatory experience for any Museum visitor who dined there. The menu of the Mitla Restaurant catered primarily to the U.S. palate, but adventurous New Yorkers could try *tamales* or *chile con carne* while lunching, and afterwards smoke imported cigars (presumably from Cuba). Apparently quite successful, this experiment in culinary diversity soon became a destination café, with a theme park mien worthy of today's Epcot.[11]

Osborn's Interest in Latin America Grows

While interest in Latin America at the Museum grew exponentially across the institution, a reflection of the impact researchers such as Putnam, Saville and Spinden had on their colleagues, credit for framing the region as a scientific "project" belongs to Henry Fairfield Osborn, the founder of the Department of Vertebrate Paleontology (Fig. 3.1). Scion of a wealthy New York family, Osborn (born in 1857) might easily have pursued the business interests of his fellow patrician Princeton graduates, or sought a career in politics, as his long-time friend Theodore Roosevelt had done. Instead, Osborn opted to devote his considerable intellectual talent to science, motivated by a deep interest in understanding the process of evolution (a curiosity said to have been fostered by his mother), as well as an appreciation of the potential for fossil evidence to elucidate the biology of extant species. Osborn created the Vertebrate Paleontology Department at the AMNH in the 1890s and rose through the ranks to become the Museum president in 1908. When he took the administrative reins that year, it was said that he brought a scientist's eye for detail to the post. More broadly, Osborn has been credited with single-handedly "turning the American Museum of Natural History from a dusty Victorian cabinet of curiosities into a leading center of scientific research".[12]

Significantly for the future direction of fieldwork at the AMNH, beginning in the 1890s Osborn publicly committed the Museum to expanding scientific research in Latin America, a choice dictated by his deeply held conviction that only the most thorough understanding of the fossil record of the Southern Hemisphere could unravel the mysteries of biological evolution. Like his contemporaries, Osborn's formative years had been shaped by the writings of Charles Darwin, whose seminal works on evolution were derived from his well-known South American travels. But it was the influence of a later Englishman, Joseph Dalton Hooker, which provided Osborn with the rationale he needed to pursue specifically South American research for answers to evolutionary puzzles.

Hooker's ideas (popular at the time Osborn began his career, and still *au courant* at the turn of the century) were based on the Englishman's belief in the existence of a large, ice-free land mass in Antarctica sometime in the past. He further hypothesized that these same lands had acted as the source of adaptive radiation for all South American animal forms (and African and Antipodean ones, as well), with

Fig. 3.1 Left, Henry Fairfield Osborn, President of the AMNH, in 1915. AMNH Asset ID 19324. Right, Frederic Lucas, Director of the AMNH, date unknown (between 1911–1924). AMNH Asset ID 17405

South America serving as the lodestone for broad evolutionary change.[13] Hooker was convinced that broad migrations of animals had moved northward from this land mass, eventually populating the Americas. But proof of Hooker's assertion was elusive, despite its wide appeal to the scientific community; the fossil evidence needed to confirm his hypothesis had yet to be uncovered when his writings caught Osborn's attention.

Once Osborn was smitten with Hooker's ideas, collecting the scientific evidence necessary to support his hypothesis became an irresistible challenge. Of course, charting a new research direction would mean underwriting costly expeditions to South America if he was to succeed. It was, therefore, a *carpe diem* moment when Osborn recognized that the "opening" of Latin American countries to U.S. interests in the 1890s (especially after the conclusion of the Spanish-American War in 1898) would help create the environment necessary to frame and finance his plans for expeditions to the Southern Hemisphere. If he prevailed in his quest, it would be AMNH scientists who would collect and amass the faunal evidence necessary to fill the lacunae in Hooker's grand scheme, and his men would finally answer lingering questions with regard to the biological development of South America.

In addition to Hooker, turn of the century evolutionary studies also witnessed the meteoric rise of the Ameghino brothers, Florentino and Carlos, two widely respected Argentine scientists. The Ameghinos insisted that the oldest mammals on the globe would be found in Patagonia, the southernmost section of their country, a suggestion which also had Osborn champing at the bit. He imagined that if he could only muster the necessary resources to mount a full-fledged expedition to Argentina he would be able to verify Hooker's ideas and/or those of the Ameghinos.[14]But Osborn's hands were tied by the Museum budget. To finance such costly paleontological fieldwork in South America there were realistically only three options: an expedition which could be directly funded (an extremely unlikely proposition given the limited budgets and competing interests within the AMNH in the 1890s); or an expedition which might be paid for by outside support (i.e., a wealthy patron); or, as a final possibility, Osborn might manage to have his men inserted into the expeditions of other institutions, as Putnam had managed to do with the Harvard-Peabody expedition to Copán, Honduras. Latin America was fast becoming Osborn's bailiwick; he was willing to stake his enviable reputation as a scientist by dedicating himself to the delineation of its evolutionary history.

Luckily for him, he found a tailor-made opportunity to advance his scheme. Not yet president of the institution, Osborn would manage to insert Barnum Brown, a new hire in his Vertebrate Paleontology Department, into the ongoing, final phase of Princeton University's exploration of southernmost Argentina. Headed by Princeton's Professor John Bell Hatcher, the areas to which this expedition was headed in late 1898 included Patagonia and Punta Arenas, where Hatcher aimed to uncover the large mammal specimens which the Ameghino brothers had insisted could be found in situ.[15] Osborn, a Princeton graduate, recognized that having someone represent the AMNH on this important mission was absolutely essential and, through his powers of persuasion and personal friendship, he succeeded in having Brown join Hatcher's Argentine team.

When the Princeton group concluded its work the following year, Professor Hatcher was slated to return to New Jersey, leaving Brown (the AMNH staffer) to either accompany his fellow expedition members home, or be left in Argentina on his own. Ambitious for his personal career, Brown considered Hatcher's departure an unexpected opportunity to remain in South America where he would now collect exclusively for the Museum, instead of for Princeton. With his own ambitions in full throttle, Osborn apparently let himself be convinced to allow Brown to continue the Argentine explorations on his own.

As it turned out, Osborn was repaid in kind. Unencumbered by the formalities of the Princeton arrangement, Brown roamed through southernmost Argentina, amassing a scientifically important collection of fossil mammal specimens which he then forwarded to the Museum.[16] Included among his finds was a substantially complete skeleton of an early glyptodont (forerunner to today's armadillos), as well as the bones of ground sloths. These fossils are still on display on the fourth floor of the Museum. A clear sense of triumph reverberated through the account of Brown's success which appeared in the Annual Report of the President for 1899. Gloating that the young researcher had sent back a remarkable total of fourteen boxes of fossils to New York, the account concluded that his contribution, when combined with other specimens already at the Museum, would now provide "a history of life upon our sister continent."[17] This was the essence of what Osborn had defined as his "Latin American mission".

Given this early level of success in Argentina, as well as on the ongoing ornithological expeditions, the Museum's collections of Latin American materials grew dramatically in the first decade and a half of the twentieth century. Research teams from the Museum explored vast swathes of the hemisphere, aiming to complete increasingly ambitious projects in their initial areas of interest, or choosing new ones. Barnum Brown, for example, shifted his attention away from Argentina and began searching for mammals in Mexico and Cuba. Other scientists scoured Nicaragua for fossil birds, reptiles and mammals, eventually pushing further southward through Central America to Venezuela. Rollo Beck was engaged to collect specimens of pelagic birds off the coasts of Peru, Chile and Argentina, while mammalogist William B. Richardson set out to collect in Colombia in late 1910 and was joined in the field by ornithologists Leo Miller, Frank Chapman, and Arthur A. Allen. Henry Crampton, a noted zoologist who had previously devoted his research to the study of *Partula* (snails) in the Pacific, trekked through British Guiana and Northern Brazil in search of invertebrate specimens. Accompanied by his Museum colleague, Frank Lutz, the pair reached the foot of Mount Roraima, where the nations of Guyana, Venezuela and Brazil meet, in 1911. The New York Times covered this expedition in detail, and the resulting article gave the Museum's Latin America agenda even greater exposure. The story that ran in the newspaper featured a sensationalist, but dubious, headline proclaiming that Crampton had "faced death in the jungle."[18] What mattered to the Museum, however, was not the veracity of the article, but that the institution had received recognition for its scientists who worked in Latin America and for its collections. This was outstanding publicity, of the kind Osborn could only have hoped for.

Yet another expedition, this time funded by Alfred M. Collins and Lee Garnett Day (the "Collins-Day Expedition"), saw George Cherrie and others traipsing through South America during 1915; this group started out in Mollendo, Peru, on the west coast, and ended up in Pará, Brazil on the east coast. In addition to the specimens they collected, the expedition members produced 1500 photographs, plus 10,000 feet of motion-picture film![19] If the public craved visual images of Latin America, the AMNH was prepared to give it to them, relying on the latest technology to accomplish this aim.

It was full steam ahead for Latin American field expeditions in the years leading up to U.S. involvement in World War I. President Osborn personally encouraged Frank M. Chapman, the Curator of Ornithology, to expand upon his successes in Colombia and head a new expedition to Bolivia, Peru, Chile, Argentina and Brazil. Joined by his department colleagues Leo E. Miller, Howarth S. Boyle and George

Cherrie, Chapman had covered most of South America by the summer of 1916. Another party, under the direction of Clarence R. Halter and L.A. Mannhardt, was beginning a collection of reptiles and fishes in Nicaragua in early 1917, just prior to the U.S. Declaration of War. Yet another distinct line of investigation, pursued by George Cherrie (who was on his own from September, 1916 to February, 1917), documented the life histories of birds and mammals in the *pantanales* (swamp lands) of upper Paraguay and Brazil.[20]

All of Osborn's plans for Latin American research seemed to be coming to fruition when, unexpectedly, the ideas on evolution which he had long embraced were dramatically, and irrevocably, challenged. Science, of course, rarely stands still; by 1910, just two years into Osborn's presidency of the AMNH, the Hooker hypothesis of the existence of an Antarctic land mass, the very rationale upon which Osborn had predicated his multiple requests for Museum funding of South American ventures, was repudiated. Well before Osborn awakened to the new reality, the scientific staff at the Museum had moved beyond Hooker's postulated existence of a south polar land mass to explain evolution toward an acceptance of competing hypotheses, ideas which now considered the role of early land bridges between continents in affecting the biome and/or the possibility that continental drift had acted as a mechanism for changes in evolutionary patterns.

The notion that migration of larger mammals had occurred in a south to north direction, the very crux of the Hooker theory, was further called into question by William Diller Matthew, a former student of Osborn's who became his colleague at the Museum. In contradicting Hooker's premise, Matthew suggested that the exact opposite was true, i.e., that mammals had actually traveled from the Holarctic North to South America, rather than the reverse. Matthew's ideas rapidly gained currency, largely bolstered by the positive reaction he received to his seminal article, "Climate and Evolution,"[21] published in 1915, which cogently laid out his north–south hypothesis. As a result, Osborn felt pressured to abandon the great southern land mass theory altogether and, after a brief flirtation with land bridge theories, ultimately came to accept Matthew's views, if for no other reason than to appear abreast of new findings in his field. Without doubt, Osborn would have appeared foolish had he clung to his previous notions.

Hooker's fall from grace became an inflection point for the new Museum President and his research agenda. As a result of having to abandon his original justification for the sponsorship of research in South America, President Osborn was now obliged to reframe the Museum's mission in the Western Hemisphere. To his credit, he managed to accomplish this transition with notable panache and an equally powerful sense of one-upmanship. Without drawing attention to what was clearly a seismic shift in his thinking, Osborn shifted the goal of AMNH Latin American expeditions away from an all out search for the proof of the existence of an Antarctic ice-free land mass, and the mammalian fossil evidence necessary to support this thesis, to an equally intense effort to underwrite the most comprehensive surveys of the flora and fauna of Latin America (including the West Indies) ever attempted by a North American natural history institution. This reformulated objective provided the underpinning, as well as the substitute rationale, for the acquisition of additional Latin American faunal collections both for study and display purposes. The same thinking was applied to cultural projects; Osborn was just as eager to add to the Museum's Latin American archaeological and ethnographic collections. But, it is worthwhile noting that even during this period of mission reorientation, Osborn remained convinced that knowledge of the history of living and fossil forms of Latin America was critical for understanding the broader direction of evolution as it played out in global context.

To implement his new agenda of assembling massive collections of Latin American biological (both living and fossil) and archaeological specimens, Osborn unabashedly approached diplomats, businessmen or anyone else working in Latin America who might be willing to finance a portion of the project; clearly, the cause of advancing broad natural science studies south of the border continued to be his passion. Anyone familiar with the Museum president recognized that he would not be satisfied until the AMNH housed the quintessential collections of Latin American specimens and artifacts, coveted not only for answering evolutionary and cultural questions and display, but for their simultaneous ability to launch the AMNH to the forefront of natural history research on Latin America within the United States.

Osborn Moves the AMNH into the Caribbean

Osborn's insistence on being able to elucidate the evolutionary history of Latin America's biomes included the Caribbean, as well. The redefined Museum mission now moved far beyond the earlier, turn of the century, studies of Antillean volcanoes and coral reefs, to coalesce into a broader scheme to document the complete natural history of this region's fauna. Parenthetically, this meant that AMNH scientists would now be conducting research in areas which were increasingly coming under the scrutiny of the United States government, and it was soon apparent that Osborn's newly defined goals for the AMNH in Latin America were a product of, and forged by, a convenient elision of politics and science. In what is obvious from hindsight, the Museum's interest in the West Indies from the turn of the century through World War I had significant political implications; possessing a scientific staff knowledgeable about the Caribbean, as a result of Museum-sponsored research to the region, would bring the institution to the attention of the State Department and was likely the critical factor in the Museum's deep dive into Washington's political orbit.

Interest in the natural history of the Greater Antilles, however, was not the exclusive purview of the AMNH. Instead, it snowballed across the country, in large part resulting from the U.S. government's political takeover of the island of Puerto Rico from Spain at the conclusion of the 1898 Spanish-American War. With an eye toward evaluating the potential of its newly acquired territory, Washington quickly authorized a systematic accounting of all of the fishes of the island, a project relegated to the United States Bureau of Fisheries. When it was completed, some two years later, the survey yielded an impressive list of 291 piscine species.

Reminiscent of the zeal with which Spanish and Portuguese naturalists had earlier assessed New World resources, Washington politicians decided to expand this canvass beyond fishes to now include all insular biological species. In turn, this updated plan morphed (c. 1910) into a comprehensive multi-year project which promised, when it was finished, to provide a register of the complete flora and fauna (both living, and importantly, fossil) of the entire island colony and the nearby Virgin Islands, which the United States had purchased from Denmark in 1917.[22] To implement the extended survey, the Bureau of Fisheries turned for assistance to the venerable New York Academy of Sciences, entrusting that highly-respected institution with supervision of future expeditions to the islands.[23]

Ever the savvy operator, Osborn (by now the Museum's president) recognized at once that Washington's expansion of the Puerto Rican survey represented a golden opportunity to augment the Caribbean collections at the Museum. But the fact that the U.S. government had chosen the New York Academy of Sciences, and not the AMNH, to conduct this canvass presented an obstacle; to insert the Museum into the government survey, Osborn would have to convince the Academy to accept the AMNH as an equal partner. Notwithstanding Osborn's considerable skills, achieving this ambitious goal while simultaneously appearing not to challenge the authority of the Academy, would require great tact. Osborn opted to rely on a surrogate to work out the delicate arrangements between the institutions; for this sensitive negotiation, he chose Henry E. Crampton, who was then serving as both Curator of Invertebrate Zoology at the AMNH and Professor of Zoology at Columbia University. Conveniently, Crampton was also the Secretary of the New York Academy of Sciences. Although Osborn might privately have wondered about his proxy's potential conflicts of interest, Crampton proved to be a tough negotiator when it came to working out the details of institutional cooperation, and his priorities, apparently, lay chiefly with the Museum.

Crampton made his feelings regarding the Puerto Rican survey abundantly clear in a letter he wrote to Museum Director Frederic A. Lucas (second only to Osborn in rank; see Fig. 3.1, right): "As I see it," Crampton averred, "we are under certain obligations to the New York Academy of Sciences through our participation so far, but entirely beyond that I feel that our department is under greater obligation to the interests of the American Museum [of Natural History]; by contributing our men at intervals, we gain an unusual amount of extraordinarily valuable material without incurring field expenses."[24] This was exactly the kind of leverage which Osborn had hoped for: cooperation with the Academy could provide the means to expand the Museum's Caribbean collections, without having to advance any funds.

Working behind the scenes in his Academy capacity, Henry Crampton was able to tweak and stretch the original Puerto Rico project far beyond the original focus (study of the island and the neighboring Virgin Islands), to one that would ultimately encompass the biological species of all the Greater Antilles, including Cuba, Jamaica, and Hispañola (the island shared by the Dominican Republic and Haiti). As soon as he recognized that a combined institutional effort could represent a potential windfall for the AMNH, Crampton cajoled colleagues from the Museum's departments of ichthyology and herpetology to join the Academy's efforts. When the widened scope of the project was presented to the New York Academy, it was practically a *fait accompli*, and the venerable institution eagerly accepted the expanded proposal, welcoming the AMNH as a partner. Largely due to Crampton's diplomatic skills, the Museum could now potentially assemble even more complete collections of Caribbean living and fossil forms than had previously been anticipated. Crampton first alerted Curator Hussakoff[25] and then Director Lucas to his thinking on the matter: "The reason why this material is so valuable, "is that it is taken from a part of the territory which is the natural field for intensive evolutionary research. The faunal relations between North and South America, Central America and the Antilles, so far as the invertebrates are involved, are the larger objects of study by our department."[26]

Crampton's reformulation of the Academy's original project not only widened its scope to consider the entire Greater Antilles, but also expanded its focus from fishes to include other vertebrates and invertebrates. As noted in the Museum's Annual Report for 1916 (which gave him full credit for the feat), "Under Curator Crampton this [the biological survey of Puerto Rico] survey has been extended to the neighboring Antillean islands, with a view to throwing light upon the whole Antillean problem, namely of the geological antiquity of these islands and of their former relations with Central and South America. Already important discoveries have been made in vertebrate paleontology, through work upon the fossils of the Porto[sic] Rican caves, which have thrown new light on the migrations from South America."[27]

The extension of the New York Academy's mandate paid out handsomely for the Museum in unexpected ways. In one instance a particularly significant collection of butterflies was gifted to the Museum by Charles T. Ramsden, the collector/manager of a large sugar estate in Guantánamo, Cuba with whom Frank Lutz had stayed when he visited that nation in 1913. Ramsden, in fact, donated his entire collection to the Museum, an assemblage which Lutz reckoned was "one of the best, if not the best, collection of Cuban butterflies that an American ever got his fingers on."[28] To demonstrate the institution's gratitude for his contribution, Crampton graciously had a lifetime membership at the Museum bestowed upon Ramsden.[29]

While Crampton's reformulation of the Caribbean survey proved successful, securing monies for expeditions which extended research beyond the U.S. government's specific interests continued to be problematic, as competing interests between Museum departments prior to World War I meant that scarce resources had to be distributed judiciously among them. Occasionally, funding was obtained from private individuals who solicited help from the AMNH in procuring specimens for their personal collections. One such collector was Bostonian Preston Clark, who petitioned the Museum to search for insects in the Dominican Republic on his behalf, advancing the generous sum of $300 to cover the cost. The ever ambitious Henry Crampton recognized Clark's request as yet another unexpected windfall for the Museum. So, too, did Director Lucas, who went so far as to suggest that the man who would eventually be sent to Santo Domingo to find sphingid moths for Mr. Clark, might also collect any mammals he found in addition to the insects, without suffering any pangs of conscience. "As we both know", he wrote conspiratorially to Crampton, "Santo Domingo mammals are rare and opportunities for getting them few."[30] For the AMNH, *carpe diem* now defined the Museum's mindset toward collecting in the West Indies and in Latin America broadly, an attitude which prevailed through the entrance of the United States into World War I.

Osborn Reaches Out for Support and Funding

As he grew more confident in his leadership, President Osborn proved to be a persuasive advocate for the AMNH. When necessary, he solicited contributions from his wide circle of personal associates and social contacts, in return for which they were often rewarded by having their names put forward for Museum trusteeship or sponsorship positions (depending on the amount of funds they provided). While private individuals were continuously solicited to contribute toward specific, earmarked projects (such as Beck's "Brewster-Sanford Expedition"), it was the trustees who brought in the big money. These wealthy men represented some of the most readily identifiable families from New York society, as well as the broader United States. Each year a slate of new trustees was drawn up and put forward for appointment and approval by the existing board. The "class" of trustees for the critical War year of 1917, for example, saw the appointment of the banker, and future governor of New York, Averill Harriman to the board, while businessmen/ philanthropists Harry Payne Whitney and Edward S. Harkness were proposed to the Nominating Committee. Not surprisingly, many of the moguls who garnered trusteeship positions were involved in Latin America in some way, either in the investment sphere or politically. As a man who enjoyed broad political connections, Osborn critically understood how the growing desire among businessmen as well as Washington politicians for closer cooperation among the American nations could work in his (and the Museum's) favor. Thus, in the same year that Harriman, Whitney and Harkness were proposed for Trusteeship, former President Theodore Roosevelt was made an Honorary Fellow of the Museum, "in recognition of the interest he has shown in…publications, his expedition through the Brazilian Wilderness, his contributions to the Roosevelt South American expedition, and his service to the cause of science."[31]

With his numerous political and business connections, Osborn was also able to secure passage for his scientists planning research trips to Latin America without spending Museum funds. To accomplish this goal, he targeted major shipping and railroad businesses with interests in the region, many of which, such as the United Fruit Company with its substantial operations in Central America, were enticed not only into underwriting the cost of travel of Museum scientists, but to offering whatever accommodations and services they could to these men.[32]

Where things did not go smoothly, as sometimes happened when the Museum attempted to secure permission for its scientists to conduct research in reluctant Latin American nations, Osborn could always fall back upon assistance from the United States State Department. Rollo H. Beck's plans to collect oceanic birds in Peru, Chile and Brazil, to cite just one example, would require the intercession of Washington on more than one occasion (see Chapter X).[33]

Ever the pragmatist, Osborn attempted to keep abreast of the Latin American scene while, at the same time, he cautiously avoided taking a public stance on political issues. A perfect example of his thinking may be seen in the way he approached events which unfolded in 1910 when Porfirio Díaz, the strong-arm president of Mexico since the 1870s, was effectively challenged for the first time by the political upstart Francisco I. Madero. Ordinarily a political kerfuffle of this type would not have mattered in the slightest to Osborn, but it had been Díaz who approved of Marshall Saville's project at Mitla, and Madero was an unknown quantity.

Political instability, therefore, threatened to interfere with archaeologist Saville's (and Herbert Spinden's) ability to collect additional material at Mitla for the Museum's new Hall of Mexico. Osborn expressed that very sentiment to Clark Wissler, Curator of Anthropology, in October 1910, a month before Madero actually crossed over the Mexican border from his hideout in Texas and began moving toward Mexico City, where he would ultimately oust Díaz. Demonstrating his uncanny ability to maneuver between political fault lines, Osborn opined: "I trust that in the course of the winter that we may reestablish our old and very cordial relations with the Mexican Government…". He went on: "It is not our duty to ask any moral questions, but to recognize the existing authorities and to do the best we can."[34] Seemingly, whether Díaz or Madero prevailed was of little consequence; what mattered to Osborn was the predisposition of the person in charge to allow the Museum's excavations to continue on schedule.

Later, when Mexico teetered on the brink of complete chaos during the Mexican Revolution (essentially, a protracted civil war which lasted from 1913 until 1917), Osborn responded negatively to a 1915 request from anthropologist Franz Boas who hoped to have his colleague hired for future Museum research in Mexico. It was unlikely that an opening for this person could be found, Osborn opined, given the fact that "prospects for science in Mexico were never worse than at the present time…".[35]

While Osborn proceeded with caution, presumably believing that as a major figure in New York social circles it was inappropriate for him to express his political opinions in public, such restrictions apparently did not extend to his wife, Lucretia. Mrs. Osborn (aka "LouLou") did not hesitate to speak her mind, even to the point of conducting writing campaigns to politicians when she thought it necessary. Was she Osborn's cat's paw, the public *persona* behind this determinedly private museum president, or merely an outspoken spouse? In any event, she certainly was no shrinking violet. Writing to the New York Senator James W. Wadsworth just months before the United States entered the Great War, she evinced no qualms in revealing her view of the rapidly deteriorating situation in Mexico. The failed U.S. invasion of that country had been unsuccessful in completing its mission to hunt down the rebel leader, Pancho Villa, and further had not helped the political future of that country. Signaling his agreement with her ideas, the Senator responded that it was imperative to "confine our attention to several problems essentially American before we attempt to lecture the nations of Europe. The spectacle of Mexico, our nearest neighbor should command our gaze…we have made a wreck of that unhappy republic."[36] These were essentially isolationist views to which both Osborn and his wife initially subscribed. But, where the AMNH president kept his opinion to himself, his wife apparently had no hesitation in expressing hers publicly.

If Osborn and the Museum needed to keep a low profile with regard to political turmoil in the Americas and Europe, it was a different matter altogether when opportunities arose to heal existing rifts between the U.S. and the sister republics of Latin America. The fact that Frank Chapman, the Curator of Ornithology, had succeeded so brilliantly in his expedition to Colombia, notwithstanding that nation's longstanding resentment of the role played by the United States in the Panama Canal debacle, was proudly celebrated in the Forty-eighth annual Report of the AMNH for the year 1916: "Not the least important feature of these expeditions from the Pan-American standpoint is the very friendly relations which our representatives, especially Dr. Chapman, have established with the officers and scientific men of the various South American governments."[37] It is clear that the AMNH understood the potential of its scientists to act as informal diplomats (a recognition which, with hindsight, can be viewed as a portent of their future political importance within the region during World War I).

It was for this same reason that Osborn embraced the Second Pan American Scientific Congress of 1915–6 with such unbridled enthusiasm. His appointment of Frank Chapman as the Museum's primary delegate to the Congress was a savvy move in light of the ornithologist's apparent ability to impress the seemingly intractable Colombians. Conversely, Chapman understood the gravity of his responsibility, as evidenced in his letter accepting the appointment: "I feel that, in attending this Congress, I can greatly strengthen our relations with South American Republics, by making known to their representatives our plans for exploration in their respective countries…I should be glad to offer to deliver to the Congress an illustrated address on our explorations in Colombia, as an example of the character of the biological survey which we propose to make of all of South America [i.e., Latin America]."[38] Gleefully, Osborn transmitted Chapman's acceptance letter to Director Lucas, adding that he thought "it very important, in connection with our forthcoming exploration in South America, that we should establish most friendly relations with the South American Republics."[39]

Chapman's successes aside, perhaps the most auspicious moment in which to mend political fissures with Latin American neighbors occurred when the former U.S. President, and Osborn's long time friend, Theodore Roosevelt, approached the Museum administrator with a plan to venture into "uncharted" Brazilian territory. Roosevelt's father, one of the founders of the AMNH, had championed the cause of natural science, and his son Theodore was equally enamored of the wilderness. Well known for the considerable time he had already spent in the American West, the former President proposed a future expedition to South America to Osborn. His scheme included helming an expedition of men who would travel in 1913 down a previously unexplored section of a Brazilian river, a plan which would be executed under the Museum's aegis.

Osborn was initially hesitant, reluctant to have his friend embark on what he personally considered an ill-conceived and potentially dangerous venture. But he eventually acquiesced to the entirety of Roosevelt's plan, agreeing not only to Museum sponsorship of the River of Doubt journey but assigning Frank Chapman to the former President as an advisor, as well as scientists Leo Miller and George K. Cherrie to accompany him downriver. In reality, the motivation behind Osborn's change of heart and his final approval of the Roosevelt expedition had little to do with its touted scientific purpose, but hinged on what the Museum President perceived as the expedition's political advantage.

Roosevelt's misadventures during his extended safari along the River of Doubt, his falling out with Colonel Candido Rondon, the Brazilian-appointed expedition head, as well as his near-death experience along the river's banks, have been documented elsewhere,[40] and need not concern us here. With respect to the interests of the AMNH in this trip, however, Osborn was focused on the unannounced portion of Roosevelt's trip, a multi-country foray which would occupy the former President for several weeks before his venture into the Brazilian interior began. During this brief, pre-adventure interlude, Roosevelt hoped to reverse the damage to U.S.-Latin American relations which resulted from the heavy-handed, "gun-boat diplomacy" he had executed earlier in the century. To accomplish this goal Roosevelt planned to visit several Latin American nations in the Southern Cone. While the former President might not have been prepared to openly apologize for his past actions in the hemisphere, Osborn recognized that this prologue to Roosevelt's trip was a thinly veiled effort at a *mea culpa.* By its association with this part of Roosevelt's project, provided that former President's pre-riverine excursion proved successful, the travel and research of AMNH scientists would benefit for years to come.

Encouraged by Osborn, Roosevelt embarked on a multi-nation speaking tour of Uruguay, Chile and Argentina[41] in the weeks preceding the October, 1913 start of his descent down the Brazilian River of Doubt. Rather pointedly, Roosevelt avoided visiting Colombia, the nation that arguably harbored the deepest resentment toward him and the United States.[42] Instead, Roosevelt made appearances in those countries which he felt had actually benefited from the increased hemispheric trade generated by the opening of the Panama Canal.

Against a background of earlier, vehemently anti-Roosevelt (and anti-U.S.) feelings in Latin America, the former President's speaking engagements across the Southern Cone were, in fact, a remarkable success. Roosevelt didn't hesitate to extol the economic benefits which the U.S. built Panama Canal had provided Latin American nations, while he studiously avoided reference to the awkward political consequences (i.e., the breakaway of Panama from the Colombian nation) which his actions had caused. His message apparently reached mostly receptive audiences, although the former President was greeted with vocal anti-United States protests when he visited Santiago de Chile. Outwardly undaunted, Roosevelt didn't hesitate to lecture the Chileans on how much that nation had actually benefited from the new Panamanian route. As Candice Millard has recently written about this incident, the former President left the people of Chile "if not convinced of Roosevelt's righteousness, at least impressed by his conviction."[43] Understandably anxious to document Roosevelt's movements during this conciliatory tour (accounts of which

Fig. 3.2 Disgruntled former President Theodore Roosevelt walking with crowd in Valparaiso, Chile, 1913. Photo by Rollo H. Beck. Image copyright California Academy of Sciences Archive, Rollo H. and Ida M. Beck collection, Box 43, with permission

would eventually be reported in the Bulletin of the Pan American Union),[44] the Museum assigned Rollo H. Beck, who happened to be conveniently in Chile at the same time as Roosevelt, to photograph the local reaction to the former President.[45] Beck's photographs of the event (Fig. 3.2) depict a rather grumpy-looking Roosevelt, clearly less than pleased by the hostile Chilean reception.

But Roosevelt's reception was quite a different matter across the Andes, where he received a mostly positive response to his speaking engagements in Argentina and Uruguay. The publicity garnered by his plans to explore the interior of Brazil simultaneously served the cause of political harmony and elevated the importance of scientific exchange in the Americas. Roosevelt was, in fact, so successful in his outreach that the head of the U.S. diplomatic delegation to Uruguay declared that the ex-President's Montevideo address had caused local skeptics to conclude that "after all these southern republics have not so very much to fear from 'the great colossus of the North.'"[46] As Roosevelt's institutional sponsor, the AMNH obviously stood to gain from the positive publicity, and it is not at all surprising that Roosevelt was made an Honorary Fellow of the Museum just a few years later.

Into the mid-1910s, Osborn's all hands on deck approach to Latin America was paying off handsomely. The administrator had taken whatever steps he thought necessary to advance his Latin American mission, including urging his staff to publish their findings from the region as quickly as possible. In addition to Chapman's magnum opus on Colombian birds, George Cherrie and Leo Miller each wrote accounts of their travels in Latin America, which were published either as a memoir, or appeared as an article in the American Museum Journal.[47] Whether scientific, or written for a popular audience, these publications signaled to the United States reading public that the AMNH was well on its way to achieving pre-eminence in the field of Latin American scientific research.

This perception was furthered by the ever increasing volume of Latin American artifacts on exhibit at the Museum. These collections were the physical proof that Osborn had been able to fulfill his goal of having the most comprehensive surveys of the flora and fauna of Latin America ever mounted by a North American natural history institution. Although the Hall of Mexico and collections of West Indian materials had their advocates, it was the exhibits of South American specimens which garnered the most praise. The American Museum Journal now boasted of having "especially valuable" South American archaeological, fossil vertebrate and mineral collections, as well as a "South American study collection of mammals" deemed "the largest in the world." Moreover, the South American bird collections were said to be the "largest in North America."[48]

In brief, up to the early months of 1917, Osborn was on target with his "Latin American project." He had already sponsored a wide spectrum of research in the region, as well as substantially broadened collections from the area. As a result, the AMNH had become the acknowledged leader in the field of Latin American natural history studies, exactly as he had planned. Nonetheless, the United States Declaration of War against Germany in April of 1917 threw a wrench into the Museum president's carefully crafted goals which even he, with his superior managerial style, had difficulty recovering from. Only by means of sheer cunning and secret deals arranged with federal agencies was Osborn able to maintain an AMNH presence in Latin America during the course of the War and to somehow keep his project on course.

Notes

1. "Collections of Birds from South and Central America", July 28, 1916, (CA), AMNH, File 370.
2. These items are reported in the Minutes of the Regular Meeting of the Board of Trustees, AMNH, November 12, 1917, p. 50. See Chapter X for a discussion of Beck's ventures.
3. Competition between natural history institutions was rampant globally. Osborn, for example, vied with the Carnegie over the exchange of *Diplodocus* fossils with European museums. The Carnegie sent casts to Berlin, but the AMNH responded by sending originals to Frankfurt. See Joe Kember, John Plunkett, Jill A. Sullivan, Popular Exhibitions, Science and Showmanship, 1840–1910, (London, Pickering and Chatto, 2012), pp. 223–224. Stanley A. Freed, Anthropology Unmasked…, Vol. 1, p. 144 points to the intense competition in the early twentieth century between the AMNH and the Field Museum in the area of anthropology.

4. See Ian Dyck, "Scientific Archaeology in Central North America: The Life and Work of W.B. Nickerson (1825–1926)," Archaeology paper, 177, Mercury Series, Canadian Museum of History and the University of Ottawa Press, 2016, p. 71.
5. See Douglas Sloan, "Science in New York City, 1867–1907," Isis, vol. 71, No. 1 (March, 1980), p. 62. By way of comparison with the AMNH, the Smithsonian, for example, listed only one expedition to Latin America (in Panama) for the year ending June 30, 1914. Cf. Annual Report of the Board of Regents of the Smithsonian Institution showing the Operations, Expenditures and condition of the Institution for the year ending June 30, 1914. (Washington, 1915).
6. Freed, Anthropology Unmasked... discusses Bandelier, Lumholtz and the Duke of Loubat in Vol. II, pp. 793–802. Dr. Hovey (Geology Department) received praise for his work on the volcanoes of Martinique and St. Vincent after his Museum expedition of 1902. Cf. "The American Museum of Natural History," Science, N.S. Vol. 7, No. 439, (May 29, 1903), p. 875.
7. *Ibid.* Loubat's material is still on display today.
8. Referenced in the Forty-eighth annual Report of the AMNH for the year 1916 (New York, May 1, 1917), pp. 29–30.
9. Freed, Anthropology Unmasked... discusses this new Hall of Mexico in Vol. I, pp. 422–425.
10. American Museum of Natural History, Annual Report of the President for the year 1902. See also the American Museum Journal, Vol. 2, No. 3 (March, 1902), pp. 25–28.
11. The plans for the Mitla restaurant were first announced in the AMNH, Executive Committee Report, arch. 20, 1908, p. 106.
12. Regal, Henry Fairfield Osborn..., p. xv. By the mid and late nineteenth century in England as well, practitioners of science began to head museums rather than trustees, as had previously been the case. See Samuel J. M. M. Alberti, "The Status of Museums: Authority, Identity, and Material Culture," in David N. Livingstone and Charles W. J. Withers, eds., Geographies of Nineteenth-Century Science, (Chicago, University of Chicago Press, 2011), pp. 51–72.
13. Rainger, An Agenda ..., p. 192. For an overview of Hooker, see also "The Travails of a Young Botanist" in Iain McCalman, Darwin's Armada: Four Voyages and the Battle for the Theory of Evolution, (New York, W.W. Norton and Company, 2009), pp. 106–126 *et passim.*
14. *Ibid.*, p. 93. The Ameghino brothers have been the subject of much recent research among Latin American historians of science. See Irina Podgorny, "Bones and Devices in the Constitution of Paleontology in Argentina at the end of the Nineteenth Century," Science in Context, Vol. 18, No. 2, (June, 2005), pp. 249–283.
15. Dingus and Norell, Barnum Brown..., pp. 62–64.
16. *Ibid.,* pp. 72–73.
17. Rainger, An Agenda..., p. 38 and Annual Report of the President for the year 1899 (New York, 1900), p. 19.
18. American Museum of Natural History, Annual Report of the President for the year 1912, p.52. The Crampton-Lutz exploration of the foot of Mount Roraima took place in 1911. See "Dr. Crampton Faced Death in the Jungle," New York Times, September 20, 1911.
19. American Museum of Natural History, Annual Report of the President for the year 1910, p.43. The Collins-Day expedition is reported on in American Museum Journal, Vol. 16, No. 1 (January, 1916), pp. 23–44. See also J. A. Allen, "List of Mammals Collected in Colombia by the American Museum of Natural History Expeditions, 1910–1915, Bulletin of the American Museum of Natural History, vol. XXXV (1916), pp. 191–238.
20. American Museum of Natural History, Annual Report of the President for the year 1916, pp. 29–30.
21. The hypothesis of William Diller Matthew, which promoted the notion of southward migration of larger mammals from the Holarctic, was probably the most prominent reason for Osborn's change of perspective. Osborn, at this point, "recapitulated his ideas on Africa and Antarctica as major regions of adaptive radiation and faunal migration." (Rainger, An Agenda..., p. 192) Matthew published his ideas in an essay entitled "Climate and Evolution," Annals of the New York Academy of Sciences, vol.24, no. 1 (February, 1915), pp. 171–318. See also Ronald Rainger, "Just Before Simpson: William Diller Matthew's Understanding of Evolution," Proceedings of the American Philosophical Society, vol. 130, no. 4, (1986),

pp. 453–474. A recent interpretation suggests that a later migration, around three to two million years ago, went both ways (North and South) through the newly emerged Isthmus of Panama, but that "something" interfered with the South-North route, leaving fewer animals to make the trip. See New York Times, "Science Times," October 27, 2020, p. D.2.

22. See Julio C. Figueroa Colón, ed., The Scientific Survey of Puerto Rico and the Virgin Islands: An Eighty-Year Reassessment of the Island's Natural History (New York, New York Academy of Sciences, 1996). Cf. also original survey in Scientific Survey of Porto Rico and the Virgin Islands, Vols. 1–10 (New York, New York Academy of Sciences, 1919).
23. The five year plan proposed by the New York Academy of Sciences is contained in J.T. Nichols, Memorandum on the Possibilities of Field-work in Porto Rico, CA, AMNH, File 1000.
24. CA file 1000, AMNH, Crampton to Lucas, May 3, 1915. Crampton was not the only AMNH curator to hold a professorial position at Columbia University. Osborn (Vertebrate Paleontology) and Clark Wissler (Anthropology) did as well. For a study of the unique relationship between Columbia and the AMNH see Douglas Sloan, "Science in New York City, 1867–1907," Isis, vol. 71, No. 1 (March, 1980), pp. 35–76.
25. Background on this exchange is contained in CA 1000, Hussakof to Lucas, Apr. 28, 1914.
26. Crampton to Lucas, May 3, 1915. CA, AMNH, file 1000.
27. Forty-eighth Annual Report of the Trustees of the American Museum of Natural History for the year 1916 (New York, 1917), p. 31. It is interesting to note that Franz Boas was also affiliated with this project, although most of his anthropological survey was conducted in Puerto Rico. See Rafael Ocasio, "Franz Boas's Trip to Puerto Rico (1915) and the Contemporary Boriken Nation," The Latin Americanist, Vol. 66, no. 1 (March, 2022), pp. 81–94.
28. Letter of Frank E. Lutz to Dr. Lucas, 8 October 1913, Santiago de Cuba, CA, AMNH, File 1000. A report by Lutz on "Collecting in Cuba," in which he compares the insects of Florida and Cuba, was published in the American Museum Journal, Vol. 14, 1914, pp. 99–106.
29. Crampton to Lucas, 2 December, 1913, CA, AMNH, file 1000.
30. CA file 1000, Lucas to Crampton, March 25, 1915.
31. As noted in the Minutes of the Regular Board Meeting of the American Museum of Natural History, November 12, 1917, p. 50.
32. See Thomas W. Cuddy, Political Identity and Archaeology in Northeast Honduras (Colorado, University Press of Colorado, 2007); p. 22 suggests that "The fruit companies also maintained modern facilities and transportation routes that archaeologists visiting the region were obliged to use".
33. Beck was able to proceed on his mission with several State Department interventions in 1912, 1913, etc. National Archives, Record Group 59, [hereafter cited NA, RG 59] Scientific expeditions….
34. Osborn to Curator Wissler, October 25, 1910, CA, File 906.
35. Osborn to Franz Boas, Oct. 21, 1915, Boas papers, American Philosophical Society, Philadelphia, Pennsylvania.
36. Letter of Senator Wadsworth to Mrs. H. Fairfield Osborn, Washington, January 27, 1917. AMNH, special collection Osborn files, 0837, Box 2, Fl. 13. Loulou's ideas are also referenced in a letter of Senator O.W. Underwood to Mrs. H. Fairfield Osborn, Washington, D.C. August 23, 1913. AMNH, special collection Osborn files, 0837, Box 2, Fl. 12.
37. Forty-eighth annual Report of the AMNH for the year 1916. (New York, May 1, 1917), p. 30.
38. Frank Chapman to Osborn, 30 June, 1915. CA File, 632.
39. Osborn to Director Lucas, 7 July, 1915, CA File 632.
40. Candice Millard, The River Of Doubt: Theodore Roosevelt's Darkest Journey (New York, Doubleday, 2005).
41. *Ibid.*, p. 64.

42. *Ibid.,* p. 65.
43. *Ibid*., p. 66.
44. Boletín de la Unión Panamericana, Vol. 40, No. 4, (Mayo, 1915), pp. 567–575.
45. The photographs of Roosevelt taken by Beck are in the AMNH collection, Negative #s 237788, 237789, 237790.
46. Patricia O'Toole, When Trumpets Call: Theodore Roosevelt After the White House (New York, Simon and Schuster, 2005), pp. 250–251.
47. George Cherrie published "To South America for Bird Study" in the April, 1917 volume of The American Museum Journal, pp. 269–273; his larger work, Dark Trails: Adventures of a Naturalist was published in New York by G.P. Putnam's Sons, in 1930. Roosevelt's Through the Brazilian Wilderness came out in 1914 (New York: Charles Scribner's Sons). Scribner's also published Leo E. Miller's In the Wilds of South America in 1918, while Chapman's work appeared in museum publications and other scientific journals.
48. American Museum Journal, Vol. 16, No. 1 (January, 1916), p. 76.

Chapter 4
Convergence: The Declaration of War, the ONI in Latin America and Osborn's Change of Heart

Worries About Germany in Latin America

The outbreak of World War I affected the Latin American research program at the American Museum of Natural History in several distinct ways. First, the neutral stance of many Latin American nations posed difficulties for U.S. researchers who hoped to work in the region, as Washington politicians were politically aligned with the Allies. After the United States Declaration of War in 1917, this situation grew even more fraught, with outright antagonism against the United States making the process of receiving Latin American research permits problematic. As the individual experiences of several AMNH scientists presented in the chapters which follow will show, those who managed to get to the region during the War encountered either hostility or decidedly pro-German attitudes from several Latin American governments.

A second impact on the AMNH was the realization by the Office of Naval Intelligence (ONI) that it lacked sufficient data on Latin America, leading to enhanced agent recruitment among academics. In fact, New York City (where the AMNH is located) itself became a center of intrigue during the war, a situation which the U.S. government recognized by opening a branch of the ONI on Wall Street. Finally, President Osborn had to overcome personal hurdles to prepare his institution for the War: he would need to reconcile his personal isolationist views with a posture of patriotism. This conundrum extended not only to the public but in dealings with his own staff as well. These three different aspects of how the War settled in on the Museum will be examined in turn.

The United States government initially took a cautious stance when World War I erupted in Europe in 1914 and waited several years before deciding that the overseas situation warranted direct involvement of the nation's troops and treasure. Yet, few observers were truly surprised when President Wilson spoke in early April, 1917, of the inevitability of American intervention in the European conflict. A few days later,

The original version of the chapter has been revised. A correction to this chapter can be found at https://doi.org/10.1007/978-3-032-18117-6_12

R. M. Delson, *Scientists or Spies?*, The Latin American Studies Book Series, https://doi.org/10.1007/978-3-032-18117-6_4

on April 7th, Congress followed Wilson's lead, and declared war against Germany, a decision which reflected the desire to protect the nation against threats within the Western hemisphere, as much as it did the necessity of reinforcing the Allied front by sending arms and men to Europe.

While it was more or less assumed that Latin America (or North America, for that matter) would never become a major theater of conflict, notwithstanding the ferocity of some of the earliest naval battles of the War fought between German and British forces off the coast of Chile and the Falkland Islands in 1914,[1] there were other troubling indicators within the Americas that gave Washington politicians pause. These included likely German disruption of international shipping across the Atlantic and the clear possibility of German submarine attacks along U.S., Caribbean or Latin American shores. Other perceived dangers emanated from the belief that the substantially large, long-standing communities of German immigrants resident in Brazil, Argentina, Chile and Central America would harbor individuals with conflicting loyalties who might evolve into what later would be called "fifth columnists" (see Fig. 4.1).

Equally concerning was the very real prospect of an international struggle over valuable Latin American mining resources. Brazilian manganese and Chilean nitrates, especially, were crucial components in the production of steel and armament, respectively. Without the ability to produce these elements or compounds on its own, the United States had long relied on India and Russia, the two major global producers of these commodities, for its supplies. The War effectively closed these markets to the United States, leaving Washington to look southward to make up for the shortfall. But even with the possibility of gaining an expanded market from the United States, the Brazilian and Chilean governments maintained postures of neutrality with regard to the European War which, in turn, rendered access to their raw materials complicated. The fact that the enemy nation of Germany, equally shut off from its traditional sources of nitrates and manganese, was competing for the same commodities in Latin America made it all the more imperative for the United States to be the first nation to lock down access to these valuable chemicals.

Initially, during the first years of the War, the goal of maintaining open trade routes with Latin America depended on convincing regional politicians to reject neutrality and support the Allies. But, following its Declaration of War against Germany in April, 1917, the United States shifted its position from cajoling southern neighbors to end cooperation with the Germans, to now outright demanding that Latin American nations join the Allied efforts to win the War. This new attitude toward Latin America implied that the region was not only expected to furnish requisite nitrates and manganese (and later petroleum and platinum, and mahogany for airplane propellers, among other goods) for the Allied forces but, additionally, to pledge explicit loyalty to the Allied cause, even to the point of sending troops to Europe if necessary.

This study has already noted the intense scrutiny Secretary of State Robert Lansing paid to Latin America and, as seen earlier, the inability to obtain reliable intelligence on the region was one of the factors driving his sponsorship of the

Fig. 4.1 Map of Latin American nations' relationships to the Allied cause in World War I, with dates of their actions (declaring war on or breaking diplomatic relations with the Axis powers, or remaining neutral). From *Latin American History: A Teaching Atlas* by Cathryn L. Lombardi and John V. Lombardi with K. Lynn Stone © 1983 by the Board of Regents of the University of Wisconsin System. Reprinted by permission of the University of Wisconsin Press

1915–1916 Second Pan American Scientific Congress. In his War Memoirs, published in 1935, Lansing went much further in expressing the anxieties he had experienced while serving as the nation's top diplomat. First, and foremost, Lansing believed that there was "little doubt that the German Government had been for some time seeking to establish submarine bases in the Caribbean Sea, from which U-boats could operate against enemy vessels approaching the Panama Canal."[2] It was this assessment which informed the U.S. decision in 1915 to send U.S. Marines to Haiti and, a year later, to the Dominican Republic, an intervention which paralleled Henry Crampton's extension of the greater Antilles biological survey to those exact areas. Lansing's justification for landing Marines in the Antilles was based on the ongoing political upheavals in both nations, which he thought rendered their respective populations vulnerable to German advances. It was Lansing who persuaded Washington politicians to purchase the Danish Virgin Islands in 1917, despite the fact that the U.S. government had waffled for decades on offers to buy these colonies from Denmark. Changing the official posture of the government, Lansing now thought it imperative to preclude these valuable pieces of real estate in the Anegada Passage (which provides the most direct access route across the Caribbean Sea to the Panama Canal) from falling into German hands. He principally feared that under Danish control these islands might not have the wherewithal to prevent the enemy from building submarine bases on their territory, a prospect considered so risky that he moved quickly to assure U.S. ownership.

But these were not Lansing's only concerns. The Secretary also insisted (although the evidence was not particularly strong) that the Germans were "intriguing to break down the Cuban Government and that German gold was being furnished to the revolutionists for that purpose."[3] A third concern involved the elimination of Colombia's antagonism toward the United States, a hatred which had festered in the wake of President Theodore Roosevelt's involvement in the breakaway independence of Colombia's region of Panama. Lansing claimed that "German agents were actively engaged there in anti-American propaganda and in intrigues to purchase islands not far from the eastern entrance of the Panama Canal."[4] Obviously, any disruptions to Canal activity would have been disastrous to the U.S. war effort, leaving Lansing no choice but to be fully committed to thwarting even the consideration of such a possibility.

Finally, the Secretary of State understood that Latin American neutrality had resulted in businessmen south of the border stubbornly refusing to end their commerce with German firms and/or the Kaiser's government. Lansing was certain that if this situation was left unchecked, not only would nitrates and manganese wind up in German hands, but petroleum might also become a commodity of contention. In this regard, the outright interference of Germans agents in the political affairs of oil-producing Mexico did not augur well for the maintenance of peace within the hemisphere.[5] Lansing suggested in his War Memoirs that the "Mexican oil fields [are]…among the largest and most productive in the world… To have them closed or their output seriously interrupted would have been a severe blow to the Allies…".[6]

If the inherent danger of a politically neutral (or worse, pro-German) Latin America had been obvious to Lansing and others in Washington, it was equally clear that it would take a yeoman's effort on the part of the United States to persuade Latin American governments to change their political positions. Years of distrust, the legacy of many heavy-handed United States interventions in the region, had produced an unwillingness to bend toward Washington. The result was that a number of diplomatic channels to Latin American nations had been effectively closed to the U.S. government over the years. Thus, in those countries it was all but impossible for the State Department, or military intelligence, to obtain reliable information about rumored German activities on the ground.

Washington and the ONI Seek Military and Mercantile Intelligence

Influenced by Lansing's concerns, throughout the course of World War I Washington administrators labored to obtain up-to-date intelligence and risk assessments with regard to Latin American nations. Interestingly, the surveillance of Latin American countries had actually begun decades before the War started. For the most part, gathering such data was a task largely given over to the U.S. military, and specifically to the Office of Naval Intelligence (ONI), the nation's oldest intelligence service.[7] In existence from the 1880s onward, the purview of the ONI was ramped up in the 1890s when it was tasked with specifically averting potential crises in the Western Hemisphere. This led to ONI naval attachés becoming involved in the pursuit of channels of communication with the Spanish government prior to the outbreak of hostilities between the colony of Cuba and Spain in 1895. They were further invested with the authority to gather intelligence on the world's navies (including those of Latin American nations) after Theodore Roosevelt became Assistant Secretary of the Navy in 1897.[8] Recognized as an official intelligence agency by Congress in 1899, the ONI gathered data about the Caribbean, Central America, and Mexico from the turn of the century onward.[9] The purpose of having ONI agents in the region was to keep tabs on all potential military threats, as well as to catalogue the merchant activities of British, German and Japanese nationals (i.e., rivals to U.S. business interests) in the Western Hemisphere.[10]

While it might seem on the surface that pursuit of commercial interests was inconsistent with the military posture of the ONI, the reality was that "[T]raditionally the Navy had permitted naval attachés and other officers to encourage American business interests overseas."[11] Where U.S. business activities seemed threatened, the Navy responded with detailed profiles on the local situation; these reports might even include plans for potential military intervention. Before World War I had begun, alarming briefs on political instability in Nicaragua surfaced at the ONI. Rather than allowing U.S. businesses in Nicaragua to fend for themselves, a 1910 "Nicaraguan War Plan"[12] was created in the event the situation escalated; this document indicated how far the U.S. government would be willing to go to protect the interests of its nationals in that Central American nation. A year later, the so-called War Plan Black, which circulated through Washington offices, contained precocious calculations of the risk of taking on the Germans throughout the hemisphere. This plan even "include[d] distance charts and sailing routes to Germany and comparative tables of both [U.S. and German] fleets".[13]

It was not just Central America and the Caribbean which attracted the early attention of the ONI. The service was also instructed to collect data about the military capacity of the southern cone nations of Argentina, Brazil and especially Chile, a consequence of the so-called "*Baltimore* Crisis" of 1891. In that year two U.S. sailors from the ship USS *Baltimore* were killed in a bar brawl in Valparaiso, Chile, and the antagonism that followed that event almost brought the two nations to the brink of war. By the second decade of the new century, with the potential of war looming large in Europe, Washington determined that misunderstandings between Latin American nations and the United States must be avoided at all costs.[14]

In addition to the ONI pursuit of "greater diplomacy," as the United States attempted to block German activities in the hemisphere prior to World War I, the acquisition of up-to-date reports on Latin American nations became increasingly critical to the mission of the service. While the United States government hesitated for several years to declare war against Germany, it definitely recognized how hamstrung the ONI's conventional fact-gathering approach had become in Latin America while war raged in Europe. This realization convinced Washington officials to expand the already existing corps of known ONI intelligence operatives with clandestine agents, many of whom were recruited from distinctly non-military backgrounds, including scholars from academe and specifically scientists from the AMNH.

Prior to this decision, in the years between the conclusion of the Spanish American War in 1898 and the entrance of the United States into the European conflict in 1917, most agents of the ONI were naval officers who received specific country assignments. But as direct U.S. involvement in World War I became inevitable, and with the situation in Latin America growing more precarious, the mission of the existing intelligence corps was broadened to include the recruitment of non-military assets, individuals who were deemed qualified simply because their previous work had been based primarily in the Caribbean or Central and South America.

In this respect the goals of the ONI closely paralleled those of the Department of State, which had similarly solicited academics for intelligence work during the Second Pan American Scientific Congress. Both the ONI and the diplomatic corps were on the lookout for unlikely, but promising, operatives who could maintain the "cover" of their real-time employment in Latin America, but who could be trusted to simultaneously (and secretly) report back on suspicious activity in the hemisphere to pre-arranged contacts in Washington or at Naval Headquarters.

Most of the academics who agreed to work for the ONI had no prior military or government service. Upon their recruitment, some of these newbies were given temporary rank within the Navy, although many were not. Regardless of their actual status within the Navy, such volunteers for espionage duty had to appear to be "in country," pursuing their customary line of work, and evincing no obvious connection to the U.S. government other than their citizenship. Anyone with previous knowledge of the region, especially of the hotspots like Cuba, the Dominican Republic, petroleum-producing Mexico, manganese-rich Brazil and nitrate-mining Chile might be considered a likely candidate for ONI service. The ideal covert informant would be someone who could blend into the surroundings without difficulty, who knew the terrain outside the major cities, spoke Spanish or Portuguese and/or German and whose normal activities required frequent travel through these nations; in short, someone who could fly under the radar. The familiarity of AMNH natural scientists with Central America, Mexico, the Southern Cone countries and the Caribbean placed them at the forefront of ideal candidates for the ONI and the State Department. Their attractive qualifications included: their ability to read, if not speak, German (the language of science at the time) and often Spanish or Portuguese; their advantage of already being armed (a necessity to bring down living specimens) to defend themselves; and the sheer improbability of scientist-scholars acting as government intelligence spies while simultaneously engaged in strenuous specimen collection. Beginning in 1916, and especially after the declaration of war in 1917, scientists from the AMNH were consequently recognized as having excellent operative potential.

Osborn Turns About-Face to Support the War

Responding to its expanded mission, the ONI extended the breadth of its intelligence services beyond Washington, D.C. to New York City, where a satellite office opened in early 1916.[15] Operating from its new headquarters at 2 Wall Street, the ONI was now able to probe the entire metropolitan area for significant new talent as well as surveil the political scene. Coincidentally, yet importantly, this office was only a short distance by subway from the American Museum of Natural History, which stood proudly on Central Park West and 77th Street.

The choice of New York City for the new ONI office had been predicated upon the growing concern Washington was voicing for the political climate of the New York metropolitan area. New York was home to approximately 750,000 German immigrants, a number which represented a bit more than a fifth of the total urban population, making New York "the world's third largest German-speaking city after Berlin and Vienna."[16] Because the United States had waited to declare war until 1917, many New York immigrant communities (and not exclusively German ones), had loudly proclaimed their doubt of, if not outright opposition to, the Allied cause.

The sentiments of the German community were echoed by other ethnic groups in the city, including residents of Irish background, who could not see themselves aligned with the British despite German aggression; and the more recently arrived Russian Jews, who considered the acceptance of the Czar as a wartime ally abhorrent, especially in light of the devastating anti-Semitic *pogroms* he had ordered against them in Eastern Europe.[17] Despite the fact that most of these individuals had fled to the United States to avoid persecution in Europe, Washington was concerned about the loyalty of all of these ethnic groups, especially if the United States determined to go to war against their "homelands."

Contemporaneously, in 1915, Theodore Roosevelt addressed a packed audience of the Knights of Columbus in New York, concluding his remarks with the rallying cry "There is no room in this country for hyphenated Americanism."[18] It was not at all unusual to hear pundits wonder out loud where New Yorkers' loyalties would lie in the event of declared hostilities. The perceived "New York situation" was also a stimulus for the creation of the "Preparedness Movement," a civilian response to the possibility of war. With New York City seen as being potentially vulnerable to foreign intrigue or worse, some prominent voices, such as J. Bernard Walker (a former editor at Scientific American), argued that the City's defenses were inadequate to protect it against attack.[19]

It's not surprising, therefore, that the Preparedness Movement first gained strength in New York City, later spreading throughout the country. Some of the "preparedness" activities spawned by the Movement, like the 1916 Flag Day rally in which 125,000 New Yorkers paraded down Fifth Avenue to give voice to their feelings about the nation,[20] foreshadowed future wartime appeals to patriotism. Other activities masked a sinister and markedly anti-immigrant posture.

The term "preparedness" was also used to characterize the plans for action being drawn up by multiple institutions across the nation as they faced the possibility of war. Museums and other cultural, educational and municipal organizations outlined their priorities in the face of such an eventuality, doing so well in advance of the April, 1917 declaration of war. By contrast, the American Museum of Natural History stalled on developing its formal response to the pending crisis until just after Congress made its decision. While sister institutions such as the Carnegie in Washington had drafted wartime procedures as early as 1915,[21] President Osborn's initial opposition to American participation in the "European" War and his reluctance to have his own staff (including *his* scientists) participate in the armed forces stalled matters at the AMNH. Ironically, it would be the staff which took matters into their own hands by developing a "preparedness" agenda which was subsequently presented as a *fait accompli* to the president of the Museum; ultimately the staff's plan was adopted across the institution.

Osborn's pessimism about the war was deep-seated, stemming from his isolationist views, an attitude echoed in his personal correspondence. As late as January 27, 1917, when others were aggressively urging the United States to become involved in the European conflict, Osborn wrote to his long-term friend E.B. Poulton, the distinguished Professor of Zoology at Oxford, expressing his hope that the U.S. would be able to stay out of the War. While he longed for a British victory, Osborn imagined that Germany could only be defeated by prolonged starvation; he thought that the allies would face food shortages as well. Unconvinced of Britain's ability to prevail in the event of attrition, as a student of science and technology he opined that, "[it] is impossible for flesh and blood to stand up against machinery."[22]

But by April 27, 1917, a mere three weeks after the U.S. declared war against Germany, and three months after writing to his English colleague, Osborn had apparently experienced a dramatic epiphany. Buoyed by his staff's preparedness plans which he was forced to acknowledge, he now expressed "his joy" to Poulton, to be "at last in the cause for which Great Britain and France are fighting."[23] Quite charitably for someone who was Theodore Roosevelt's lifelong friend, Osborn unexpectedly declared that T.R.'s political foe, President Woodrow Wilson, had redeemed himself in early April by urging the United States Congress to declare war against Germany.

For Osborn, this was a profound about-face. By late 1917, and into the next year, Osborn morphed beyond his initial endorsement of U.S. participation in the hostilities into a full-throated supporter of the Great War. Once a confirmed skeptic, Osborn emerged from his isolationist chrysalis as a newly minted super-patriot. By contrast to his previously whispered misgivings, he now loudly encouraged the participation of his museum staff in the war effort at every level, even approving of his own family's involvement, as well. In a June 11, 1918 letter to his cousin Henry, Osborn proudly boasted that many members of the Osborn family (including his own sons and daughter) were actively engaged in important war-time activities.[24]

Osborn's new, wartime vision for the Museum was implemented after the April 7th Declaration. Many of his "pet" projects echoed the prevailing thinking of the Progressive Movement, an intellectual current which had been attempting to change the habits of the American public for the better. Among these was the "Food Values and Economies" exhibition, an ambitious show mounted in the massive hall at the Museum's 77th Street entrance (where the Northwest Coast Canoe used to hang) in the Spring of 1917. This exhibition, perhaps the first of its kind in the nation, and certainly the only one held in a natural history institution, presented the public with displays on how to manage food resources during wartime. From the outset, it drew large crowds, resulting in high praise for the Museum and its war efforts and ironically, given his initial reluctance, for its "prescient" president. Especially noteworthy was the exhibition's emphasis on the conservation of existing food supplies and the attention it called to "the serious nature of the food crisis which the world is facing." This exhibition singularly tackled the role that "science" could play in "increased production, elimination of waste and intelligent selection of an economical dietary [sic]."[25]

"Food Values and Economies" opened a scant month after the declaration of War, in May of 1917 (Fig. 4.2). Visitors were enticed by its superbly rendered, artificial models of food (displayed with indications of their caloric and nutritional values; Fig. 4.3) which simultaneously furnished guidelines on how best to deal with potential food shortages. The exhibition made its mark by emphasizing previously overlooked, but locally available, food sources, such as kelp, shellfish, corn (not widely consumed in the northern portions of the United States at the time) and cottage cheese, all foods which could provide inexpensive alternatives to the consumption of meat which the government had rationed to feed troops slated for Europe.

The month before "Food Values and Economies" opened, the Museum launched a complementary exhibition on Military Hygiene. Precise and graphic in its displays, this exhibition focused on such issues as exactly how many calories (calculated at 4199) a soldier should consume to operate at peak performance; it

Fig. 4.2 American Museum of Natural History Food Values and Economies exhibition, view of central area, 1917. AMNH Asset ID 36698

Fig. 4.3 American Museum of Natural History Food Values and Economies exhibition, proposed dinner of 1087 cal, with models of soup, lettuce, meat, bread, butter and buckwheat cakes. AMNH Asset ID 36483

also showcased items found in standard mess kits, including gas masks and packets of sterile dressings, paraphernalia of vital interest to a public whose sons were soon going off to war (Fig. 4.4). Helmets, shoes and other necessities including water purification methods developed to prevent the horrors of cholera, dysentery and typhoid, were displayed. As stated in the American Museum Journal, the exhibit clearly presented the "needs and requirements of the modern soldier, and the part that science plays in supplying these needs."[26] In conjunction with the exhibition, the Museum published a small, inexpensive Handbook of Health in War and Peace, which was made available to the public to use as a guide to keeping healthy and maintaining food safety during wartime. Distributed to New Yorkers at the urging of President Osborn,[27] the Handbook contained much useful information including, but not limited to, instructions for dealing with the common household fly in an effort to eradicate typhoid and other diseases.[28] By October of 1917, the public's reaction to the pamphlet had been so positive that a revised edition was issued to be sold at local newsstands for "a nominal price."[29] Although it did not bring in much revenue, it advanced the public profile of the Museum, which had been Osborn's goal all along.

Osborn also authorized the creation of a War Relief Association, composed of thirty-one women who volunteered from across the staff, including the wife of Director Lucas.[30] According to the Association's guidelines, the trustees would grant three work hours per week to each woman employee to spend in a war-relief workroom furnished with sewing machines and fabrics; this group was also associated with the Red Cross, as Auxiliary No. 117. Just a few months into the effort, the auxiliary took pride in announcing that it already had provided socks, hats and dressings for soldiers and sailors. According to its records, sewing was confined to designated workshops, whereas knitting could be done in spare time on subways, in restaurants and even at the movies. In mid-January, 1918, First Lieutenant Harold Anthony, a Museum mammalogist stationed at Fort Dix, New Jersey, wrote to the committee to thank them for the scarves and hats they had knitted for his outfit.[31] By end of the War, the War Relief Association announced

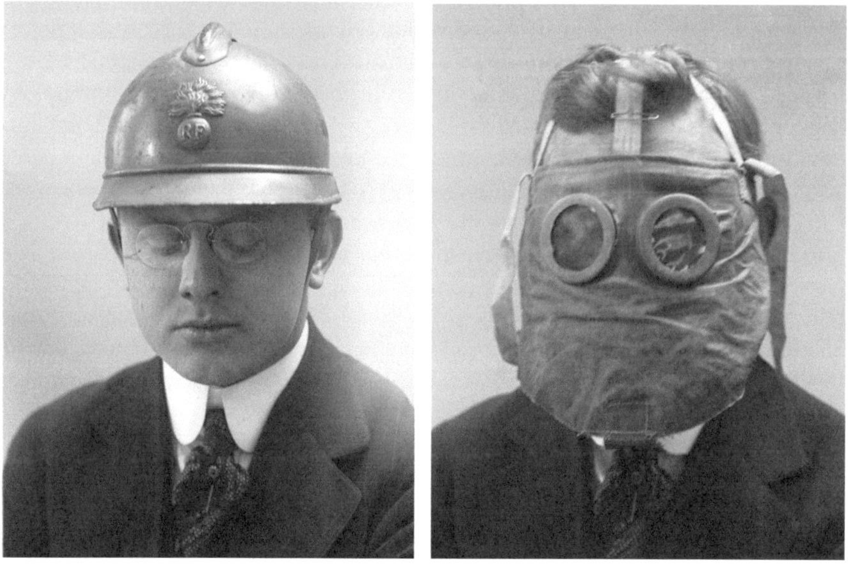

Fig. 4.4 Photographs from the American Museum of Natural History Military Hygiene Exhibition, 1917: left, a French steel helmet; right, a "primitive gas mask"

they had completed and sent "nearly 30,000 surgical dressings, 688 hospital and refugee garments, and over 900 knitted articles," with most of these designated for "Museum men" in service.[32]

In a totally different realm, Osborn commissioned the Head of Museum Construction to draft plans for a collapsible military field hospital. Although the archives do not confirm if the project ever proceeded beyond the planning stages, an illustration of his model hospital may be found in the Museum archives; it was later published in the Museum Journal.[33]

But that was not all. Osborn encouraged his staff to conduct safety drills in the center of the Philippine Hall, a routine which lasted for several months during 1917. The staff used old Springfield rifles to practice, weapons ironically left over from the Philippine War which were, for some unexplained reason, in the Museum's possession.[34] It is unclear if there was serious concern about the Museum becoming a target of aggression, but certainly greater New York City was considered vulnerable. There was much serious discussion among the staff as to which men would be required to drill and who might be excused. It was decided that those men who were over the age of fifty-five, "or who feel that they are not in proper physical condition" should not be expected to join the effort. Older men could participate in the exercises if they liked, but painters, roofers and those involved in Museum construction might be excused. When the drills ended after some months, some of the same rifles were lent to Columbia University at the urging of Invertebrate Zoology Curator Henry Crampton, who considered them necessary for alumni practice.[35]

Finally, Osborn took the extraordinary step of allowing the War Provost Marshal in New York City to set up a recruiting station for soldiers within the bowels of the Museum. Dozens of potential soldiers from the borough of Manhattan passed through the Museum doors wherein they were interviewed, measured, and processed before being sent to boot camp.[36] Curiously, the statistical material obtained in this manner remained at the AMNH, where it is housed in the archives of the Anthropology Division. To this day, an elaborate plaque is found on the wall at the West 77th Street entrance to the Museum commemorating "the men in the World War who entered the armed service of their country during 1917–1918 from this building under the auspices of Local Board Division N° 129, New York City."

But even all of these stepped-up war efforts had not been undertaken with sufficient alacrity to suit the newly patriotic Osborn. In a letter to Director Lucas in mid-June, 1917, he lamented that he had hoped "a strong initiative would be taken by the Museum independently, with my advice and counsel. I must confess," he continued, "that I am very much disappointed at the result. Thus far the American Museum has not begun to respond as other institutions are doing, such as Columbia University. Our men do not seem to have awakened to the fact that every man, woman and child in the United States has got to take part in the this great movement, and the sooner the better for the life and health of all concerned."[37] If Osborn was guilty of hypocrisy, he certainly did his best to conceal it, or was incapable of recognizing it.

One sticking point remained: what was to be done about the progress of science? Notwithstanding his newly avowed affirmation of volunteerism, Osborn faced serious quandaries when it came to deciding whether or not to authorize the continuation of research abroad during wartime. As it happened, Osborn would find himself having little say in this regard; events that had transpired by mid-April, 1917 spiraled far beyond Osborn's ability to influence them, ironically leading to the intersection of the seemingly disparate trajectories of the science-oriented American Museum of Natural History and the United States government, with its urgent need for up-to-date intelligence on Latin America. The manner in which protocols for research would change within this new environment, and the growing influence of the ONI, the State Department, the Council of National Defense and other government agencies which sought to recruit intelligence assets from the staff scientists of the AMNH, led to an extraordinary merging of political and scientific agendas. This new elision and its implications for the future of the Museum are discussed in the mini-histories which follow.

Notes

1. On November 1, 1914, off the coast of Chile, German Admiral Von Spee tricked a British squadron into attacking his fleet with disastrous consequences for the British forces. British losses after the attack totaled almost 1600, while the Germans sustained little damage. About a month later the two forces faced each other again off the Falkland Islands, but this time the British prevailed. Despite the ominous tone of these encounters, they were the only real naval battles in the Americas during World War I. See "The Battle of the Falklands" in Stephen R. Brown, A Most Damnable Invention: Dynamite, Nitrates and the Making of the Modern World (New York, Thomas Dunne Books, 2005), pp. 185–206. See also Chapter X on the activities of Rollo Beck along the Chilean coast.
2. Robert Lansing, War Memoirs of Robert Lansing, Secretary of State, (Indianapolis, Bobbs-Merrill Company, 1935), p. 310. See also Robert Lansing, "Drama of the Virgin Islands Purchase," New York Times Magazine, July 19, 1931, pp. 4–5.
3. Lansing, War..., p. 311.
4. *Ibid*, p. 308.
5. The role of the German government in advancing the Zimmermann telegram in Mexico, which purported to promise a return to Mexico of territory taken by the U.S. in the Mexican–American War in return for official Mexican support, is well known.
6. *Ibid*, p. 315.
7. See Jeffery M. Dorwart, The Office of Naval Intelligence: The Birth of America's First Intelligence Agency, 1865–1918 (Annapolis, U.S. Naval Institute, 1979).
8. Captain Wyman H. Packard, A Century of U.S. Naval Intelligence (Washington, Department of the Navy, 1996), p. 8.
9. *Ibid*.
10. Dorwart, The Office..., p. 91.
11. *Ibid*., p. 92.
12. *Ibid*., p. 89.
13. *Ibid*., p. 91. War Plan Orange considered the threat of the Japanese in the hemisphere.
14. Packard, A Century...p. 7. See also Mark Rice, "Transnational Business and U.S. Diplomacy in Late Nineteenth Century South America: W. R. Grace & Co. and the Chilean Crisis of 1891," in Journal of Latin American Studies, Vol. 44, No. 4 (2012), 765–792.
15. Packard, A Century..., p. 249.
16. See Steven Jaffe, New York At War: Four Centuries of Combat, Fear, and Intrigue in Gotham, (New York, Basic Books, 2012), p. 178.
17. Jaffe, New York..., pp. 184–186.
18. *Ibid*., p. 186.

19. Walker wrote a book entitled America Fallen! which was published in 1915. See Jaffe, New York…, p. 186.
20. Jaffe, New York…, p. 190.
21. See Castañeda, "The Carnegie Missions…", pp. 50–51. The Carnegie Institution was involved directly in the war effort as early as 1915.
22. Osborn to E. B. Poulton, January 27, 1917, New York Historical Society, Diaries of Henry Fairfield Osborn (hereafter cited as NYHS/HFO), Box IX, folder 1917.
23. Osborn to E. B. Poulton, April 27, 1917. NYHS/HFO, Box IX, folder 1917.
24. Osborn to Henry Osborn, June 11, 1918, NYHS/HFO, Box IX, folder 1918.
25. Announcement on "Special Exhibit on Food Conservation and Military Hygiene." June, 1917, Central Archives (CA) file 1069. Couched in the language of the Progressive Movement, the exhibition clearly stretched the boundary of traditional natural history museum offerings. For a discussion of Progressivism, see Helen Zoe Veit, Modern Food, Moral Food: Self-Control, Science, and the Rise of Modern American Eating in the Early Twentieth Century, (Chapel Hill, 2013). See also Roberta Marx Delson, "'The Bullet and the Crust': A World War I Exhibition on nutrition and food conservation at the American Museum of Natural History," Museum History Journal, vol. 11 (2018), pp. 1–23.
26. American Museum Journal (April, 1917), p. 274.
27. American Museum Journal (May, 1917), p. 355.
28. American Museum Journal (November, 1918), p. 623.
29. American Museum Journal (October, 1917), p. 420.
30. See "list of American Museum War Relief Association Workers for Regiment 15." n.d., CA, file 1069.
31. Letter of H.E. Anthony to Miss Hood, War Relief Committee, American Museum of Natural History, January 18, 1918, CA, file 1069.
32. Report of Lindsey M. Sterling, President, The American Museum War Relief Association, December 30, 1918, CA, file 1069.
33. "Model of Liberty Field Hospital Ward" designed by President Osborn and Superintendent of Construction de Beers. Negative no. 36756, CA, iconographic files.
34. Minutes of Preparedness Committee Meetings, April 19, 1917, CA, file 1069. These drills lasted until late June, 1917 when "those really interested have gone or will go to drill in earnest." Frederic Lucas to President Osborn, June 26, 1917, CA, file 1069.
35. Henry Crampton, the Curator of Invertebrate Zoology at the AMNH, but also Professor of Zoology at Columbia University and the Chief of Staff of the Columbia University committee on mobilization, asked Director Frederic Lucas even before the war declaration for permission to borrow some of the rifles for alumni training. Crampton to Lucas, April 2, 1917, CA, file 1069.
36. Osborn directed his construction staff to prepare a large sign for the west face of the Museum, which could be illuminated at night to direct recruits to the building. Osborn to Lucas, June 11, 1917, CA, 1069.
37. *Ibid.*

Chapter 5
On a War Footing: What to Do About Research in the Face of Aggression

Osborn was quite adamant when he reproached Director Lucas with his expectation that the staff must confront wartime exigencies with promptness. But his exhortation referred to officially sanctioned activities, such as the drills Osborn had authorized on the premises, and work for the Red Cross, or allowing his general (non-scientific) staff to enlist. When it came to the subject of research, Osborn was strangely reticent.

Faced with his silence and no clear direction on the matter, several months before the President of the Museum publicly acknowledged that scientific research outside the country might be impacted by the War, his staff of scientists had taken it upon themselves to consider the thorny questions of what kind and where to conduct studies during wartime. Discussions amongst the scientific staff began early in 1917, conducted principally during what was referred to as "faculty" meetings. In this regard, the simple truth was that Osborn's scientific staff was far ahead of "the boss" in planning for the inevitability of United States involvement in the War. With their several months' head start, these staff scientists were able to arrive at practical and timely solutions to potential research problems before the United States Congress actually declared war on Germany.

Bashford Dean, the retired Curator of Ichthyology (see Fig. 8.1), was the first to broach the subject of potential wartime involvement to his fellow Museum scientists. He did this by placing a formal resolution before a committee of his peers at their February 15, 1917 meeting. Taking the floor, and expressing the sentiment of those of his fellow scientists with whom he had previously consulted, he resolved "that the American Museum of Natural History, prepare a roll of its Staff, Scientific and technical, in the interest of National Defense, and that this shall include a classified subject index of the specific fields or subjects in which these men may be of service to the State; also that copies of this classified list be presented to the Council of National Defense and to similar organizations as the President of this Museum shall find expedient."[1] Assistant Curator Willard G. Van Name of Invertebrate Zoology seconded Dean, adding a fulsome warning that the Museum's staff should "respectfully express their anxiety at the delay on the part of

R. M. Delson, *Scientists or Spies?*, The Latin American Studies Book Series,
https://doi.org/10.1007/978-3-032-18117-6_5

Congress in taking measures for the preparation of this country for the possibility of war… in such a time this country should adopt prompt measures to place itself in the strongest position in respect to naval and military strength and *scientific preparedness* [my italics]."[2] The motion passed; apparently, neither Dean nor Van Name were privy to President Osborn's contemporaneous pessimism about the futility of war.

Once Osborn got wind of the staff proceedings, however, he was obliged to take action. Recognizing that the Museum scientists had, in effect, challenged his authority on this important issue, he knew that he had to act quickly in order to reassert control. Therefore the next time the staff faculty met, on March 6th, Osborn had a ready response, although initially he chose not to show his hand out of concern for the optics of what might be perceived as him taking over. Instead Osborn relied on Curator George Sherwood (the Secretary of the Museum) to put his resolution before the body. Acting as Osborn's proxy, and without actually mentioning the possibility of the U.S. entering the War, Sherwood resolved, as amended by President Osborn, that a new committee of seven or more members (with President Osborn present in an *ex officio* status) be created and empowered to "formulate plans and make recommendations as to how the employees as individuals or as a body may render aid to our government."[3] Osborn had effectively pounced at precisely the right moment to seize the Museum's readiness response from under the noses of his scientists. From this point on, it would not be the scientific staff which would decide how to face the challenges of war, but rather President Osborn, via his newly constituted committee. Not surprisingly, thus presented with a *fait accompli*, no one at the staff meeting raised an objection to Sherwood's resolution, and it passed.

Still wavering in his convictions about the War, Osborn stalled in setting a date for a meeting of this "new committee." The official "Preparedness Committee," as reconstituted by Osborn, was not convened until April 9, a few days after President Wilson's and Congress's declaration of war against Germany, and a full month after the vote was taken to formalize it. When the previously selected staff members of this new committee finally met, it was President Osborn who spoke first. By taking the floor before the others could express their opinions, he was clearly asserting the prerogative of his office; moreover, he continued to chair the session, an action which allowed him to shape the direction of the debate as he saw fit. This was scarcely the *ex officio* status which Sherwood had initially suggested.

When it came to the issue of research, Osborn had to tread carefully. While he personally favored the continuation of Museum fieldwork abroad despite the ensuing war, he did not, or dared not, say this publicly for fear of having his priorities (not to mention his patriotism) questioned. To deal with this conundrum Osborn resorted to a classic bluff: instead of revealing his actual preference to continue research outside the country at all costs, Osborn reported that "it has been suggested by a member of the Museum staff" (without naming the individual) "that, unless, there are some special reasons, we should not suspend our ordinary fieldwork this year."[4] This clever ruse allowed him to play the devil's advocate: Osborn appeared to be challenging the wisdom of this position as well as the continuation of overseas fieldwork. He reinforced this impression by pointing out that this decision would make it impossible for Museum men needed for military service to respond, especially if they happened to be out of the country pursuing research. While he personally hoped that someone would take the opposite point of view, all of the staff present at the session silently agreed that this situation would be problematic for the AMNH. Sensing defeat, Osborn moved to deflect debate at this point, suggesting that discussion be tabled until the following day when the full Scientific Staff was scheduled to meet. In deferring the discussion until the next day's larger caucus, Osborn appeared to be magnanimous and willing to take more opinions into consideration than he actually was. He seemed to be suggesting that the subject would now be broached in a manner in which he averred everyone could "express an opinion." The reality was, however, that the Museum president continued to hope to outmaneuver his staff.

On the next day, April 10, 1917, the Scientific staff held its own "Preparedness Meeting", which President Osborn attended and which, again, he also took the liberty of chairing. Clearly anticipating dissension, Osborn labored to give his men the false impression that the Museum was already in the throes of planning a wartime research agenda. He began by outlining a specific project he was championing that would involve future exhibitions on health and food. Dr. Spinden, from the Anthropology Department, interjected that an amplification of this exhibit could include an examination of the relationship between the native food supply and the broader food supply for the entire nation. The staff unhesitatingly gave its approval to this proposal and, as discussed earlier, an innovative exhibition on "Food Values and Economies" opened to the public in late May, 1917.

Having the staff in his thrall, Osborn then deftly shifted the discussion to the one issue that clearly was paramount in his mind. Masking his own opinion once again, he coyly asked the faculty body, "Would it be right for us to go right on with our fieldwork as certain members may be called upon to do work for which they are best qualified. In other words, should we be prepared to make some sacrifice of our scientific plans under these unusual conditions? The institution is hardly deserving of the name American if it is not prepared to make some sacrifice in the regular progress of scientific work."[5]

From the context it is clear that Osborn personally did not believe in the proposal. In essence, Osborn's question was yet another clever obfuscation of his own decision to continue AMNH expeditions regardless of his staff's opposition. However, in front of his scientists, Osborn naturally felt it was necessary to portray himself as patriotic; privately he worried what the net result would be for the Museum if it suspended all its initiatives, especially the important evolutionary work in Latin America. Thinking he was helping Osborn, Curator W. D. Matthew of Vertebrate Paleontology took what he thought was the President's cue and expanded upon the theme. He reported that the Vertebrate Paleontology Department thought that all field expeditions should be postponed "in view of the possible curtailment of Museum funds or staff and desirability to have new men in readiness for possible national call."[6] Anthropology Curator Pliny Goddard echoed Matthews, chiming in that the "Aztec" work his department had presently been conducting would be put on hold for another year.[7]

At this point in the discussion it was clear that the "faculty" saw the issue of research abroad in two distinct ways. Although most were likely willing to contribute their expertise to any branch of government (although intuitively this might entail working in a foreign country), at the same time they expressed caution about conducting any personal research outside of the country, fearing that such activities might place them in harm's way. Foreshadowing the sentiments which his fellow anthropologist Franz Boas would raise in his 1919 article in The Nation, Anthropology Curator Goddard opined that "We feel that expeditions outside of this country would involve too great difficulty to undertake and would place certain responsibilities upon our government. Personally, I had in mind a trip to British Columbia. *However, a stranger would be a subject of suspicion there and it might be of some difficulty to undertake it.* [italics mine] Therefore, we should not have any expeditions outside of the United States."[8] Speaking for the Department of Invertebrate Zoology, Assistant Curator Van Name was clearly of the same mind when he proclaimed that "It seems to me that any expeditions outside the United States Territory would be almost out of the question. The matter of passports and suspicion aroused by strangers would be serious obstacles."[9]

The Science Faculty discussion reflected its very real concerns that Museum personnel attached to field expeditions in other countries might not only be mistaken for spies but, by extension, their actions might reflect negatively on the Museum. They were hardly envisioning the scenario that was developing behind the scenes in which several of their colleagues might voluntarily engage in secretive spy work for the United States government while they ostensibly pursued their fieldwork abroad. Without hesitation, the Science Faculty voted nay on the issue of future fieldwork outside the United States, a recommendation that carried for the entire period that the country was at war. When the meeting finally concluded, their decision was forwarded to the members of the Museum's Executive Committee for their consideration.[10]

This was clearly not the outcome Osborn had wanted; his grand Latin American scheme, among other global projects, was now in real jeopardy. A more empathetic, or for that matter cautious, museum president might have accepted this outcome as the end of the matter. But the truth was that it really didn't matter to Osborn what his Science Faculty thought. No matter how sincere their consternation, the scientific staff did not, and would not, have the final say. The final decision could only be determined by the President meeting in closed session with the Executive Committee.

The EC, however, was not scheduled to meet until later in the year. When this committee was finally convened on September 19th, 1917, President Osborn sidestepped the earlier recommendations of his Scientific Staff, proposing instead that the Executive Committee fund a new expedition to New Guinea under the direction of the Swedish naturalist Dr. Eric Mjöberg. President Osborn argued that such an expedition would not represent a financial hardship to the AMNH as the American Geographical Society had already contributed $5000 toward the project, while the New York banker and philanthropist, Felix Warburg, was prepared to add an additional one thousand dollars to that sum once the expedition was authorized.[11] Apparently, Osborn thought that the Swedish citizenship of the expedition leader, and the fact that he was not a salaried Museum employee, would render this project palatable to the EC.

He was greatly mistaken. The EC rejected his proposal flat out, ruling decisively that "it was inadvisable for the Museum to undertake expeditions until after the war is over."[12] This recommendation, consistent with that of his scientific staff, must have taken President Osborn by surprise. Osborn had likely hoped that by bypassing the recommendation of his staff, and leaving the Executive Committee to make the final recommendation he could avoid the proposed moratorium on foreign expeditions. But the Executive board stood firm and vetoed the New Guinea fieldwork notwithstanding Osborn's comforting reassurances.[13] Conducting research overseas in a time of war was dangerous business; the higher-ups at the Museum, including the Executive Committee, had no desire to risk the lives of their highly esteemed scientists. The Executive Committee decision remained the official position of the American Museum of Natural History throughout the War, and seemingly that was the end of the discussion.

Osborn's Secret Support of Intelligence Gathering

However, unbeknownst to most Museum personnel, by this point Osborn had already managed to outmaneuver both his scientific staff and the Executive Committee. Even before the Science Faculty "Preparedness meeting" was convened on April 10th, Osborn had given his authorization to staff archaeologist Herbert Spinden to travel to Central America, ostensibly to do fieldwork for the Museum but simultaneously allowing him to offer his service to the government as a secret agent. The day after the Science Faculty meeting, in fact, the archaeologist left New York for Washington D.C., where he was sworn in to a joint commission as a naval officer and Office of Naval Intelligence agent #56 (see Chaps. 6 and 7). From there he proceeded to Central America where he met up with his Carnegie colleague, Sylvanus Morley, who was already in Central America working for the ONI. Theoretically in the region for several weeks to briefly look at Mayan inscriptions, in fact the two men would work for the next two years ferreting out intelligence on German activities and infiltrating networks in the region.[14]

Neither Spinden's fellow scientists, nor his direct superior, Curator Wissler of the Anthropology Department, were aware of the subterfuge involved in Spinden's acceptance of an ONI assignment. But Osborn not only understood that Spinden had been tasked with spying in Latin America, he gave this deceit his explicit approval. As he wrote to the doubtful Director Lucas, "Dr. Spinden and I think there are especially good reasons why he should go with Dr. Morley at the present time."[15] Had the Science Faculty (or the EC for that matter) been made aware of the true nature of Spinden's trip, there might have been an uproar. Osborn circuitously avoided a dustup by simply not informing anyone of what Spinden was up to.

This proved to be an inflection point for the AMNH: in the months to come, before the Executive Committee even met, Osborn would engage in the same order of subterfuge for a number of other staff scientists who were soon acting as government intelligence agents while they ostensibly conducted Latin American field work. As he saw it, the felicitous convergence of the interests of the AMNH with those of the United States government (especially the ONI and Department of State) in Latin America was "a marriage of convenience" in which both parties stood to benefit. If he had experienced any qualms about going behind his Museum colleagues' backs, he certainly didn't demonstrate them. If needed, he could always fall back upon the justification that a working relationship with Washington politicians was nothing new for the AMNH, since up to this point, the Museum had often relied on the government's diplomatic corps to facilitate entrée into less than friendly nations in Latin America for its scientists, including Chapman (as noted in Chap. 2).[16]

Indeed, similar negotiations between the State Department and the Museum had also occurred when Rollo H. Beck's expedition stalled unexpectedly in January of 1913, in Lima, Peru. Beck initially tried, by himself, to obtain the necessary permits from the Peruvian government when he began his expedition to collect oceanic and shore birds.[17] Only after repeated delays, and with the intervention of the State Department, was the collector allowed to proceed, and then only with the proviso that he not collect or shoot any of the shore birds.[18]

But such peripheral interactions by the Department of State in Museum affairs were one thing. Osborn's decision to sanction the direct involvement of his scientific staff in government espionage was quite another matter. Perhaps he rationalized his actions to himself by the fact that his scientists had volunteered for such assignments, each one knowing full well the limits imposed by the Museum research moratorium. Then again, perhaps Osborn recognized that the Museum had not always been forthright about its actions in Latin America; there had already been a certain degree of skullduggery involved in the AMNH expeditions to that highly competitive research arena. In early 1916, for example, when Spinden was sent on a secret expedition to Venezuela to collect artifacts, Curator Clark Wissler informed Osborn that "we are letting it be understood that he (Spinden) is going to Porto Rico (sic) as a blind to his movements."[19] In point of fact, when it came to Museum expeditions in Latin America, subterfuge, intrigue, lack of transparency, as well as suspicion of colleagues from other institutions, went along with the territory. The objective had always been to forward collections of important Latin American specimens and artifacts home to New York before a rival institution could get to them.

Alternatively, perhaps Osborn imagined that there would not be too great a difference in their assignments if AMNH scientists were to gather intelligence for the ONI (or other government bureaus) as long as they continued to collect for the Museum. There is something to be said for this argument. Viewed in broad perspective, the goals of locating, retrieving and forwarding requested data to safe hands was an identical mission for both the Museum and the government. In short, the very same skill set that a researcher would employ on site would serve him well for ferreting out data of a very different kind. In either situation, as a spy or as a natural history researcher, the scholar involved functioned as a detective who would "specialize in examining evidence and making informed judgments in cases where data is poor and evidence is ambiguous."[20] Perhaps Osborn rationalized that the difference between the jobs of Museum scientist and that of government secret agent were not as great as one might imagine.

It was a measure of the competence of the AMNH scientists who would eventually participate in this deception that they not only met the political and physical challenges presented by working in Latin America in wartime, but they were savvy enough to competently keep their *alter personas* hidden as they appeared to conduct their day-to-day scientific research. Provided that their home institution would allow them to participate in such "extracurricular" activities, such AMNH scientists familiar with Latin America, and who spoke Spanish (and/or Portuguese) and German,[21] could provide the key to collecting vital, wartime intelligence which otherwise might be unobtainable.

Importantly, even if he personally did experience qualms about his actions, President Osborn was more than willing to ignore them as the plans to secretly "loan" his men to the ONI or other governmental agencies provided the singular means to circumvent the AMNH moratorium. As the War played out, if there were any ethical considerations that actually bothered him regarding his decision to allow his men to engage in espionage, it was not evident from his actions. Just how involved in spying scientists from the American Museum of Natural History became in Latin America during World War I, as well as the consequences of their participation, has remained a secret to the present day. One fact is irrefutable: the boundaries between science and politics steadily dissolved as the War was being prosecuted and as these "borrowed scientists" became integral to the Latin American intelligence and political scene. This essentially unacknowledged story is revealed through the rest of this book.

Notes

1. "Resolution of Curator Emeritus Dean", in Resolutions on Preparedness, Faculty Minutes, 15 February, 1917, CA, File 1069.
2. "Resolution of Assistant Curator Van Name" in Resolutions on Preparedness, Faculty Minutes, 15 February, 1917, CA, File 1069.
3. "Resolution of Curator Sherwood as Amended by President Osborn", Faculty Minutes, 6 March, 1917, contained in Resolutions on Preparedness, Faculty Minutes, 15 February, 1917, p. 2. CA, File 1069.
4. "Minutes of the Preparedness Committee Meeting", 9 April, 1917, p. 2, CA File 1069.
5. Preparedness Meeting, 10 April, 1917, p. 3, CA File 1069.
6. Preparedness Meeting, 10 April, 1917, p. 3–4, CA File 1069.
7. Preparedness Meeting, 10 April, 1917, p. 6, CA File 1069.
8. Preparedness Meeting, 10 April, 1917, p. 5, CA File 1069.
9. Preparedness Meeting, 10 April, 1917, p. 6, CA File 1069.
10. This decision was reported to the Executive Committee of the AMNH on April 18, 1917. Minutes of meetings, Executive Committee of the American Museum of Natural History, unpublished mimeograph (hereafter cited as EC AMNH).
11. EC AMNH, September 19, 1917, pp. 39–40.
12. *Ibid.*
13. *Ibid.*

14. See Chapter 7 on Herbert Spinden's war service.
15. Osborn to Lucas, New York, 4 April, 1917. CA, AMNH, File 906.
16. See correspondence between Acting Secretary of State, Alvey A. Adee to F. A. Lucas, Director AMNH, October 24, 1912 and Adee to Leland Harrison, October 24, 1912. NA, RG 59, box 291, folder 2.
17. Osborn's letter to Professor John Bassett Moore of May 8, 1915 talks about his request to the State Department to enable Beck, who was then stalled in Lima, to proceed with his collecting. NA, RG 59, Box 291, folder 1. There is a lengthy correspondence in the State Department records concerning Beck; the Museum was billed $9.78 to cover the cost of a telegram to the American Legation at Lima to assist him.
18. A permit was issued from Lima on May 13, but a telegram to the Secretary of State explains that Beck cannot collect guano birds as the (Peruvian) "Foreign Office believes Minister of Hacienda cannot relax law against their destruction." Guano birds produced the nitrous fertilizer that was used globally and had accounted for substantial Peruvian revenues in the nineteenth century. Telegram from Howard to Secretary of State, May 14, 1913. NA, RG 59, box 291, folder 1.
19. Clark Wissler to Osborn, 6 January 1916, Division of Anthropology, AMNH File 224. See Chap. 7.
20. This was certainly the case for the academics who engaged in intelligence gathering during World War II, as Elyse Graham indicates in Book and Dagger: How Scholars and Librarians Became the Unlikely Spies of World War II, (New York, Ecco, 2024), p. 185.
21. President Osborn, himself, was fluent in German as numerous letters in his file written in German indicate. See, for example, the letter of Ernest Haeckel to Osborn, Jena, November 6, 1908, CA. file 811. Similarly, a letter from F. Deverman to Dr. William D. Matthew (AMNH Vertebrate Paleontology), Frankfurt, April 28, 1915, CA. file 818, is written entirely in German.

Part II
The Actors: Scientists or Spies?

Chapter 6
For Museum and Country: ONI Agent 270, Henry E. Crampton, the "Fixer"

If a drum roll had ordered the AMNH scientists recruited by the ONI to "fall in," they would have responded as follows: Agent 56, Herbert J. ("Joe") Spinden; Agent 139, Charles R. Eastman; Agent 241, Roy Chapman Andrews; Agent 245, George K. Cherrie; and finally Agent 270, Henry E. Crampton.[1] In addition to these *bona fide* intelligence assets, other scientists associated with the AMNH, including paleontologist Barnum Brown (for a short while Agent 233), mammalogist Harold Anthony and bird collector Rollo H. Beck, reported to government officials on the current political situation in Latin America at various times during the War, and immediately after. Other members of the Museum's scientific staff worked directly within branches of the Federal government during the War, including Frank M. Chapman, a distinguished commissioner for the American Red Cross who was subsequently dispatched to Latin America on the organization's behalf in the waning days of the War; Bashford Dean, Curator of Ichthyology, who served as a major in the Army Ordnance Department in Washington, D.C.; and the already very political Henry E. Crampton, who was appointed to both the high-level Council of National Defense and the ONI. Finally, a number of the Museum's staff (scientific and not) were drafted to serve in the military.

With the exception of those who were drafted outright into the armed forces, all of the scientists cited above who knowingly crossed the line between politics and science had done so with the express approval of the AMNH president, Henry Fairfield Osborn. Moreover, those who participated in government sponsored subterfuge by suggesting that they were engaged in their usual research while they were in actuality involved in developing intelligence did so with the understanding that they had willingly violated the Museum's overseas research moratorium. Sworn to secrecy, the scientist/agents who worked simultaneously for the Museum and the U.S. government were instructed by Washington that it was of the utmost importance that they prevent their colleagues at home from knowing the real nature of their activities. It is somewhat ironic that it was likely just as wise for these individuals to not share their assignments with their fellow scientists because they

R. M. Delson, *Scientists or Spies?*, The Latin American Studies Book Series,
https://doi.org/10.1007/978-3-032-18117-6_6

continued to collect their salaries as Museum scientists while they were in Latin America, along with additional monies provided by the U.S. government.

Yet there is little doubt that the recruitment of the AMNH scientific staff for espionage activities was undertaken with the express acquiescence of President Osborn and that of a very few other individuals (including Director Lucas) who represented the highest levels of the Museum's administration. By the end of hostilities in 1918, when it was possible to acknowledge that museum scientists had played a multitude of roles during the conflict, the Museum openly took pride in such efforts. It probably came as no surprise to the Museum community when the mid-1919 publication of the Annual Report of the Trustees of the American Museum of Natural History proudly boasted "[T]he Scientific Staff for the year 1918 numbered 54. Of these, 15 members have been directly engaged in the Federal Service, while many of the others have indirectly contributed to the winning of the war by furnishing expert information."[2] A masterful stroke of understatement, this acknowledgement purposely failed to disclose the often dangerous situations in which such "expert information" had been collected. Even at the conclusion of the War, discretion was the order of the day; not only did the U.S. government insist upon such secrecy at the time but, astonishingly, the work of these agents remained officially classified by the relevant government authorities until the 1990s.

All of the men involved in the secret provision of intelligence to Washington had one thing in common, i.e., the service they rendered to the U.S. government was either conducted in Latin America and/or related to Latin America in some way. The one exception to this rule was the service of paleontologist Roy Chapman Andrews, ONI Agent 245, who worked in China. The expertise of those who were recruited to work in Latin America ranged from anthropology to ornithology, ichthyology, paleontology and mammalogy (living as well as fossil forms). Each individual was an expert in some field of natural science relating to Latin America. Chapman, Cherrie and Beck studied birds of South America; Crampton oversaw evolutionary studies of living and fossil forms of invertebrates in the West Indies and British Guiana; Eastman was sent to Brazil to collect specimens of fossil and living fishes; Spinden delved into ancient Mayan cultures while Brown traversed Argentina, Mexico and Cuba in search of fossil mammals and, in the last year of the War, petroleum.

It was soon obvious that what the U.S. government could expect from these volunteers who spied in Latin America was highly qualified, intelligent observations of conditions on the ground. What was not completely apparent from the "special" arrangement between the U.S. government and the AMNH was that such agreements functioned as a *quid pro quo*. Hypothetically, at least, the American Museum of Natural History stood to benefit as much as, if not more than, the U.S. Government from having its scientists engaged in Latin American wartime activities. In Osborn's transactional framing of these arrangements, the delineation of Latin America's evolutionary history was at least as important as the Government's assignments, leading him to insist that his men take every opportunity to broaden the Museum's scientific collections and engage in *bona fide* research during the intervals in which they were otherwise engaged in intelligence gathering. He was also keenly aware that having his scientists in Latin America during wartime gave the AMNH an advantage over those rival institutions which had chosen to restrict expeditions abroad. Consequently, and without consideration of the difficulties each would face in maintaining the delicate balance between scientific and political interests, Osborn (a man not known for compromise) demanded that each scientific staff member who worked for the government in Latin America also pursue their specific research in earnest. From the Osborn's vantage point, in addition to sleuthing, these scientists were in Latin America to do "the Museum's work," to excavate and do fieldwork, or to examine local collections and, above all, to establish rapport with other local scientists under the rubric of "scientific Pan Americanism." Osborn viewed this latter objective as distinct from all other goals, one which could potentially be exploited after the conclusion of the War. Had they known about the secret espionage assignments of these AMNH scientists, no doubt their Latin American hosts and colleagues would have seen things quite differently.

In short, irrespective of the government agencies to which they were assigned, at no time were the AMNH staff scientists who operated as secret agents in Latin America unaccountable to the Museum administration. On the contrary, Osborn fully appreciated the irony inherent in his men undertaking double duty as "spies", a designation which could simultaneously provide cover for the territorial ambitions of the AMNH. Thus service to the nation proceeded in tandem with responsibility to the Museum. Each scientist was on his own to navigate a mutually beneficial détente between the two entities; each, as will be seen in the coming chapters, found the task exceedingly difficult.

Crampton's Multiple Roles in the War Effort

Arguably, the most significant AMNH scientist/informant on loan to Washington during the course of World War I was ONI Agent 270, Curator of Invertebrate Zoology Henry Edward Crampton (Fig. 6.1). Considered a brilliant zoologist by his peers, Crampton was also Professor of Zoology at Columbia University and a specialist in a wide range of mainly marine invertebrates (including, but not limited to, snails of the genus *Partula*)[3]; on *terra firma* he specialized in butterflies. Most of his early fieldwork had been conducted in Polynesia (where he would return when the War concluded), but from 1911 to 1915 Crampton shifted gears, overseeing several important expeditions to the West Indies on behalf of the AMNH and in conjunction with the New York Academy of Sciences.[4] As seen earlier, Crampton was quite savvy when it came to mediating potentially difficult situations; he evinced few qualms when it came to juggling conflicting loyalties. This particular "skill set" was evident in the masterful way he connived to have the New York Academy of Sciences extend the Puerto Rican survey to eventually cover the entire Greater Antilles, much to the benefit of the AMNH (see Chap. 3).

Fig. 6.1 Henry Crampton, date uncertain

This extension of the Academy's (and the Museum's) mission simultaneously allowed Crampton to familiarize himself with the politically and strategically sensitive West Indies. His experience in the Antilles, coupled with the time he had spent with Frank Lutz on a previous investigation of Mount Roraima in South America, and the wide network of personal contacts he had cultivated during the course of his scholarly career (including those he made at the 1915–6 Second Pan American Scientific Congress in Washington, D.C.), rendered him an "old Latin American hand" relative to the other, less experienced AMNH staff. Additionally, he was on close terms with some of the most distinguished scientific minds in Latin America and in the United States. It was not surprising, therefore, that months before his other Museum colleagues would volunteer for secret government work, Crampton was tapped for an important post in the newly created Council of National Defense; he also obtained agent status within the ONI. The "experienced" zoologist would subsequently, but unofficially, assume the mantle of "control" for those of his Museum colleagues who later joined him at the ONI as undercover agents, or who were dispatched on other intelligence-gathering missions south of the border.

Crampton's appointment to the prestigious Council of National Defense began in December, 1916. The main objective of the Council, to coordinate and oversee a national academic/university level and business response to the impending war, required appointees to not only work at the Council's headquarters in the nation's capital but to keep their assignments close to the vest. To fulfill the Council's mandate, only the most distinguished academic scholars and businessmen across the country were asked to join this advisory body[5]; Crampton landed an especially plum position as Vice Chairman of the Committee of Science and Research, which included Engineering and Education (later the name was changed to the Committee on Engineering and Education). His long-time associate, the Philadelphia engineer Dr. Hollis Godfrey,[6] was named Chair of the Committee, effectively becoming Crampton's boss.

Crampton served in this capacity at the Council until late 1917, during which time he was obliged to relocate from New York to Washington part time. Ensconced at the Council's headquarters, the zoologist acted as a conduit through which his fellow Museum agents could periodically report observations from their Latin American field assignments and/or petition him to ask for favors with regard to government service. Despite the fact that Crampton was a sworn agent of the ONI, he does not seem to have held a military commission, and he was not assigned active espionage fieldwork until the waning days of the War.

The Council as a whole, and the Engineering and Education Committee in particular, was committed to recruiting other suitable academics for future government work, a process which began in the uneasy months before the United States entered the War, and intensified when the country entered the fray for real, starting in April of 1917. In keeping with the theme of readiness, as well as the precocious understanding that this would be a "technical war", the Council was especially anxious to link scientists to the war effort, looking to recruit those who were capable of bringing rigorous training and methodology to government service. As Crampton would later explain to Osborn in one of his periodic progress reports: "It may be said that it was recognized at the outset that the War was a technical war and therefore requires the services of technically trained men and Institutions to continue the production of technically trained men."[7]

Crampton's appointment to the Council of National Defense came at the moment that the United States government had finally, but unofficially, recognized that its own participation in the Great War was unavoidable. As required, Crampton dutifully attempted to keep his appointment secret from his Museum colleagues, but he also had to overcome Osborn's reluctance to see him absent from his museum duties while he served in Washington. While not discussing his exact mission, the zoologist made it very clear to Osborn that his work was ultra-important, explaining to his boss that "the nature of the work in the particular field is such that no documents will be issued, certainly at the present time. I can only say, that the investigations have been effective on account of the comprehensive scientific methods which only can bring about the desired solution."[8] Crampton, apparently, quickly adopted Washington doublespeak.

Of course, to have one of its scientists chosen to be in this front line of national defense bestowed enormous prestige on the AMNH (and simultaneously on Columbia University and the New York Academy of Sciences). Osborn was understandably thrilled by the zoologist's appointment, but failed to keep his excitement checked; the scientific staff was soon aware that Crampton had become their designated contact in D.C. Seemingly unaware of the familiar adage, "Loose lips sink ships," Osborn responded without hesitation to a question put to him by the Museum's Preparedness Committee in April, 1917, concerning which agency might be approached for information on shaping future war policies. He suggested that they need only contact the Council of National Defense through their own colleague, Curator Crampton. "(Although) he has been very modest about it,"

Osborn confided to the committee "…he has been playing a very important part for the last eight months"[9] (in actuality, it had only been less than five). Not only did Osborn openly demonstrate pride in his curator but, in ignoring the inherent secrecy constraints both he and Crampton were under, he demonstrated his willingness to mine the curator's newly acquired position for all the prestige it was worth, or could potentially be worth, to the Museum, anonymity be damned.

To be fair to Osborn, keeping Crampton's new position a secret before the declaration of War would have been difficult, if not impossible, given the rarified, intimate world of the Museum's scientific staff. There is no doubt that when former Curator Dean suggested at the February meeting of the staff that a classified list of museum personnel with special talents be passed on the "National Defense Council" [i.e., Council of National Defense], he intended that such information be directed to Crampton.[10] Already deep into the Washington orbit, Crampton had become the "go-to" man for the AMNH for any matters having to do with the government.

During the period in which Crampton served as Vice Chairman of the Engineering and Education Committee (December, 1916 to September, 1917), he worked alongside Hollis Godfrey, enjoyed dining at the prestigious Cosmos Club (the preferred watering hole of the nation's most distinguished scientists) and only occasionally put in time at the Museum in New York. Crampton's long-term absence eventually began to weigh on Osborn, who viewed the time the curator was spending in Washington as a significant departure from the normal work hours curators usually spent at the AMNH. Above all, President Osborn worried that his actions might set a negative precedent for the rest of the staff. Further compounding this fear, Osborn was also fully aware that Crampton had previously demonstrated a chameleon-like ability to change loyalties, a facility which he had personally witnessed in the zoologist's dealings with the New York Academy of Sciences. Perhaps Osborn felt threatened.

In any event, by June, 1917, Osborn decided that Crampton's frequent absences from the Museum required action, a conclusion which subsequently devolved into an acrimonious dispute between the two men. Osborn balked at continuing the Museum's portion of Crampton's pay (the remainder was paid by Columbia University), on the grounds that the curator had not put in the number of hours required of a curator at the Museum. In an attempt to allay the President's doubts, Crampton informed Osborn that he had been in the Museum "at least" nine days during the month of June, 1917 and that, furthermore, he had spoken by telephone with Museum staff either from other parts of the city, or from Washington, whenever necessary. Characteristically sure of himself, Crampton boldly suggested that Osborn request one of the trustees provide additional funding so that the museum portion of his salary might be perpetuated without interruption. He reckoned that such a move would ease Osborn's sense of fairness to the remainder of his staff, while allowing an "important government official" [meaning himself] to continue his War work.[11]

Evidently caught off guard, Osborn eventually acquiesced to Crampton's suggestion and solicited the additional funds which would allow the curator to remain at his prestigious Washington assignment. But before the details were mutually agreed upon, he took the opportunity to throw a gentle barb at his man in Washington. It appeared that while working at the Council, Crampton listed his affiliation on the official Committee on Engineering and Education letterhead as Professor of Zoology at Columbia University, failing to make any mention of his AMNH curatorship. Viewing this omission as an unforgivable affront, Osborn cautioned Crampton that, "It seems important that the Museum should receive recognition in your Committee work, as well as Columbia University."[12] If Crampton was engaging in important work which might reflect positively on the AMNH, Osborn would insist that his institution get its due.

On the question of wages, Osborn conceded victory to Crampton by managing to convince trustee Cleveland Dodge (whose business dealings included Latin American interests) to direct non-appropriated ("NAFI") funding to compensate for Crampton's reduced curatorial salary.[13] But the reality was that both men had accomplished their aims: with the additional NAFI funds Crampton continued to receive the same total salary[14] that he previously had gotten from the Museum, his government duties and frequent absences from New York notwithstanding. At the same time, Osborn continued to bask in the reflected glory of Crampton's glamorous government service.

During his time on the Committee on Engineering, Crampton's tasks expanded far beyond his initial assignment of helping to recruit scientists deemed necessary for the War effort. In retrospect, the new assignments he subsequently received might be considered downright bizarre for a man whose line of work principally involved the study of snails. One example, in particular, stands out: within a short period of time (there is no date on these papers), Crampton was tasked by the Council with evaluating a chart which detailed the amount of armament needed to equip an "army of 1,000,000 men." By being asked to give his opinion on the survey, this assignment surely signaled that the zoologist had been given access to highly classified data. A copy of that chart (essentially a spreadsheet full of painstaking details; see Fig. 6.2 and Table 6.1) appears in Crampton's personal papers at Columbia University, an indication that following his time in Washington he spirited it away from the Council. This important chart diagrammed numerous details about the number of pistols, grenades, machine guns and other equipment which was either readily available to the U.S. government at the moment or, alternatively, still needed to be produced. Crampton's hand-written notes accompany each of the categories.[15]

It may be somewhat unnerving (or even alarming) to imagine that a zoologist quite unschooled in warfare might have played a crucial role in estimating the amount of equipment needed by the U.S. military not only to defend the homeland, but to carry the fight abroad. Nonetheless, Crampton was apparently considered enough of a materiel expert, or at least he passed himself off as one, to be given this assignment. Undoubtedly his intelligence, and the thoroughness of his research methods, not to mention the organizational skills provided by his previous work in classification, made up for his lacunae in the military arts. It was in this newly assumed identity as "arms expert" that Crampton would also become deeply involved in the search for manganese from Brazil, as well as other minerals equally critical to the manufacture of weaponry, including the Chilean nitrates essential to the manufacture of explosives.

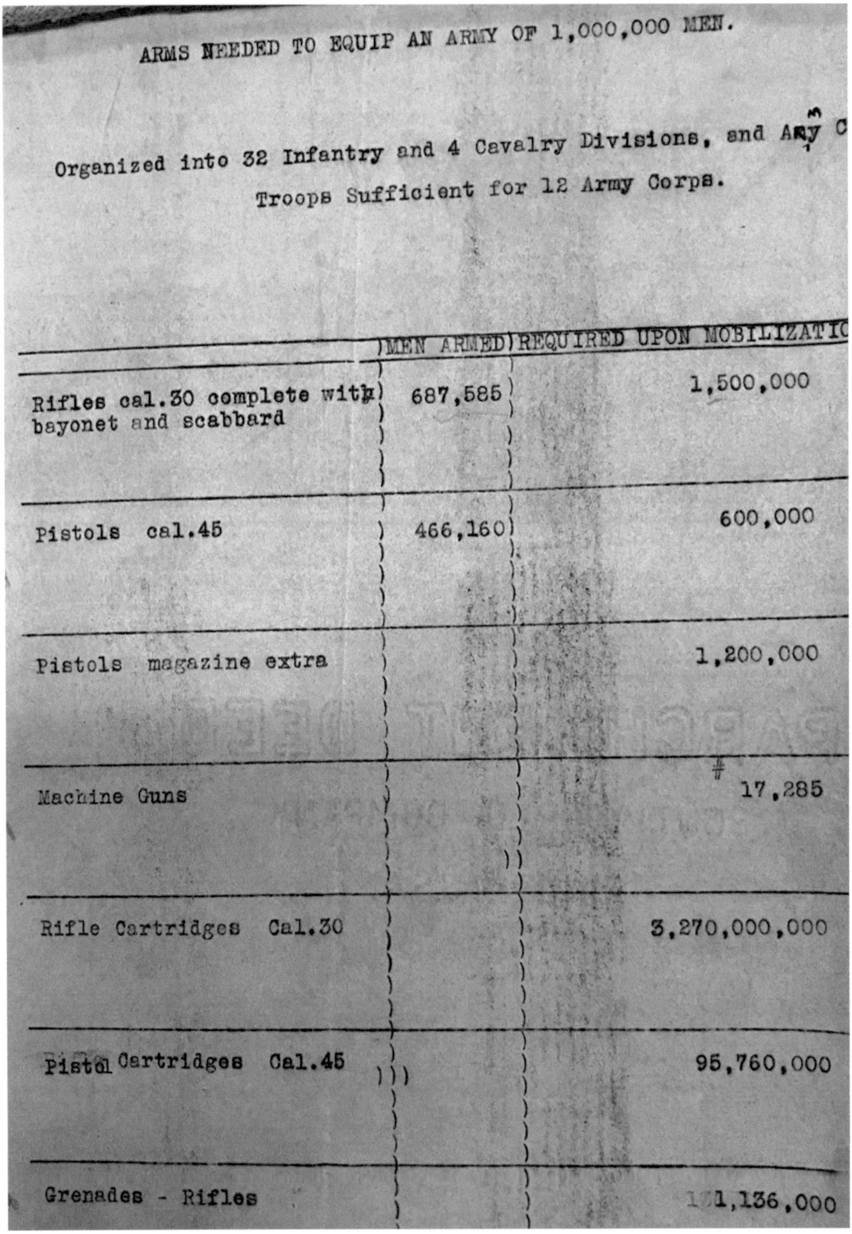

ARMS NEEDED TO EQUIP AN ARMY OF 1,000,000 MEN.

Organized into 32 Infantry and 4 Cavalry Divisions, and Army C

Troops Sufficient for 12 Army Corps.

	MEN ARMED	REQUIRED UPON MOBILIZATIC
Rifles cal.30 complete with bayonet and scabbard	687,585	1,500,000
Pistols cal.45	466,160	600,000
Pistols magazine extra		1,200,000
Machine Guns		# 17,285
Rifle Cartridges Cal.30		3,270,000,000
Pistol Cartridges Cal.45		95,760,000
Grenades - Rifles		1[illegible]1,136,000

Fig. 6.2 Photograph of first page (of 4) of typewritten table of data regarding military materiel needed to equip an army of one million men, prepared by Henry Crampton in 1917 for the Council of National Defense. Full dataset in Table 6.1. Original in Columbia University Rare Book and Manuscript Library, Special Manuscript Collections, Henry E. Crampton papers, Box 3. Public domain after expiration of copyright

Table 6.1 Arms needed to equip an army of 1,000,000 men. Organized into 32 infantry and 4 cavalry divisions, and army corps troops sufficient for 12 army corps

	Men armed	Required upon mobilization	In hands of troops and in store	Under manufacture	To be supplied	Additional to be supplied for four months' warfare	Total to be supplied	Remarks
Rifles cal. 30 complete with bayonet and scabbard	587,585	1,500,000	615,000	75,000	810,000	250,000	1,060,000	75,000 are to be completed by October, 1917
Pistols cal. 45	466,160	600,000	100,000	65,000	455,000	100,000	555,000	
Pistols magazine extra		1,200,000	200,000	130,000	870,000	200,000	1,070,000	
Machine guns		17,285	1306	4125	11,852	5761	17,613	4000 to be completed by Nov. 16, 1917. 125 to be delivered by May 1, 1917. Money has been appropriated for 600 but is held by Sec. of War to await results of tests by Board May 1, 1917
Rifle cartridges cal. 30		3,270,000,000	181,000,000	140,000,000	2,949,000,000	2,250,000,000	5,199,000,000	140,000,000 to be completed by Sept. 30, '17

(continued)

Table 6.1 (continued)

	Men armed	Required upon mobilization	In hands of troops and in store	Under manufacture	To be supplied	Additional to be supplied for four months’ warfare	Total to be supplied	Remarks
Pistol cartridges cal. 45		95,760,000	5,000,000	30,000,000	62,760,000	90,000,000	152,760,000	15,000,000 being made and 15,000,000 being purchased
Grenades—rifles		1,136,000	1000	75,148	1,058,852	1,136,000	2,194,852	13,648 can be completed April 1, 1917
Grenades—hand		1,136,000		105,000	1,081,000	1,136,000	2,167,000	105,000 to be completed by Oct. 31, 1917
3 inch field gun batteries		449	135	80	236	652	888	20 batt. to be completed April 1, 1917 20 batt. to be completed April 18, 1917 40 batt. to be completed June 30, 1918
3.8 inch Howitzer batteries		211	7	8	196	306	552	Batt. 8 to be completed June 30, 1918
4.7 inch Howitzer batteries		25	25	2	–	–	–	To be completed April 1, 1917

(continued)

Table 6.1 (continued)

	Men armed	Required upon mobilization	In hands of troops and in store	Under manufacture	To be supplied	Additional to be supplied for four months' warfare	Total to be supplied	Remarks
4.7 inch gun batteries		105	10	13	82	215	297	4 batt. to be completed April 1, 1917 9 batt. to be completed June 30, 1918
6 inch Howitzer batteries		105	8	16	82	215	297	2 batt. to be completed June 30, 1917 15 batt. to be completed June 30, 1918
9.5 inch Howitzer batteries		105	0	23	82	152	254	10 batt. to be completed Dec. 30, 1918 13 batt. to be completed April 30, 1919
3 inch field gun ammunition		3,250,332	631,871	909,249	1,709,212	21,552,000	23,861,212	
3.8 inch Howitzer ammunition		1,620,803	29,546	41,816	1,549,441	10,128,000	11,677,441	

Keeping these resources out of German hands became one of his and the government's highest priorities.[16]

Another equally surprising document in Crampton's personal papers reveals that Hollis Godfrey also requested the zoologist to critique a plan for "aeroplanes," which were destined for wartime use. Instead of evaluating munitions, as in the chart cited above, for this assignment Crampton was required to comment on the efficacy of flying vehicles, including "battleplanes," "bombcarriers," "high speed pursuit type" machines or "heliocopters" [sic]. These devices (listed in another chart in Crampton's papers) were divided into "land types", "convertible water types, hydro-aero-planes" as well as water types which were non-convertible, i.e., "large flying boats."[17]

Exactly what Crampton knew about such flying apparatus is unclear, nor is there any evidence that Crampton was ever "airborne" during this part of his lifetime. On the other hand, the Council of National Defense routinely assigned unlikely candidates to extremely important committees, so Crampton's assignment was not necessarily out of the ordinary. To cite just one example, the Council appointed the Wall Street financier Bernard Baruch[18] to a position of an advisor to its Committee on Raw Materials; subsequently, Baruch reported directly to the General Munitions Board. Obviously, the qualifications for participation in such tasks varied; a particular candidate might have singular experience or insight that made him (or her) uniquely suited, without necessarily having a background in defense. Parenthetically, it was becoming patently clear that in the United States, as it had been in England before that, on the whole, politicians and military officers were often "completely ignorant of science."[19] Crampton's specific contribution to the nation's development of armament or flying plans is unclear, but at least he could apply a logical, scientific mind to the issues before him. He could understand the evidence presented, look for patterns inherent in the data and come to a logical conclusion. Godfrey, for one, valued Crampton's opinion enough to believe that he could be trusted with obviously highly classified data. It is likely that Crampton's contemporaneous status as an ONI agent acted as a guarantee of his discretion.

The high esteem in which Crampton was held in the nation's capital reverberated outside Washington. Despite the obvious need to keep quiet about his Washington role, one after another of his colleagues came to appreciate Crampton's new-found authority, or connived to take advantage of the curator's apparent access to the halls of power by beseeching him for favors. These appeals suggest that Crampton hardly took pains to disguise his Washington work. On the contrary, convinced of the zoologist's importance, several of his peers wrote to him as supplicants, begging the curator to apply his hallowed status as a Washington insider to secure commissions for them in the military,[20] or to be allowed to enter government service at any level.[21] Amusingly, Nicholas Murray Butler, the President of Columbia University, demanded that Crampton have some of his "secret service people" [there is no evidence that Crampton had any] keep an eye on a crank who had been sending him pro-German poetry.[22] If it is unclear from hindsight exactly how much authority Crampton actually had during World War I, there were apparently many who were deeply convinced of his importance and treated him accordingly.

Within the AMNH community, Crampton was instrumental in helping Bashford Dean (the Curator Emeritus of Ichthyology) secure war work in Washington. Dean's studies on fossil armored fishes were well known to the scientific community, but the curator was also something of an expert on medieval armament. Today he is as well known for the still highly respected installation of medieval armor at the Metropolitan Museum in New York, a project which he oversaw, as he is for his work in ichthyology. Dean implored Crampton to put in a good word for him when he interviewed for a position at the War Department in the capital. "I have heard so much of the splendid service you have been doing for the Government," Dean flattered Crampton in June of 1917,[23] clearly unaware (or perhaps not caring), that Osborn was having second thoughts about Crampton's dual appointment at exactly this time. The zoologist didn't hesitate to help his colleague and as a result Dean landed a position in the Ordnance Department of the Army. Given the rank of Major, the fish specialist was set to work comparing modern armament to medieval prototypes, and his duties included designing helmets for the troops.

With regard to Museum personnel, however, it was probably Barnum Brown to whom Crampton ultimately proved most helpful. Like most of his colleagues, the paleontologist was eager to do his patriotic duty and apparently volunteered his services to the ONI. Brown clearly had the appropriate Latin American credentials for such service and Crampton championed him to the Naval Intelligence brass. In his "Memoirs" Brown would later claim that the ONI had indeed tried to recruit him, but that he had ultimately refused the Navy's offer of a job in "foreign service."[24] Crampton saw the situation differently and related the incident in a personal letter to State Department official Frank L. Polk. According to him, the ONI had successfully recruited Brown for intelligence work in Mexico, which the paleontologist would have pursued were it not for the fact that Osborn had (inadvertently?) revealed his secret appointment in a letter to a government official.[25] It is hard to imagine that Osborn could be so naïve as to publicly reveal that one of his men was also an ONI agent. Perhaps it better served Osborn's, and the Museum's, interests to have Brown out of the direct spying business.

His cover blown, the ONI dropped its plans to place Brown in Mexico. But, as Crampton complained to Polk, Brown should have been working for the government in Latin America because he had substantial prior experience in Mexico (and Argentina), as well as fluency in Spanish and familiarity with Latin American people. He was therefore, according to Crampton, quite capable of working in any Latin American country[26] and cultivating the hearts and minds of important U.S. neighbors. Apparently Crampton did manage to influence some channels, and Brown was ultimately sent to Cuba by the U.S. Treasury Department in the Fall of 1917 (see Chap. 10) to pursue a secret mission investigating that nation's petroleum drilling potential.[27]

Crampton monitored the work of other ONI agents from the Museum including Spinden and Eastman. He also supervised the participation of AMNH mammalogist Harold Anthony in the expanded West Indian survey. Anthony had substantial Latin American experience prior to the War, including a stint at the Canal Zone in Panama,[28] and was therefore a likely addition to the Antillean survey when Crampton arranged for him to visit Puerto Rico; he had already begun work on the island. In June of 1916, after his unsuccessful Venezuelan adventure, Joe Spinden was assigned to work with him. This would turn out to be a disastrous partnership. Grumbling about the situation in a letter to Crampton, Anthony wrote that "the short time that Spinden and I had together pretty thoroughly exploded the theory of cooperation along two such very different lines of work [archaeology vs. mammalogy]."[29]

It is likely that later on Crampton, already at his Washington post, also agreed to Anthony's follow-up visit to Cuba in the early months of 1917. This was framed as an extension to his earlier Puerto Rican excavations, but Anthony found much of his research time hampered by the actions of "revolutionaries", an assessment that he likely passed on to Crampton. Interestingly, Anthony's insightful observations on these Cuban disturbances were not kept secret but were forwarded to the Museum and released to the morning papers by the AMNH Public Information Committee.[30]

A curious document found in Anthony's file in the Mammalogy Department, dated April 14, 1917 from San Juan, suggests that Anthony was also reporting directly to Washington. In what was clearly a personal letter, one Earle Vendig informed Anthony that he had already tipped off Herbert Spinden of the Anthropology Department (who was at the time initiating his stint in the ONI) to the existence of an ancient "stone collar" kept by the Manati (Cuba) Chief of Police in his office.[31] One suspects this letter was code for suspected German activities in the region, as United States Marines stationed in eastern Cuba had already voiced their suspicions of the German (or pro-German Cuban) managers of the nearby Manati sugar mills.[32] In any event, no stone collar from Manati was collected by the AMNH from this locale, making it very likely this was indeed coded information that Anthony could pass on to the U.S. government.

As a Washington insider, Crampton's importance to the Museum lay beyond the prestige of his position on the Council. His monitoring of staff scientists Anthony, Brown, Dean, Spinden, and Eastman effectively coordinated the Museum's response to the War in the early stages of the conflict. But, unfortunately for the zoologist, while serving as the primary advisor to all of these Museum assets, and after having successfully negotiated a rapprochement with President Osborn, Crampton's cherished position as an academic/scientific advisor to the United States government came to a surprising and abrupt end, when a reorganization of the Council of National Defense in late September, 1917 resulted in the loss of his vaunted position. Crampton must surely have panicked when he realized that his lofty role with the government was finished, notwithstanding the continuation of his ONI credentials. In anticipation of his return to New York City, Crampton was clearly concerned about protecting his curatorship. To avoid a potential dustup, he thus took the liberty of extracting some highly flattering portions of a review of his work, from "one who is high in the councils of the administration," which he then had forwarded to Osborn and the trustees. This "confidential" document commended Crampton in glowing terms for "the highest imaginable service to the country-service which is on a plane of its own because in the very nature of the case it cannot be properly appreciated or linked with his name." If Osborn, or the Trustees, had seriously entertained thoughts of dismissing the vainglorious

Crampton from his curatorship of Invertebrate Zoology at the AMNH, this assessment surely militated against that possibility, or at least insulated the zoologist against an awkward homecoming.[33]

Crampton reluctantly resumed his posts at the Museum and Columbia University, but he never relinquished his hope of being reinstated as a member of Washington's inner circle. He attributed the loss of his Council position to his association with Godfrey who, he claimed, was "in such disfavor that his former associates are also non grata,"[34] rather than to himself. Presumably unbeknownst to the AMNH administration, Crampton would pursue multiple leads for months to come in the futile hope of securing another equally important government position.

He began this search while still in Washington by soliciting the Bureau of Censorship and Information, to which he offered not only the benefits of his personal expertise, but his ability to contact a large number of "patriotic, able and high class workers for the cause." These were individuals whom he deemed capable of establishing close relationships with "intelligence groups of the army and navy" and who could help with censorship[35]; presumably he was referring to Museum personnel. The very fact that his letter to the Bureau of Censorship was written on AMNH letterhead suggests that Crampton had few reservations about drawing the Museum and/or his colleagues into his schemes. The apparent lack of response from the Bureau of Censorship in Crampton's files, on the other hand, is a good indication that the agency summarily dismissed his offer.

Still not discouraged, the wily curator had another strategy up his sleeve. Reworking a theme which he credited to Charles Tilden (the new secretary of the General Engineering Section of the reorganized Council of National Defense), Crampton expressed his growing concern about the patriotism of the resident alien population of New York City. Writing to C. F. De Woody, the head of the New York branch of the Bureau of Investigation (part of the Department of Justice and the forerunner of today's FBI), Crampton proposed that he and two of his friends (one was the attorney Archibald Ewing Stevenson, the other unnamed) offer their services to spy on "certain Germans and pro-Germans" resident in New York City.[36] Crampton took the liberty of attaching to his letter an embarrassingly biased sample roster of the sort of names he could ferret out once he was "onboard".[37] Once again, as in his missive to the Bureau of Censorship and Information, Crampton wrote on Museum letterhead to solicit a position with the government, apparently without any regard to how his action might reflect upon the institution.

Reading his characterizations, it is clear that the individuals on Crampton's sample list of German "evil operators" earned their epithets on seemingly the flimsiest of grounds; in several cases warnings of "suspicious behavior" were applied to the spouses or family of suspected individuals for having expressed pro-German sentiments. One Harold Schubart, for example, who was "doing war work in Cable Censor's office, New York" (was this a cheap attempt to get back at the censors?) is cited because his grandmother, mother and wife "are open pro-Germans, although he himself is regarded as loyal." Crampton expressed the view that Schubart's family "are in a position to secure information inimical to our national interests."[38]

The Department of Justice remained unimpressed, steadfastly rejecting Crampton's attempt to obtain "special agent" status for the threesome for ferreting out potential "evil operators". Instead, the Department suggested to Stevenson that the three friends continue to transmit these types of reports "without the formality of these official appointments."[39] This was a gentle rebuff, only nominally rescinded in the summer of 1918 when Crampton finally received special agent status at the Department of Justice, but without compensation.[40] Clearly a second class arrangement, his previous government service as well as his ONI connections appear not to have mattered at all to the Department of Justice. But whatever work Crampton eventually did for the Bureau, it was limited to simply submitting names of suspect individuals. Thankfully, he does not appear to have been aligned with the New York branch of the American Protective League (APL), an organization of privately financed thugs who intimidated and harassed citizens, while the Department of Justice looked the other way.[41]

It is obvious that none of Crampton's proposals for assisting the Bureau furthered the interests of the AMNH, nor did they advance the Museum's Latin America agenda. On the contrary, it would seem that Crampton was rapidly losing interest in the Western Hemisphere and was instead planning to return to his earlier research on snails in Polynesia. As proof of his change of heart, before he accepted his appointment to the Council of National Defense in 1916, Crampton applied for,[42] and received,[43] a promise of funding from the Carnegie Institution for future work in the Pacific Ocean Society Islands and Marquesas Islands. This proposal was described as a "continuation of research in Polynesia upon the variation, distribution and evolution of land-snails belonging to the genus *Partula*" and was subsequently funded. His ongoing "war work" at the Council, however, precluded him from simultaneously accepting the monies which the Carnegie had awarded him.

In January, 1918, Carnegie Institution President Robert Simpson Woodward unexpectedly informed Crampton that the $2000 grant which he had been given for future research in the Pacific would now revert back to the Institution for "other and pressing purposes."[44] This was the last straw: Crampton's one remaining possibility to work outside the New York area was about to evaporate. No introvert, Crampton literally begged the Carnegie to reconsider, suggesting that he would be willing to take a reduction in funding from the full $2000 to just $600–$750. He even magnanimously (ironically?) offered to return the remaining monies to the institution for its "pressing war work."[45] His strategy seemingly worked, and the Carnegie eventually agreed to support his now substantially reduced budget.[46]

Even with the promise of funding for overseas research, as the War dragged on Crampton continued to harbor fantasies of his own self-importance in the intelligence-gathering business. Seeking to impress the president of the Carnegie Institution, Crampton disingenuously informed him that "I am still called to conferences on the war organization at Washington…certainly I would not plan to be away from America during the next few months unless I had the assurance of Colonel House, for the administration."[47] Colonel Edward Mandell House was a chief advisor to President Wilson, who (together with geographer Isaiah Bowman) was responsible for organizing the "Inquiry," the ultra secret think tank which operated out of New York City. While Crampton probably knew House, it is unlikely that the zoologist was on intimate terms with the presidential advisor. In point of fact, in October of 1917, shortly after his return to the Museum, Crampton had unsuccessfully tried to join the staff at the "Inquiry" by directly petitioning House. Unconvinced of Crampton's usefulness, House replied noncommittally to the curator's request, and offered little hope for a future appointment.[48]

Crampton was at last able to return to the field in the late summer of 1918 when he headed back to British Guiana and Brazil (especially Mount Roraima where, in 1911, he had previously collected with Lutz).[49] Surprisingly, in light of his failures to secure any government position, in the waning months of the War, and already in his late forties, Crampton appears to have taken the plunge: he volunteered to work as an active undercover agent for the ONI in the field, if an opportunity presented itself. Throwing caution to the wind, he applied for what could have easily been the most dangerous assignment of his career. On August 24th, 1918, ONI administrators wrote to Commander L. R. Sargent, head of U.S. Naval forces in the Canal Zone in Panama and the contact for ONI agents in the field, to offer the services of "an able scientific man who collects marine invertebrates and paleontological material experienced in Latin American countries. Answer quick if you want him."[50] We cannot be sure that the agent the ONI was alluding to was Crampton, but the description certainly fits that of the invertebrate zoology curator, as does the chronology. It also coincides with the time frame of Crampton's collection trip to Latin America during the late summer of 1918, supposedly undertaken on behalf of the Museum. What government work was done during this escapade is unknown, but following his stint in South America, Crampton calmly returned to his AMNH curatorial role and professorship at Columbia University until the termination of the War a few months later.

With his reputation as a Washington insider curiously still intact in New York, Crampton continued to be besieged with personal requests to intervene on government matters, even after the War concluded. In December of 1918, for example, David Eugene Smith, a professor of mathematics at Teachers College, Columbia University, and a colleague of Crampton's, appealed to the once powerful zoologist to help remove his name from a list of German sympathizers.[51] There is no way to know whether Crampton obliged.

Aside from the occasional request, Crampton's stellar intelligence career seemingly ended with a whimper. But the ever resourceful scientist, who seemingly was always capable of re-inventing himself, was not yet finished. Unlike other agents at the Museum who ended their affiliations with intelligence services, Crampton chose to maintain his "Agent 270" ONI status even after the War concluded, and he had returned to his "daytime" job as a distinguished academic at Columbia University and the Curator of Invertebrate Zoology at the AMNH. Two years later, he was able to finalize his plans for a return trip to the Pacific where he would pursue research using the funds previously granted by the Carnegie Institution. But his venue had changed; he would now pursue specimens in the Northern Marianas Islands and not in the Society Islands or the Marquesas as he had originally intended. It is unclear whether Crampton chose to work in the Northern Marianas or whether he picked that location because the ONI urged him to take on the additional task of assessing the extent of Japanese penetration of the region.[52] Held by the German government before World War I, the Northern Mariana Islands territories became part of a larger South Pacific Mandate under the jurisdiction of the League of Nations; they were subsequently ceded to Japan when the War ended. Under Japanese authority these holdings were located sufficiently close to United States-held Guam and the Philippines to warrant Washington's heightened concern.

On assignment for the ONI during August of 1920, the ever-intrepid Crampton set out for the Pacific ostensibly to resume collecting land snails (genus *Partula*) with the unutilized grant from the Carnegie Institution.[53] Prior to 1920, Crampton's research in this region had concentrated on snail specimens from Tahiti and Moorea, but his current itinerary now took him to Guam, and thence to Saipan, in the northern Marianas. Getting to this destination required the agreement of the unsuspecting Japanese Foreign Office, which Crampton was able to obtain for himself and his son Henry Jr., who accompanied him on what was clearly a Washington-sanctioned spying reconnaissance. Crampton was "authorized to obtain information for the Office of Naval Intelligence, concerning which he received *verbal instructions* (emphasis mine)." No receipts would be required of him, but he was instructed to send back information through consular officers.[54]

Exactly what intelligence Crampton gathered on the Japanese is not known but, to his credit, he did manage to collect enough material to eventually publish Vol. II of his Studies on the Variation, Distribution, and Evolution of the Genus *Partula*: The Species of the Mariana Islands, Guam and Saipan, in 1925.[55] Although he remained a Professor at Columbia, Crampton left his formal curatorship at the American Museum of Natural History between late 1920 and 1921 and was given an honorary appointment as Curator Emeritus. Perhaps the Museum had had more than enough of their peripatetic spy, and he of them.

In retrospect, Crampton's stewardship of the projects promoted by the United States Government in Puerto Rico in association with the New York Academy of Sciences gave the AMNH a definitive research edge in the Greater Antilles. The dominant role of the AMNH in Caribbean evolutionary studies was almost exclusively the result of his finagling. To the extent that the curator supervised his fellow ONI scientist colleagues all over Latin America, the long reach of the AMNH throughout the hemisphere was similarly expanded.

The importance of Crampton to the development of Latin American research at the Museum should not be underestimated. As a founding member of the Council of National Defense, and an ONI agent, he served as the principal facilitator for those of his colleagues who wished to serve in Latin America in undercover capacities. However insignificant his position was when balanced against the more daring contributions of other AMNH agents during the War, and notwithstanding the often acerbic relationship he endured with President Osborn, for the scientific departments at the American Museum of Natural History, Agent 270, Invertebrate Zoology Curator Henry Crampton, was not only their man in Washington but, more critically, their conduit to Latin America. The fact that he was capable of serving several entities at the same time with equanimity rendered him invaluable to the

Museum and a paragon of the elision between politics and science which was concurrently occurring on a national and global scale.

Notes

1. List of ONI agents, provided in "Office of Naval Intelligence Designations", in Harris and Sadler, The Archaeologist…, pp. 370–380. The National Archives (NARA) RG 38, E 122 "Examples of Lists from Secret Lists of Agents," also includes Barnum Brown as ONI Agent 233 who operated out of Mazatlán, Mexico using the name Walter Palmer Brown. It would seem that he left the service shortly after joining, and later prospected for oil in Cuba for the Treasury Department.
2. Fiftieth Annual Report of the Trustees of the American Museum of Natural History for the year 1918 (New York, 1919), p. 31.
3. By 1916, he had already published his monumental study on the Polynesian snails: Henry E. Crampton, Studies on the Variation, Distribution and Evolution of the Genus *Partula*: The Species Inhabiting Tahiti. Carnegie Institution of Washington, Monograph 228, 1916.
4. A biographical sketch of Henry E. Crampton is included in the survey of Department of Zoology of Columbia University, 1892–1942 (New York, 1942), p. 54. Crampton was born in New York on January 5, 1875 and died in New York on February 26, 1956. See also Stephen Jay Gould, "Crampton, Henry Edward." Complete Dictionary of Scientific Biography. Encyclopedia.com, viewed 13 December, 2018 https://www.encyclopedia.com/science/dictionaries-thesauruses-pictures-and-press-releases/crampton-henry-edward. Crampton's experiences in South America are described in Forty-third Annual Report of the Trustees of the American Museum of Natural History for the year 1911 (New York, 1912), p. 53.
5. The Council of National Defense was created by order of President Wilson on October 11, 1916. See New York Times, October 12, 1916, p. 10. Despite the publicity, the workings of the Council were not made available to the public.
6. Hollis Godfrey was President of the Drexel Institute in Philadelphia when he was appointed. He was a distinguished engineering graduate of MIT ('98). See Jack Nunn, "MIT: A University's Contribution to National Defense," Military Affairs, vol. 43 (October, 1979), p. 121.
7. Crampton to Osborn, 29 June, 1917, New York, p. 2, CA, AMNH, File 1069.
8. Crampton to Osborn, 29 June, 1917, New York, p. 3, CA, AMNH, File 1069.
9. Minutes of the Preparedness Committee Meeting, 9 April, 1917, p. 8, CA, AMNH, File 1069.
10. "Resolution of Curator Emeritus Dean," presented in Resolutions on Preparedness Meeting, AMNH, 15 February, 1917. It should be noted that when it came to those scientists involved in covert government activities Osborn was far more secretive.
11. Crampton to Osborn, Washington, D.C., 15 July, 1917, CA, AMNH, File 1069.
12. Osborn to Crampton, New York, 16 July, 1917, CA, AMNH, File 1069.
13. Osborn to Cleveland Dodge, New York, 16 July, 1917, CA, AMNH, File 1069.
14. Crampton to Osborn, New York, 25 September, 1917, CA, AMNH, File 1069.
15. "Arms needed to Equip an Army of 1,000,000 men", n.d. Columbia University Special Manuscript Collections [hereafter cited as CUSMC], Crampton, Box 3. Crampton's notations to the chart are in his handwriting.
16. Not only was Crampton interested in manganese for armament production, but a clipping in his newspaper file on the U-Boat menace to Chilean nitrate shipping also demonstrates that he was concerned about other materiel from Latin America essential to the war effort. See article from Philadelphia Public Ledger, 23 March, 1917 in CUSMC, Crampton, Box 4.
17. Hollis Godfrey to Crampton, Philadelphia, 6 November, 1916, CUSMC, Crampton, Box 2.
18. The "Minutes and Reports of the Committee of the Council of National Defense," shows that Bernard M. Baruch had a position on the Advisory Committee on Raw Material, including minerals and metals. In CUSMC, Crampton, Box 3.

19. Bowler and Rhys Morus, Making Modern Science…, p. 467.
20. Herbert Winans to Crampton, Little Silver, New Jersey, 11 June 1918, CUSMC, Crampton, Box 2.
21. Jessup to Crampton, New York, 20 December, 1916, CUSMC, Crampton, Box 2. Jessup, a New York attorney, asked for Crampton's help in doing something for the war effort, since he was much older than recruitment age.
22. Nicholas Murray Butler to Crampton, New York, 21 February, 1918, CUSMC, Crampton, Box 1.
23. Bashford Dean to Crampton, 15 June, 1917 and 22 June, 1917, CUSMC, Crampton, Box 2. Bashford Dean's work on fishes included studies on fossil armored fish. Amusingly, Dean believed that he could contribute this knowledge of armor to the government. See his "Notes on the Ventral Armoring of *Dinichthys*," New York, New York Academy of Sciences, 1896. With regard to his World War I work see, "Bashford Dean and the Development of Helmets and Body Armor during World War I," by Donald J. La Rocca, In Heilbrunn Timeline of Art History. New York, The Metropolitan Museum of Art, 2017. https://www.metmuseum.org/toah/hd/bash/hd__bash.htm.
24. Notation of April 21, 1917 in Barnum Brown's "Memoirs" [handwritten] "Itinerary from letters and records." In Barnum Brown file, Vertebrate Paleontology Collection, AMNH. As noted before in note 1, the National Archives (NARA) RG 38, E 122 "Examples of Lists from Secret Lists of Agents," lists Barnum Brown, referred to as Walter Palmer Brown, as ONI Agent 233 in Mazatlán, It is possible that Brown purposely reported in his notes that he refused the offer of "foreign service" in keeping with the need for secrecy.
25. Crampton to Hon. Frank L. Polk, Washington, 14 July 1917, CUSMC, Crampton, Box 1. Again, NARA has an ONI listing for Barnum Brown, where he is referred to as Warren Palmer Brown, Agent 233 working in Mazatlán, Mexico.
26. *Ibid.*
27. See entries for October, 1917 in Barnum Brown's "Memoir", VP collection, AMNH.
28. George Goethals to H. Anthony, Culebra, Canal Zone, 10 February, 1914. Harold Anthony papers, Dept. of Mammalogy, AMNH.
29. Harold Anthony to Crampton, Utuado, Porto Rico [sic], 25 June, 1916. Harold Anthony papers, Dept. of Mammalogy, AMNH.
30. "Cuban Situation Serious", Public Information Committee, American Museum of Natural History, New York 20 March 1917. Harold Anthony papers, Dept. of Mammalogy, AMNH.
31. Earle Vendig to Anthony, San Juan, Puerto Rico, 14 April 1917. Harold Anthony papers, Dept. of Mammalogy, AMNH. Harold Anthony is also listed as having a file in NARA "Formerly Confidential Correspondence", File 20953-67, along with other confidential agents for the Navy.
32. See "Intelligence Report" from 1st Lieutenant Marc M. Ducote to Commanding Officer, 1st Battalion, Camaguey, Cuba, 5 December 1917, United States Marine Corps Intelligence Reports, National Archives, Record Group 127, Entry 38, C 298 Cuba. At the conclusion of the War, Anthony again passed on information to the War Department, this time on Ecuador (see Chap. 11).
33. Crampton to Osborn, New York, 25 September, 1917, CA, AMNH, File 1069.
34. Crampton to Frederick K. Keppel, New York, 25 February 1918, CUSMC, Crampton, Box 1.
35. Crampton to Dean Guy Ford, Bureau of Censorship and Information, New York, 23 February 1918, CUSMC, Crampton, Box 3.
36. Charles J. Tilden to Crampton, Washington, 16 October 1917, CUSMC, Crampton, Box 2.
37. Crampton to Charles Dewoody, New York, 26 February, 1918, CUSMC, Crampton, Box 2.
38. List dated 25 February, 1918, enclosure in Crampton to Charles Dewoody, New York, 26 February, 1918, CUSMC, Crampton, Box 2.
39. Letter of John Lord O'Brian, Special Assistant to the Attorney General, Department of Justice to A.E. Stevenson, Esq., Washington, D.C., 7 June, 1918, enclosed in Stevenson to Crampton 12 June, 1918, CUSMC, Crampton, Box 2.

40. On 24 July, 1918 Crampton wrote to the Attorney General of the U.S. Department of Justice thanking him for his appointment as Special Agent without compensation. CUSMC, Crampton, Box 3.
41. See Jaffe, New York At War…, p. 205 for a discussion of the APL.
42. Abstract of Application of Henry Crampton to Carnegie Institution for research trip to Polynesia. See Crampton Files, Carnegie Institution, Washington, D.C.
43. On December 15, 1916 the Carnegie Institution granted the sum of $2000 to Crampton to continue his Polynesian research. Crampton Files, Carnegie Institution, Washington, D.C.
44. R.S. Woodward to Crampton, Washington, D.C., 23 January, 1918, Crampton Files, Carnegie Institution, Washington, D.C.
45. Crampton to Woodward, New York, 5 February 1918. Crampton Files, Carnegie Institution, Washington, D.C.
46. Woodward to Crampton, Washington, D.C., 19 February 1918, Crampton Files, Carnegie Institution, Washington, D.C.
47. Crampton to Woodward, New York, 5 February 1918. Crampton Files, Carnegie Institution, Washington, D.C.
48. Edward House to Crampton, New York, 3 October 1917. CUSMC, Crampton, Box 2. For a discussion of the Inquiry see Gelfand, The Inquiry…
49. On July 30, 1918, Crampton wrote to the Institute of Arts and Sciences in Brooklyn from Latin America offering to give a talk on his current research in November of that year, Crampton to Institute of Arts and Sciences in Brooklyn, 30 July 1918, CUSMC, Crampton, Box 3.
50. Commandant 15th Naval District to Naïveté, 24 August, 1918, National Archives, RG 38, 21115, ONI as cited in Harris and Sadler, The Archaeologist…, p. 268.
51. David Eugene Smith to Crampton, 11 December 1918, CUSMC, Crampton, Box 2.
52. ONI files, NARA Record Group 38. E123, 30 April, 1920.
53. See endnotes 43–47, above.
54. See ONI files, NARA Record Group 38. E123, 30 April, 1920. On Crampton's passage to Saipan, see letter from Commanding Officer of U.S.S. *Bittern* to the Governor of Guam, August 3, 1920. ONI records, NARA file 9351-1895.
55. Henry C. Crampton, Studies on the Variation, Distribution, and Evolution of the Genus Partula: Vol. II, The Species of the Mariana Islands, Guam and Saipan, Carnegie Institution of Washington, Monograph 228b, 1925.

Chapter 7
Herbert "Joe" Spinden in Central America: ONI Agent 56

Herbert Joseph Spinden's "war" was very different from that of Henry E. Crampton. Although both men were AMNH scientists with Latin American research interests, and both had been sworn in as *bona fide* agents of the Office of Naval Intelligence, their service to the nation could not have been more dissimilar. Crampton's patriotic contribution consisted of time spent either safely ensconced in Washington or machinating in New York; his sole intelligence-gathering experience in Latin America was limited to the summer of 1918. By contrast, Spinden was actively engaged in what proved to be potentially dangerous espionage in Central (and later South) America from early April 1917 throughout the course of the War; he would not return to the Museum until the Spring of 1919.

At the time of his ONI recruitment, "Joe" (his preferred moniker) Spinden was an Assistant Curator in the Anthropology Department of the AMNH, where his specialty was Mayan archaeology (Fig. 7.1). Unlike the politically savvy Henry Crampton who, as seen in the last chapter, was comfortable wearing several hats, Spinden was primarily interested in his academic career, which focused on Mexican research for the AMNH. One of the most respected scientists on the Museum staff, he was nationally recognized for his determination to pinpoint the full extent to which Mayan civilization had expanded from its Mexican origin to influence the development of greater Central American culture. Something of a *wunderkind*, Spinden completed his doctorate in anthropology at Harvard University only three years after receiving his bachelor's degree[1] and was shortly thereafter hired by the AMNH. He began his tenure in 1909 working alongside the more established staff archaeologist Marshall Saville. Like his colleague, Spinden was quickly drawn into the planning of the American Museum's Hall of Mexico (one of President Osborn's pet projects) and was the beneficiary of the senior administrator's interest in all things Latin American. So it should come as no surprise that on the eve of Mexico's descent into what would devolve into a full-blown civil war (1910), Spinden was in the midst of important negotiations for a transfer of artifacts from that nation to the AMNH.[2] As the years

R. M. Delson, *Scientists or Spies?*, The Latin American Studies Book Series,
https://doi.org/10.1007/978-3-032-18117-6_7

Fig. 7.1 Herbert "Joe" Spinden, 1916. AMNH Asset ID 36108

progressed, Spinden would demonstrate an unerring knack for being in diplomatic hotspots at critical moments, not all of which were coincidental.

As notable as his excavating skills were in the field, it was his superb scholarship which caught the attention of the Museum's higher-ups. It was on the basis of his promising publishing career in Mexican archaeological studies that Spinden was chosen to represent the AMNH as the official alternate delegate to the Second Pan American Scientific Congress in Washington, ranking him second only to the primary representative, the esteemed ornithologist Frank M. Chapman. For a relative newcomer to the Museum (he had been there only six years at the time he was chosen for the Congress), Spinden's appointment represented a great honor as well as validation of his personal progress. He dutifully attended the sessions of the Congress from late December of 1915 to early January, 1916, where he delivered two important papers which demonstrated the breadth of his knowledge of Central

American archaeology. But the significance of Spinden's presentations went beyond their demonstration of his high degree of intellectual mastery. "On the Origin and Distribution of Agriculture in America" and "Recent Progress in the Study of Maya Art"[3] coincidentally revealed Spinden's familiarity with the countries and landscape of Central America. This included his knowledge of the nation of El Salvador which, given its pro-German sentiments, was a growing concern for the U.S. government. As global tensions increased, Spinden, along with other academics who were familiar with Latin America, would increasingly draw the attention of Washington politicians; as noted earlier, the Second Pan American Scientific Congress had served as a conduit for channeling potential academics into government work. In April of 1917, Spinden became one of the first nationally chosen scientists to be commissioned in the ONI and, not surprisingly, he was assigned intelligence and reconnaissance work in the Central American region with which he was most familiar.

Spinden's research interests had shifted to Central America when the bloody civil war in Mexico (1913 through 1917) precluded further fieldwork in that country. Unable to work at Mitla and other Mexican sites during the fighting, Spinden turned his attention to the nations of El Salvador and Honduras (exclusive of Copán), where he hoped to conduct a highly original survey of the southernmost limits of Mayan culture. As Anthropology Curator Clark Wissler explained to Director Lucas, "Doctor Spinden's field-work in Central America revealed to us an important field in San [sic] Salvador and Honduras which we have looked forward to as the main parts of our future work in that country."[4] This exciting new project was not only ambitious, it had the potential of putting the Museum at the forefront of Latin American archaeological expertise in the United States. Of course, Spinden's future success was completely dependent on his being able to work in the region.

With the assistance of Wissler, Spinden developed a five year research plan for El Salvador and Honduras, with 1915 as the target starting date. Its purpose was to "determine the southern limits of Maya culture; …to obtain evidence bearing on the origin of agriculture; and to determine the stratigraphic relations of human art."[5] The project outline clearly recognized the necessity of obtaining the right to excavate Mayan relics in both of these nations; perhaps naïvely, AMNH administrators anticipated that the governments of El Salvador and Honduras would readily agree to grant them the right to excavate artifacts, much as the Mexican government had done earlier in Mitla. In the minds of these administrators such finds would henceforth be considered the "exclusive" domain of the Museum,[6] and the agreement they proposed to these Central American governments would simultaneously serve to exclude other researchers and institutions from working in the country.

Fulfilling such a broad ambition would not come cheap; Wissler estimated that Spinden's cost, along with that of a student assistant to accompany him, would be about $5000, to which $10,000 would need to be added to cover the cost of excavation, purchase and transportation of artifacts.[7] The total cost of the project would therefore be $15,000, a substantial sum. When spread out over the five-year duration of the plan, this would break down to an annual budget of $3000, and the Museum was more than willing to defray the cost of this effort. Without referencing the exclusivity issue, Osborn laid out his expectations for this undertaking in a letter he sent to the government of El Salvador, anticipating nothing less than an enthusiastic response in return. "What we have in mind," he wrote, "is a complete systematic survey of the whole of San [sic] Salvador. All ruins will be uncovered, mapped, and described, such excavation as may be necessary, made, etc., and the scientific aspects of the data thus obtained be made the basis of a series of memoirs to be published by this institution."[8] Reading between the lines, Osborn made it abundantly clear that involvement of other institutions in this project would not be necessary.

At the very minimum, such long range planning would necessitate the written approval of El Salvador, if not a formalized agreement. In his capacity as the project's designated leader, Spinden was selected to function as the logical intermediary between the Museum and the government of El Salvador. If the Salvadorans agreed to have the well-regarded Spinden as the AMNH representative, then his approval would go a long way toward securing permission for the broader project.[9]

But there was a hitch to this project which Osborn had not foreseen. At the same time that Spinden pitched his proposal, El Salvador's standing vis-à-vis the United States government was rapidly deteriorating, and its political posture had turned decidedly pro-German. To avoid the possibility of Spinden's project being rejected out of hand because of the rift between the nations, the Museum decided to circumvent the usual diplomatic channels and, instead, solicited the intervention of former Salvadoran President, Dr. Alfonso Quinones. Not only was Quinones himself an archaeologist but, conveniently, he was traveling in the United States during the time Spinden's plan was conceived. A flurry of letters between the Salvadoran consulate and the Museum resulted in an invitation for the ex-president to visit the AMNH, in June of 1915.[10] Unfortunately for the Museum, Dr. Quinones contracted "a severe attack of pneumonia" while staying in Philadelphia, his last stop before reaching New York.[11] At this point, without any explanation, Dr. Quinones abruptly changed his travel plans and headed west to San Francisco rather than New York City. The Consulate assured worried Museum officials that at the end of July the ex-President would indeed visit the AMNH in New York,[12] but there is nothing in the files to confirm that such a meeting actually took place. A hand-written note to a Museum official on the Consul's letterhead suggests that the matter was referred to Spinden for "his information." The Quinones ploy, apparently, had come to a dead end.

Spinden was left twiddling his thumbs. Not only did El Salvador not break its ties with Germany, as Washington had requested, but it refused to honor the prohibitions imposed by the U.S. on trading with the "enemy" (e.g., Germany). Even worse from Washington's perspective, the rhetoric from government offices in the capital city of San Salvador was increasingly marked by anti-American sentiments, a situation exacerbated by the newly concluded Bryan-Chamorro treaty (1916) between the United States and Nicaragua. That treaty had guaranteed the United States the "exclusive right to build an inter-oceanic canal across her [Nicaragua's] territory, and lease[d] to the United States for forty-nine years a naval base on the Gulf of Fonseca [in which El Salvador also had territory] and also the Great Corn and Little Corn Islands as coaling stations."[13] In response El Salvador and Costa Rica mounted a sovereignty challenge to the treaty in the Central American Court of Justice. Although El Salvador prevailed, the United States essentially ignored the decision. Thus, the treaty stood in non-legal limbo, infuriating the Salvadoran government, and it became clear that if Spinden hoped to secure approval for his project in El Salvador, the diplomatic climate would have to improve dramatically. While Spinden waited for relations to make any headway, negotiations for his project stalled. Tensions between the two nations would later increase in the months following Wilson's 1917 War Declaration, given El Salvador's preference for the German cause over that of the Allies.

With his project stalled, Spinden was sent by his superiors at the AMNH in early 1916 on a brief, but hushed-up, reconnaissance trip to Venezuela. This hastily organized expedition was a solo venture which the Museum hoped to keep under wraps, lest the archaeological world learn that Spinden was looking for evidence of early indigenous village sites in that country. To cover Spinden's tracks, Curator Wissler let it be known that Spinden was going to Puerto Rico (not Venezuela) which, in Wissler's own words, was a "blind" to Spinden's real movements. Notwithstanding his prior Central American focus, having Spinden on assignment in the Caribbean would have made sense to any outside observer and, as the Museum had several ongoing projects in the region undertaken in conjunction with the expanded Puerto Rican survey, Spinden would have been a logical addition to any survey team. At the same time, Wissler requested (and received) a $2500 appropriation for his younger colleague so that Spinden might have sufficient monies with which to purchase any artifacts he came across while in Venezuela.[14]

The results of this secretive trip, however, were largely disappointing. According to the report Spinden forwarded to Wissler on February 21, 1916, the archaeologist found little evidence of old village sites in Venezuela, although he did recover some bones (human?) from caves. Meanwhile, he whined about the "poor 'accomodations' [sic], a bout with fever, the lack of cleanliness and the expense" (as the "European armies have depleted the mule trains"). He reckoned that there might yet be some good ethnographic material on indigenous populations,[15] a paltry attempt to justify the cost of his misadventure. Spinden had also become familiar enough with the region that he was able to sketch a map of it which he sent to Wissler,[16] although it is unclear what, if anything, Wissler did with this data. However, Spinden's Venezuelan venture would later be viewed as an invaluable experience by the United States government, which would return Spinden to South America in 1918 as a Naval Intelligence officer in search of platinum.

Fresh off his failure in Venezuela, Spinden managed to save face by sending Wissler a small collection of pots, while he headed north to Puerto Rico, giving credence to the lie that he was already participating in the Greater Antilles survey. Once in Puerto Rico, Spinden joined the ongoing expedition under the direction of Henry Crampton and Franz Boas, which represented the combined interests of the Museum, Columbia University and the New York Academy of Sciences. It was expected that Spinden would prove an important addition to this expedition and the Curator of Ethnology, Pliny E. Goddard, confirmed with Boas that the Museum intended Spinden to join the party in either April or May of 1916.[17] According to the survey's protocol, the New York Academy needed to approve Spinden's appointment before the Museum could send him to join the others already in Puerto Rico. Boas confidently informed Crampton that Spinden had the votes necessary at the Academy to allow for his participation and that the New York institution would also help defray the cost of his work at the dig.[18]

As discussed in Chap. 3, AMNH participation in the Puerto Rican expedition provided Osborn with the ability to have his men conduct the faunal surveys he considered so vital to his Latin American evolutionary project without much expenditure. Although this kind of work was hardly Spinden's forte, the Museum evidently thought having him involved in the survey's work would be advantageous. While on the island, Wissler instructed the archaeologist not to interfere with the site Boas was already working with John Mason but, instead, to assist Harold Anthony, one of the Museum's mammalogists, in excavating faunal deposits from caves. Wissler also begged Spinden to bring back a sample of cassava (*Manihot esculenta*) from the island so that the Museum could display it in a special exhibit on food plants of the New World.[19]

Despite Crampton's and Wissler's insistence that they work as partners, as discussed earlier, Anthony and Spinden did not get on. Nonetheless, Spinden worked assiduously in Puerto Rico on the extended faunal survey through June of 1916, amassing a number of bones and shells unearthed from middens to augment the Museum's heretofore limited Caribbean collection. Although Wissler thought Spinden's finds important from a stratigraphic point of view,[20] the archaeologist himself was far less enthusiastic about his work in Puerto Rico, preferring to have begun his long-planned overview of Mayan civilization in Central America. In the meantime, he thought he might placate his superiors (after his contentious stint with Anthony) by preparing a report on his Venezuelan trip. Obviously pleased with the result, Wissler forwarded Spinden's findings to Osborn, and asked for his permission to publish them in the Proceedings of the National Academy of Sciences. Evidently the secrecy under which Spinden had been operating during his Venezuelan venture was no longer necessary, and the AMNH was now ready for the scientific world to know Spinden had "planted the flag" for the Museum in the area. Presumably the AMNH would claim it had rights to future treasures that might be unearthed in that South American nation.[21]

When he returned to New York from Puerto Rico, Spinden once again took up his duties at the Museum, hoping to pursue his Central American expedition in the very near future, even though El Salvador had evinced little change in attitude. After New Year's in 1917, the archaeologist busied himself in helping Osborn obtain a set of important photographs of the Yucatan, which the president had considered essential for the "development of the Mexican Hall."[22] The El Salvador project remained on hold through March of 1917. Yet, even if arrangements had been completed in time for Spinden to initiate his ambitious Mayan research during the first half of the year, such plans would inevitably have been superseded by global events and especially by the entrance of the United States into the War. So it is triply ironic that given the United States declaration against Germany in early April, 1917, El Salvador's contemptuous attitude toward the U.S., and the Museum's self-imposed moratorium on fieldwork outside the country, Spinden actually found himself in Central America later that same month, exactly as he had hoped he would.

Spinden Moves into Intelligence Work

The circumstances under which Spinden made his way south of the border, however, were not what Museum officials might have imagined, nor were they in any way connected with the archaeologist's erstwhile Salvadoran project. Instead, Spinden's colleague at the Carnegie Institution in Washington, Dr. Sylvanus Morley, had secretly recruited the AMNH archaeologist to pursue intelligence work in Central America as a formally commissioned officer of the Office of Naval Intelligence.

During the several month interval in early 1917 in which the AMNH scientists were slowly acclimating themselves to the prospect of war, Sylvanus Morley had begun to enlist fellow scientists for the war effort. This was, in effect, an extension of the 1916 initiative of the Council of National Defense which had begun recruiting scientists, such as Invertebrate Zoology Curator Henry Crampton, to work on wartime priorities in Washington even before the formal Declaration of War.[23] In March of 1917, about a month before the U.S. formally entered the war, Morley gave the Office of Naval Intelligence (with which he was now associated), a list of researchers familiar with Mexico and Central America who he thought might agree to work secretly for the U.S. government in that region. Included in his list of potential intelligence assets (a compilation which encompassed anthropologists, archaeologists and naturalists)[24] was Morley's colleague and fellow Mexicanist, Joe Spinden.

Spinden promptly accepted Morley's invitation but wisely did not broadcast his decision at the Museum. So when he nonchalantly let it slip during the April 9th meeting of the AMNH War Preparedness Committee that he would be soon be leaving New York for Washington on a previously unannounced trip, no one imagined its ulterior purpose and no one raised any objections.[25] Spinden neither revealed the details of his intelligence assignment to his fellow scientists, many of whom were still reeling from the impact of the Declaration of War, nor the fact that he had already been commissioned as an ensign in the Naval Coastal Defense Reserve. Despite the overseas research moratorium which the Science Faculty hinted at on the 9th and which was adopted with full-throated approval at their meeting the next day, Spinden proceeded to Washington, D.C. where he had been ordered to report for intake at the Office of Naval Intelligence. That there was no reaction among his colleagues when Spinden announced his unexpected departure suggests that the purpose of his departure had not been revealed to them. Perhaps more to the point, even had he wanted to reveal his new position, ONI instructions expressly prohibited him or any of its recruits from discussing their future intelligence assignments.[26]

Exactly what his colleagues might have said had they known that Spinden had already been appointed Agent 56 of the Office of Naval Intelligence, and that his previously unannounced "fieldwork" would actually be used as a cover for intelligence gathering in Central America, can only be imagined. Although the mood among scientists working at the AMNH was unquestionably patriotic, the staff was as apprehensive about any fieldwork undertaken abroad during wartime as it was toward the prospect of putting themselves in harm's way. As seen earlier, when the staff took a vote to reject any research outside the United States for the duration of the War, they thought the matter was done and finished[27]; so, for that matter, did the more powerful Executive Committee which upheld the Faculty's wartime research moratorium.[28]

And yet, notwithstanding the obvious conflict of interest between official Museum policy and the goals of the Museum president, Spinden was on his way to pursue a highly secretive intelligence gathering mission by mid-April under the guise of AMNH research. Curator Wissler had quite innocently set the process in motion when he sent a letter to Director Lucas on April 4, 1917, explaining that Sylvanus Morley was going to Central America to do "new work on inscriptions" and wanted Curator Spinden to accompany him. This sounded to Wissler like a legitimate request, as he likely was unaware of Morley's or Spinden's already assumed double identities. As Wissler pointed out to Lucas, "[t]he trip will take about six weeks and as Mr. Morley is about ready to set out, the war can hardly settle down upon us in that time."[29] In retrospect, Wissler was greatly mistaken about the amount of time the archaeologist would spend in Central America. Spinden not only met Morley for the agreed-upon joint venture, but he would move back and forth among El Salvador, Honduras, Costa Rica, Venezuela and Colombia for the next two years, not returning to the Museum until the Spring of 1919.

Director Lucas responded ambiguously to Wissler's request. "Under ordinary conditions," he wrote to the Anthropology curator, he would have no hesitation about sending Spinden to work with Morley. "As matters are, while I approve the recommendation, yet I cannot help questioning a little the desirability of undertaking fieldwork outside of the United States, the more, as our recent experiences in South America, Cuba and Nicaragua have been somewhat unpleasant." He neglected to mention the outright recalcitrance on the part of El Salvador's government toward the Museum's (and Spinden's) Mayan project and the scientific staff's self-imposed research moratorium. Nonetheless, he gave Spinden the go-ahead for the expedition contingent upon Osborn's approval.[30]

Osborn's response to Lucas when he learned of Spinden's trip was short, cryptic and hand-written, so that no carbon copy would be kept. He informed the Director that he was withdrawing any objections that he might otherwise have had about Spinden working outside the country. "Dr. Spinden and I think there are especially good reasons why he should go with Dr. Morley at the present time." As there is no memorialization of what transpired at the meeting between the two men, it is unclear how specific their conversation was. Did they discuss whether there were dangers involved? Were expectations laid out for what Spinden might collect on the

Museum's behalf? Was the issue of patriotism brought up, and did Osborn evince any pride in his younger colleague?

Clearly, Spinden had informed the President of the real nature of his trip (although it is unclear if Lucas understood the full implications of his travel at this time). While that knowledge should have raised flags in light of the agreed upon moratorium, Osborn intimated that he saw a golden opportunity for the Museum through Spinden's intelligence gathering. Embellishing upon the pretense, Osborn informed Lucas that "It appears he [Spinden] has had a conference with President Woodward and the Carnegie Institution will do no collecting…We shall get the benefits of the work of two men." Osborn speedily allotted the sum of $900 to Spinden and gave the archaeologist his blessing. But a curious handwritten postscript to this letter, which was not necessarily in Osborn's hand, belies the president's optimism; it reads, "Doubt if any collections are made."[31] There is no written reaction from Director Lucas recorded in the files and, for the time being Curator Wissler, Spinden's immediate superior, apparently remained in the dark.

Osborn's approval raises serious questions about how this obvious breach of the scientific staff's official policy on research overseas could have been allowed. Was this hypocrisy, or merely dissembling on Osborn's part? The answer to this question is that Osborn's actions were probably a bit of both and were based on his astute calculation of what allowing Spinden to leave the country could mean to the AMNH and its project in Latin America. Perhaps believing that the location of Spinden's assignment in Central America might actually work to the Museum's advantage, and being required by the ONI to maintain secrecy, he chose not to reveal the real purpose of Spinden's travel to the rest of the staff. At any rate, for Osborn, it seemed obvious that Spinden could not possibly engage in "spying" all of the time he was in Central America. In return for receiving "official" permission to join the ONI, Osborn likely suggested to the archaeologist that he might, in his "off time," pursue some casual collecting for the Museum, especially as he would be in the exact region of his long-thwarted Salvadoran project. Whatever Spinden managed to ferret out of the country, it would be more than the Museum currently possessed, nor would he likely obtain much, given the hostilities between the two countries. The AMNH would thus become the hidden beneficiary of what was clearly a duplicitous arrangement, especially as other natural history institutions in the U.S. had simultaneously put their overseas collecting on hold. While it goes without saying that any misstep which occurred in the pursuit of Osborn's gamble could easily have put the archaeologist's life in jeopardy, provided that Spinden managed to succeed on both fronts, this dual assignment could potentially work to both Washington's and the Museum's advantage. The timing of Spinden's recruitment was, therefore, essentially a gift to Osborn and his ambitious Latin American research dreams.

Were any ethical issues discussed between the Museum president and his underling? Unfortunately, the written record is also silent on this point, so we have no idea. But, in retrospect, neither Spinden nor Osborn should be considered innocent in this venture as both men had been party to the staff's decision to curtail any further research outside the country during wartime. Compounding this infraction, as if to counter any conceivable objection to his trip, Spinden was allowed to proceed to Washington armed with a general letter of introduction from President Osborn (presumably to be kept on his person at all times) which speciously announced that the archaeologist was "planning a journey of exploration through Central America for the purpose of collecting ethnological and archaeological specimens and information in the interests of The American Museum of Natural History."[32] Given the fact that Spinden's five year plan for evaluating the Mayan cultural horizon in Central America was well known among the U.S. archaeological community, Osborn's loosely-defined AMNH research "cover" in El Salvador carried a high degree of validity. Yet this was surely complicity born of expediency; from Osborn's point of view there was no need to obfuscate the matter further by drawing attention to Spinden's primary spying mission as opposed to the Museum's goal of collecting; in a very real sense, at least for the Museum president, the two objectives had become indistinguishable. Moreover, if Osborn had experienced any hesitation about what he had just authorized, ONI regulations required that he, and Spinden, remain quiet about the assignment.

By contrast, the Government's instructions to Spinden were anything but obscure. He was informed in an ominous situation report that "Reliable information indicates German activity in Central America," and further that "Reports have reached this Office that bases for the supply of enemy submarines and raiders have been established on the Coast of Central America. It is also rumored that marked pro-German activity exists among the inhabitants and that German agents are operating." For his assignment, Spinden was "to thoroughly investigate the coast of Central America for evidences of enemy bases, and endeavor to organize an information service covering this coast."[33] There was no obfuscation here; Spinden had just been ordered to hit the ground running.

The immediate orders that the archaeologist received required that he proceed from Washington to Central America via the port of New Orleans. Once there he was instructed to waste no time and book passage on the "first available steamer sailing" for the region. When he had arrived safely, he was to cable the ONI with his address. And then, in classic spy-craft tradition, Spinden was to destroy his orders as soon as he had memorized their content. Further, Spinden was ordered to communicate with other agents by using codes (the "A 24" and the "B 24"), using the key given to him to send messages in cipher.[34] According to Harris and Sadler's monograph, Morley's instructions were not nearly as specific as those given to Spinden.[35] Perhaps Washington, or the Navy, had determined that Spinden's assignment was the more important one or, perhaps, they recognized greater spy potential in Spinden. Notwithstanding such confidence in his abilities, the reality was that Spinden's "training" for this mission was limited to non-existent: after his processing in Washington, the archaeologist was sent on his way in a matter of days, provided with scant "equipment" other than a code key to a cipher, and the coded name of a contact in the New York City office.

While he pursued his ONI assignment and simultaneously aimed to maintain his cover, Spinden faced the burdensome reality that relations between several Central American nations and the United States government had continued to deteriorate.

During the interim between the formulation of Spinden's five year Mayan project in 1915 and the archaeologist's actual departure for Central America, in April, 1917, the antagonism which El Salvador had displayed toward Washington had expanded to other nations in the region. The result was that by the time Spinden arrived in Central America, the region was divided in its loyalties with regard to the ongoing hostilities in World War I Europe: El Salvador, Honduras, Costa Rica and Mexico were openly sympathetic to the Germans, while Nicaragua and Guatemala had fallen under the sway of the United States. Adding insult to injury, by late April, 1917, the Mexican government had already provided El Salvador with aircraft to bolster its defenses, while Mexican President José Venustiano Carranza had personally gifted a German-built radio station to the nation, allowing the capital city, San Salvador, to communicate directly with Mexico City.[36] As a consequence of this turn of events, placing "agents," like Spinden, in Central America had become a matter of intense urgency for Washington.

For the next few months Spinden and Morley, as well as Samuel K. Lothrop (an archaeologist from the Harvard Peabody Museum) and John Held Jr. (an artist), traipsed through Central America, sometimes as a team, sometimes independently, looking for evidence of German infiltration. Morley's escapades have been well documented by Harris and Sadler,[37] but Spinden's contributions to intelligence gathering may have been equal to Morley's, yet they are much less well known. This remained the case, notwithstanding Franz Boas's scorching critique of Spinden in *The Nation* referenced at the beginning of this study. As it turned out, as a sleuth Spinden would do an excellent job in gathering and forwarding data on German activities in those Central American countries to which he was assigned, even as he took advantage of whatever limited opportunities presented themselves to enhance the collections of the AMNH. He was constantly on the lookout for possible future sites for excavation (although he certainly could not undertake an excavation at the moment), as well as any already existing collections of Pre-Columbian artifacts he might encounter and purchase for the Museum.

The movements of Agent 56 in Central America can be tracked from slightly before June 1, 1917, through the end of 1918. On that first date, the United States consular representative in El Salvador, Boaz Long, was able to report to the Secretary of State that Spinden had traveled via horseback from Zacapa, Guatemala to Santa Anna, in El Salvador, accompanied by a Mr. Hebard, whom Long identified as an American engineer. Long also confirmed that Spinden had checked in with the consulate upon his arrival in country and was given an order to proceed to the estate of Gustavo Brinckmann, a German national, where Spinden would, ostensibly, "see some ruins." But, as Long explained, Spinden "very kindly consented while on this journey to look into the rumor that a small wireless plant is to be installed in that section." Whatever information the archaeologist found about the transmitter, Long explained, would be forwarded to the Consul and thence to the State Department.[38]

Unfortunately, there is no report in his files which documents what Spinden actually found while he hunted for the "ruins" on Brinckmann's estate, nor it is clear if the wireless plant which he was instructed to locate was closer to Santa Ana, in El Salvador or to Copán in Honduras. Either way, stopping communication with the Germans (even if communications were transmitted second hand through Mexico) was a high priority for the Allies. Although Spinden's involvement in what transpired is not clear, at exactly the time that the AMNH archaeologist was assigned to the district, another German planter, identified in the records as "Bach", was arrested, convicted and sent to prison in Guatemala for the crime of possessing a wireless transmitter. Was this the equipment Spinden had been tasked to find? It would seem possible given that Spinden had been in that very region when Bach was apprehended.[39]

By June 5th, Spinden had left El Salvador for Tegucigalpa, Honduras,[40] where he began his next assignment: an investigation of German trading houses operating in Fonseca Bay (referred to, at the time, as the Gulf of Fonseca). This horseshoe-shaped bay was a likely location for enemy activity; its deep waters could provide safe harbor for shipping activity and, given its extensive breadth, it allowed for the possibility of illicit trade to proceed without detection. The difficulty for the United States was that the waters of the Bay touched the shores of the German-leaning nations of El Salvador and Honduras as well as those of the friendly country of Nicaragua. With the Bryan-Chamorro Treaty just approved in 1916, and the successful challenge to its legitimacy by El Salvador, Costa Rica and Honduras in the Central American Court of Justice essentially ignored by the Nicaraguan (and the United States) government, it is not surprising that the tiny nation of El Salvador was considered most likely to afford accommodation to German interests, although Honduras was also suspect. There was deep concern that German submarine activity might be centered within the Salvadoran portion of the bay or, alternatively, in the Honduran port of Amapala with its numerous German trading houses. Spinden's task was to assess the potential danger of all of these possibilities.

But no sooner had he begun his investigations, than the first of several horrific temblors struck El Salvador. Spinden raced to the capital city of San Salvador to meet up with Morley and Held who were already there.[41] They entered the besieged city after a series of powerful earthquakes on June 7th had wreaked massive devastation; the three ONI agents were astonished by the level of destruction and debris strewn in the quakes' path. Everywhere houses were leveled, the drinking water was limited and people were fleeing the capital as quickly as they could. In a three page, single-spaced account written to Secretary Sherwood at the Museum on June 15th, Spinden described the damage in detail and told his superior that he expected to stay in the capital for some time while he aided in the relief effort. He offered his opinion that the United States government should grant generous aid to the suffering nation, which he considered "an excellent political move, even if all questions of humanity were left out," adding that the German colony in the country had already contributed ten thousand pesos to the relief effort.[42] Morley, who wrote separately to the Secretary of State, agreed with Spinden's assessment that this was an "unequalled opportunity for creating favorable feeling towards United States Government."[43]

A month and a half later, in late July, Sherwood sent a copy of Spinden's eyewitness account of the Salvadoran earthquake to the editor of Scientific American, inviting that journal to publish it, "provided due credit is given to the Museum." Spinden is described in this letter as having been "engaged in archaeological work in San

Salvador for the American Museum of Natural History."[44] Apparently the Museum was not only willing to substantiate Spinden's "cover," they were prepared to go to great lengths to protect the subterfuge. By then, perhaps, they had the good sense to finally realize that Spinden's life depended on it.

On June 18th, 1917, the State Department sent a coded message to the U.S. Consulate in San Salvador which was intended for Spinden.[45] Without the benefit of the deciphering key he possessed, it is impossible to interpret its contents, but it is likely that Spinden had been ordered to return to Fonseca Bay. Conditions in the capital were such, however, that Spinden could not proceed right away. Another urgent, and this time uncoded, request arrived from the State Department on July 3rd, which unequivocally ordered him to check out the rumor that on June 25 Germans had been spotted using small boats to supply a "mysterious steamship off port of Amapala,"[46] on the Honduran side of the Gulf. Spinden was to "investigate and cable report," and this time the archaeologist left immediately. On July 5th, Consul Boaz Long informed Washington that Spinden had left the capital of San Salvador for Fonseca Bay that day and that the American Legation in Tegucigalpa had been advised of that fact.[47]

On July 13th, Spinden sent his assessment to "Gustav Koch" in New York, the code name for his contact officer at the ONI Wall Street offices. Spinden's note was not written in cipher, but it contained a disguised message that gave the impression that Spinden was a businessman working in Central America. The archaeologist wrote to his contact as if he had been expecting to close a deal but was unable to seal the contract. The "account is doubtful," he opined, concluding that, "[T]he matter can remain in abeyance…market quiet; expect reaction soon, quotations are too high…machinery is out of date."[48] Was he implying that the threat of German activity in the Bay had been overstated? Or was he still attempting to ferret out data? The message leaves the impression that Washington need not be overly concerned and that, in any event, no urgency need be attached to the matter.

This task completed, several days later Morley cabled ONI for the next set of orders for Spinden, Held and himself. Unfortunately, the response was sent via the Salvadoran telegraph office and Morley believed that "as far as Salvador is concerned, my alibi [i.e. "cover"] and that of Held and probably that of Spinden as well, is gone."[49] But Morley's pessimistic assessment seems not to have panned out, since Spinden continued to operate in and out of El Salvador for almost a year following this telegram, notwithstanding a brief hiatus at the end of July, 1917. Moreover, in April of 1918, Spinden was appointed the Head of ONI Section 3 in Central America, a territory which included El Salvador and the west coast of Honduras and Nicaragua,[50] which suggested that his cover was very much intact.

To be on the safe side, Held, Morley and Spinden left El Salvador on July 22, 1917, and headed to Fonseca Bay. For now, Spinden's most important contribution to the War effort had been to alert Washington to the activities of the German trading houses of the J. Rossner and Theodore Kohncke (sometimes identified in the correspondence as Kohneke) firms, which were headquartered in Amapala.[51] On Spinden's counsel, these companies (which coincidentally performed banking functions) would continue to be watched by the U.S. government throughout the duration of the War. This caution held notwithstanding the sophistry involved when the Kohneke firm changed its name to Compañia Nacional de Agencias to avoid appearing "too Germanic." In fact the cable censor's New York office issued an alert on January 18, 1918, warning that Kohncke and other firms in Amapala were dangerous and should now be considered "code red." "[J]udging both by the cables and the mails," the censor office reported, these firms were judged "very active." Mail from this firm was to be carefully scrutinized and "[U]nless the Amapala correspondent is well known to the Censor, no mistake would be made in holding the message up for a time."[52]

From Fonseca Bay, the trio of Held, Morley and Spinden headed to Tegucigalpa, the capital of Honduras. Unfortunately for the team, Spinden came under suspicion for being too interested in Germans; to protect himself, he soon departed for the safety of Bluefields on the North coast of Nicaragua.[53] He would make that port his base of operations for the next few months, during which he found time to visit San José, Costa Rica to meet up with Lothrop and his wife Rachel, both of whom were also ONI agents.[54] By December, 1917, Spinden apparently felt secure enough about his cover to return to El Salvador, and he continued to travel within the Salvadoran-Honduran-Nicaraguan axis during the early months of 1918.[55]

But what was happening to the Museum's interests during all of Spinden's meandering through Central America? Had Spinden been able to fulfill his end of the bargain to the AMNH? The answer to that is initially yes; somehow, by September of 1917, Spinden had managed not only to correct the galleys of his forthcoming book on Ancient Civilizations of Mexico and Central America, but arranged to have a copy sent to President Osborn. The latter graciously praised Spinden's new volume, calling it "important" but not hesitating to add his own recommendations for making it "stronger".[56] At the very least, Spinden was able to give his boss the impression that he remained committed to his Museum position by publishing important research with the Museum imprimatur, even while he simultaneously tackled the thorny business of intelligence surveillance. Apparently Spinden could "walk and chew gum" at the same time.

As further evidence of his ongoing dedication to his Museum responsibilities, Spinden looked for opportunities to acquire quality archaeological artifacts. On June 19th, 1917, in the midst of the chaos engendered by the earthquake, Spinden wrote to Wissler describing what appeared to him to be a first rate collection that had recently become available for purchase. The Anthropology curator did not respond until July 11, just a few days before Spinden returned to his assignment in Fonseca Bay. Presumably Wissler's delay had been caused by the U.S. censor's examination of mail (even from Spinden) before it was forwarded to the Museum. In any event, Wissler advised Spinden to obtain an option to buy the collection, but to wait before committing actual sums to the owner. Of course, any future plans to purchase collections would be contingent on Spinden getting permission to export materials to the United States. "What may confront us next year, we cannot forecast," Wissler opined. It is possible that by this point the Anthropology curator understood what Spinden was really doing in Central America, even if he had not been in the loop in April. But if he did appreciate that Spinden was in Central America on a spying mission, that knowledge did not preclude him from melodramatically chiding his underling, "I suppose you have not had a chance to try for chronology yet… I hope that before I die someone will have a hole down through Maya culture—I am sure the story is all there somewhere."[57]

In December of 1917, Spinden successfully shipped six cases of "archaeological objects" from the port of Corinto, Nicaragua to New York, via the May and Griffith Company.[58] But in February of the following year his conscience compelled him to confess to Dr. Wissler that "the calls upon my time have been so many that comparatively little work of direct interest to this Museum has been done by me in the last two months. The time in the field has run beyond what I expected but although I have tried to get away, I am always given something more to do…[however] I feel that the time has been well spent."[59] Uncertain of how the administration would respond, and unable to get back to New York by February as he had originally anticipated, Spinden pleaded that he hoped that the "Trustees will not put me on the shelf when they meet at the Annual Session."[60]

Perhaps to make up for what he recognized might be perceived as a failure to pursue fieldwork, Spinden eventually was able to secure a spectacular collection of "painted vases in the Maya style" from one Justino Armas, a Salvadoran collector, using five hundred dollars of his own money as a down payment for the purchase.[61] In March of 1918, Spinden reported to Wissler about his preparations for shipping this collection,[62] informing the curator that, in addition to Mayan pottery, the Armas collection also contained fine representative archaic material of clay and stone. Perhaps more to the point, Armas was a dealer in antiquities, and on good terms with the President of El Salvador so that obtaining permission to export the collection would not present a problem. Spinden hoped that the AMNH would reciprocate by sending material in exchange to the Salvadoran Museum, as they had done during their previous attempts to convince that government to permit outside researchers to work in their country. In the meantime, Spinden already reproduced several drawings from this collection which appeared in his "Notes on the Archaeology of Salvador" [sic]. In his opinion, "its duplicate could [not] be obtained by any original exploration on the part of the Museum…[while] several pieces show connections with Mexico and…Nicaragua and Costa Rica."[63] This was precisely the raison d'être of the ill-fated five year Museum project which Spinden had proposed in 1915 and which had aimed to establish the relationships among various Mayan pottery horizons from Mexico through Central America.

The AMNH subsequently purchased the entire Armas collection which, including freight and consular services, cost the institution the substantial sum of $2561.20; packed up in mid-May,[64] the shipment arrived at the Museum on July 18, 1918.[65] George Pindar, the Museum Registrar, recorded it as accession number 21007, the 47th such acquisition of 1918. According to the memorandum Osborn sent to Wissler,[66] at least part of the collection was paid for by a gift from George B. Hopkins.

Notwithstanding his lack of personal time to pursue actual fieldwork on behalf of the Museum, Spinden had, incredibly, managed to please his superiors at the AMNH with the acquisition of a smashing new collection. Even before the materials arrived in New York, Osborn wrote to Spinden on March 18, 1918, informing him that, after consulting with Director Lucas and Curator Wissler, his position would be renewed at his current salary of $2200 per year and he would be allowed to "continue…travels, researches and assemblages of collections in Central America."[67] Referencing the themes which had dominated the Second Pan American Scientific Congress of 1915–6 to which Spinden had been an official delegate, Osborn's letter requested that the archaeologist aid the Museum in "increasing our knowledge of the peoples, of the geography, of the climate, the ethnology and archaeology of this most interesting country [El Salvador] south of the Mexican Republic."[68] Additionally, Spinden was reminded in the administrator's letter of the Museum's overarching goal in "establishing and maintaining friendly relations of every kind…in pursuing these researches and exploration in Central and South America."[69] Obviously, the preeminent role of the AMNH in advancing "scientific Pan Americanism" had continued to dominate Osborn's thinking.

In service to the fact that Osborn was prohibited from publicly mentioning Spinden's continuing espionage in the region, Osborn's letter to the archaeologist was essentially another crafted "cover letter" for his future ONI business in Central America. In the interim, while Spinden had been angling for the Armas collection, the ONI switched Spinden's assignment from Central America to Colombia, where he was now instructed to aid in an ongoing search for platinum. Like manganese, platinum was classified as "vital war materiel," which was "needed for the production of magnetos, sulfuric acid, amplifiers, shell primers and electrical equipment, as well as for the fixation of nitrogen."[70] Although there were several ONI agents already operating in Colombia, as well as numerous British agents,[71] Spinden had proven so effective in Central America in his tracking of German commercial activities, probable involvement in locating a German wireless transmitter and helping to increase production of mahogany for airplane propellers,[72] that he was deemed an essential component of this new intelligence nexus in the South American "theater."

Years later, in his 15th Harvard class report of 1921, Spinden diverged from most of his classmates in stating under "war record" that he served from April 1917 to March 1919 as "Ensign, Office of Naval Intelligence, (reporting on conditions and war supplies of Central and South America)." He went on to report that "Much of my work during 1917 and 1918 was concerned with mahogany along the Atlantic Coast of Central America, this wood being in great demand for aeroplane propellers. In 1918 I went to Colombia to report particularly upon the platinum situation."[73] Not only does this clarify his activities, it indicates that he was proud of his work and more than willing to announce it widely soon after the War ended.

In a letter he sent to Wissler on February 2, 1918, Spinden hinted at his change of venue to Colombia, saying that if he were to be sent elsewhere "in connection with the present arrangement," it would perhaps be better to "try" South America.[74] Wissler evidently shared Spinden's letter with his superiors in early March, 1918, causing Osborn and Director Lucas to together demand that Wissler ask for "direct official information" as to why it was "desirable" that Spinden go to Colombia.[75] Whatever explanation the Museum received in response to this question was evidently satisfactory, as evidenced by Spinden's letter to Wissler of May 17, 1918, in which he could now blithely remark that he would make his way to Panama and Colombia after June 20th, apparently without fearing any repercussions.[76]

By the end of the war, Spinden had made some preliminary contacts with other collectors in Central America but, for the most part, these were limited to those archaeologist/spies with whom he had been working, or to local collectors of *objets d'art*, rather than with the host country's scientific community. Whatever arrangements he made for future museum exchanges between the AMNH and El Salvador were most likely arranged through Justino Armas. A realistic overall assessment of the time Spinden spent in Central America would conclude that he achieved little in the way of advancing "scientific Pan Americanism," even as his cover work as "an archaeologist" at least promoted AMNH name recognition in Central America.

The ONI agreed to release Spinden from active duty upon the conclusion of the War in November of 1918, but he was not formally detached from service until February 8, 1919.[77] With the nation's thanks for "faithful and patriotic services…rendered to your country in the war with Germany",[78] Spinden returned to New York in late winter and to his waiting job at the American Museum of Natural History.

It is unclear whether the War created a distinct restlessness in the man, or whether he had determined that a move to another institution would be beneficial to his career and prove perhaps less demanding than his work at the AMNH. He also might have felt more comfortable returning to his alma mater. But similar to Crampton, Spinden left the Museum in 1921 and took up a position as Curator of Mexican Archaeology and Ethnology at the Peabody Museum of Harvard University.[79] This break with his past spilled over to his foray into intelligence work, and it would seem that his sleuthing career ended with his mustering out in 1919. Unlike Crampton, and some of the other scientist/spies at the Museum, there is nothing in Spinden's files at the National Archives to suggest that he continued spying for the United States government after World War I ended.

Notes

1. Herbert J. Spinden was born in South Dakota, August, 1879, and died in New York in October, 1967. He worked with railroad survey parties before entering Harvard where he received a B.A. in 1906, a master's degree in 1908 and a Ph.D. in anthropology in 1909. From 1909 to 1921 he held the position of assistant curator of anthropology at the AMNH. Biographical material from "Official biography of Dr. Herbert Joseph Spinden," compiled by Robin Smith and John C. Pallister, AMNH vertical biography files.
2. Osborn to Curator Wissler, 25 October 1910, CA, AMNH, File 906. In this note Osborn thanks Wissler for enclosing a letter from Spinden reporting "his progress in regard to the archaeology of Mexico, as well as the geographic distribution of the principal monuments and the sources of information which we must draw upon for materials and designs for the Great Mexican Hall."

3. James Brown Scott, Second Pan American Scientific Congress: Final Act and Interpretive Commentary Thereon. (Washington, 1916), p. 474.
4. Curator Wissler to Lucas, New York, 27 April 1915. CA, AMNH, File 906.
5. *Ibid.*
6. *Ibid.*
7. Curator Wissler to Lucas, New York 14 May, 1915. CA, AMNH, File 906.
8. The outline for this work was presented to the Government of El Salvador for preliminary approval. Osborn to the Minister of Salvador, New York, 14 May 1915, CA, AMNH, File 906.
9. Curator Wissler to Director Lucas, 14 May 1915, CA, AMNH, File 906.
10. Secretary of La Delegación de El Salvador to AMNH, New York, 19 May, 1915. CA, AMNH, File 906.
11. Director Lucas to José Alfaro Morán, Consul of El Salvador, New York, 25 July 1915. CA, AMNH, File 906.
12. Consul of El Salvador in New York to Director Lucas, 30 July 1915. CA, AMNH, File 906.
13. Perhaps the best account of the actions of the El Salvador government at this time was written almost immediately after World War I concluded. See Percy Alvin Martin's monumental, Latin America and the War: The Albert Shaw Lecture on Diplomatic History, 1921 (Baltimore, Johns Hopkins University Press, 1925), pp. 499 and 510–511 *et passim.*
14. Curator Wissler to President Osborn, New York, 6 January, 1916. DAA, AMNH, File 224. Curiously, several months after Spinden had left Venezuela, Wissler suggested to Osborn that a report on his work be offered to the Proceedings of the National Academy of Sciences. Apparently the need for secrecy vanished after Spinden had surveyed the archaeological landscape in Venezuela. See Wissler to Osborn, New York, 9 May 1916, DAA, AMNH, File 224.
15. Spinden to Curator Wissler, Valera, Venezuela, 21 February 1916. DAA, AMNH, File 224.
16. *Ibid.* Spinden mapped the area from Maracaibo in Western Venezuela to the Orinoco Delta in the east, tracing a potential collection trip up-river to Ciudad Bolívar. Later on Spinden would write a small paper on his reconnaissance which he called "Travel Notes in Western Venezuela." Published in the American Museum Journal, Vol. XVII, no. 1 (January 1917), (pp. 15–23), Spinden made it seem that there was a large expedition sent to Venezuela, rather than his solo, secretive adventure.
17. Goddard to Boas, 3 February 1916. New York, DAA, File 75.
18. Boas to Crampton, 4 February 1916, American Philosophical Society Boas files, 92. Four days later the vote went in Spinden's favor, 8 February 1916, Boas files, 98.
19. Wissler to Spinden, New York 3 March, 1916. In this letter Wissler informed Spinden that he would be working with Anthony on cave fauna and that he was to contact the mammalogist as soon as he arrives. DAA, AMNH, File 224. The exhibit in question morphed into the Food Values and Economies Exhibition which opened in May of 1917.
20. Wissler to Crampton, New York 11 October, 1916. DAA, AMNH, file 495.
21. Wissler to Osborn, New York, 9 May, 1916. DAA, AMNH, File 224.
22. Osborn to Spinden, New York, 26 February 1917. CA, AMNH, File 906.
23. See the preceding discussion of Henry Crampton and his work on the Council of National Defense.
24. Morley's list is cited in Harris and Sadler, The Archaeologist…, p. 46.
25. "Minutes of the Preparedness Committee Meeting," 9 April, 1917. CA, AMNH, file 1069.
26. Instruction to No. 56, 11:40 April 16, 1917. Confidential Files, Office of Naval Intelligence, NARA, 20578. Harris and Sadler (The Archaeologist… p. 49) suggest that Spinden left the Museum for his ONI post on April 7. That is impossible, however, as Spinden spoke up several times at the April 9th meeting of the Preparedness Committee in the Museum. It is likely that Spinden went to Washington on April 7, 1917 where he took the oath of office as an Ensign, class 4, in the United States Naval Reserve Force (ONI division), and then returned to New York to finish his preparations. See Spinden File, ONI 20317, April 5, 1917, Spinden's swearing in papers and physical. See also details of ONI general instructions to agents in Chap. 11 (p. 216).

27. Preparedness Meeting, 10 April, 1917, CA AMNH, File 1069. See discussion in Chap. 3.
28. EC AMNH, September 19, 1917, pp. 39–40. See discussion in Chap. 3.
29. Clark Wissler to Director Lucas, New York, 4 April, 1917. CA, AMNH, File 906.
30. Director Lucas to Wissler, New York, 4 April, 1917. CA, AMNH, File 906.
31. Osborn to Lucas, New York, 4 April, 1917. CA, AMNH, File 906. Woodward was the President of the Carnegie Institution.
32. Osborn "To Whom It May Concern", New York, 12 April 1917, CA, AMNH, File 906.
33. Instruction to No. 56, 11:40 April 16, 1917. Confidential Files, Office of Naval Intelligence, NARA, 20578.
34. On April 16th, 1917 Spinden reported directly to ONI offices in Washington where he acknowledged the receipt of Codes nos. A24 and B24 as well as the key and ability to "send or receive a message in cipher." Confidential files, Office of Naval Intelligence, NARA, 20578.
35. See Harris and Sadler, The Archaeologist… p. 49.
36. Bisher, The Intelligence War… p. 126.
37. As are the activities of other scientist/spies such as Held, Lathrop and Mason.
38. Boaz W. Long to Secretary of State, San Salvador, 1 June 1917. NA, RG 59, 031.11 C 21/70.
39. Bisher, The Intelligence War…, p. 123 and Harris and Sadler, The Archaeologist…, p. 77 also mention the arrest of a German whose name is given as Bach; perhaps this is a pseudonym for Brinkmann. For further discussion of German wireless activity in Central America during WWI see David McCreery, "Wireless Empire: The United States and Radio Communication in Central America and the Caribbean, 1904–1926," South Eastern Latin Americanist, Summer 1993, pp. 23–41.
40. Spinden mentioned his location to Secretary Sherwood, San Salvador, 15 June 1917. CA, AMNH, File 906.
41. *Ibid.*
42. *Ibid.*
43. Morley to Secretary of State, Washington, June 15, 1917. A carbon copy of this transmission is in AMNH, CA 906.
44. Assistant Secretary (Sherwood) to Charles A. Munn, New York, 24 July, 1917. CA, AMNH, File 906.
45. Telegram of Secretary of State Lansing to AMLEGATION San Salvador, Washington, 18 June, 1917. NARA, RG 031.11 C21/72.
46. Secretary of State Lansing to AMLEGATION San Salvador, Washington, 3 July, 1917. NARA, RG 031.11 C21/72.
47. Boaz Long to Lansing, San Salvador, 5 July 1917. NARA, RG 031 C21/72.
48. Spinden telegram sent by Minister Ewing to Secretary of State, Tegucigalpa, 13 July 1917. NARA, RG 031.11 C21/72.
49. Harris and Sadler, The Archaeologist…, pp. 80–81.
50. Division of Central America into Sections overseen by ONI agents. This list is reproduced in Harris and Sadler, The Archaeologist…, pp. 179–180.
51. Harris and Sadler, The Archaeologist… p. 81.
52. "The Daily Bulletin", issued in the Office of the Cable Censor, New York, January 18, 1918, #53. NARA. RG 38, entry 226, Box 1.
53. Harris and Sadler, The Archaeologist…, p. 106.
54. *Ibid.* See also https://www.doaks.org/newsletter/news-archives/2015/digger-curator-soldier-spy-samuel-kirkland-lothrop.
55. Harris and Sadler, The Archaeologist…, p. 194. See also note 50, above.
56. Osborn recommended inserting a paragraph on chronology to make the book read better. However, he also admitted that the book was a "really…important contribution" to the field. Osborn to Spinden, New York, September 18, 1917. CA file 827. The book was published in 1917 as AMNH Handbook Series number 3 and reprinted and revised several times afterward.
57. Wissler to Spinden, New York 11 July, 1917. DAA, File 224.

58. May and Griffith, Corinto, Nicaragua, to American Museum of Natural History, December 15, 1917, DAA, Spinden file.
59. Spinden to Wissler, Tela, Honduras, February 2, 1918, DAA, Spinden file.
60. *Ibid.*
61. *Ibid.*
62. Spinden to Wissler, March 20, 1918, DAA, Spinden file.
63. *Ibid.*
64. Spinden to Wissler, May 17, 1918, San Salvador, DAA, Spinden file.
65. Accession Record, American Museum of Natural History, George Pindar, Registrar, July 18, 1918, DAA, Spinden file.
66. Osborn to Wissler, June 11, 1918, DAA, Spinden file.
67. Osborn to Spinden, March 18, 1918. CA, file 906.
68. *Ibid.*
69. *Ibid.*
70. Richard L. Lael and Linda Killen, "The Pressure of Shortage: Platinum Policy and the Wilson Administration During World War I," Business History Review, Vol. LVI, No. 4 (Winter, 1982), p. 546.
71. Bisher, The Intelligence War…, pp. 232–234.
72. Harris and Sadler, The Archaeologist… p. 109. In a technical report about the use of "old" wood to repair the Capitol after the activity of January 6, 2021, the authors wrote: "One common use of mahogany was for airplane propellers, …which must have dimensional stability, high strength, clear grain, and consistent properties throughout the laminated structure." (R. Bruce Allison, James R. Anderson, Amy L. Androff, Robert J. Ross, C. Adam Senalik, Jeffrey Hagan, Mary L. Oehrlein, Nathan Kamprath and Keith York, "Lumber from World War I Research at the Forest Products Laboratory Helps Repair the U.S. Capitol Building," General Technical Report FPL-GTR-296 (Forest Products Laboratory, U.S. Department of Agriculture Forest Service), (December 2022), p. 4.)
73. Harvard College Class of 1906. Fifteenth Anniversary Report. Printed for the Class at the University Press, 1921, p. 314.
74. Spinden to Wissler, February 2, 1918, DAA, Spinden file.
75. Director Lucas to Wissler, March 7, 1918, CA, file 906.
76. Spinden to Wissler, May 17, 1918, San Salvador, DAA, Spinden file.
77. Navy Department, Bureau of Navigation, Washington, February 8, 1919. Spinden file, NARA, declassified May 31, 1995, Authority NND750041.
78. *Ibid.*
79. "Official biography of Dr. Herbert Joseph Spinden," compiled by Robin Smith and John C. Pallister, AMNH vertical biography files.

Chapter 8
Brazil, Where the Manganese Comes from*: The Strange Journey of Charles R. Eastman, ONI Agent 139

Surely the most unlikely member of the AMNH intelligence-gathering cadre was the scholarly, middle-aged ichthyologist, Charles R. Eastman (Fig. 8.1, left). A Harvard-trained undergraduate who specialized in fossil fishes (especially early Devonian sharks), Eastman earned his doctorate in Germany, rather than in the United States. He was an accomplished linguist who spoke German fluently, as well as Spanish and French and, after his stay in Brazil from June, 1917 to April, 1918, apparently a decent Portuguese. In contradistinction to C. P. Snow's premise which began this book, Eastman was also a humanist and widely admired as a Greek and Roman classics scholar. Eastman held a number of prestigious positions, including an instructorship at his *alma mater*, and a post at the Carnegie Museum in Pittsburgh, before he initiated his career at the AMNH.[1] Well thought-of as a scientist, Eastman was recruited by Bashford Dean (Fig. 8.1, right), Curator of Ichthyology at the AMNH, in 1914, to take charge of the curator's important project, a monumental bibliography of all fossil and living fishes.

Eastman began working at the AMNH at a propitious moment for Latin American research, although he had previously expressed scant interest in the region. Nonetheless, the enthusiasm generated by successful Museum expeditions south of the border would eventually push Eastman in that direction as well. Among these were the spectacular return of Theodore Roosevelt, Leo Miller and George Cherrie (ONI Agent 245) from their perilous, yet ultimately triumphant, exploration of the Brazilian River of Doubt; Rollo Beck's reports and collected specimens, both of which arrived periodically from South America; Frank Chapman's substantial and highly acclaimed work on Colombian birds; and Joe Spinden's grand (but unrealized) five-year plan to study the extension of Mayan culture beyond the Mexican border. Other scientists from the Museum were actively involved in collecting birds, mammals and reptiles from Latin America at

*With credit (and apologies) to the line, "Brazil, where the nuts come from," in "Charley's Aunt," an 1892 farce by British writer Brandon Thomas.

R. M. Delson, *Scientists or Spies?*, The Latin American Studies Book Series,
https://doi.org/10.1007/978-3-032-18117-6_8

Fig. 8.1 Left, Charles R. Eastman, date uncertain; right, Bashford Dean in his uniform of Army Major, date uncertain, AMNH Asset ID 312546

the time of Eastman's hiring. Moreover, Museum president Henry Fairfield Osborn, who was constantly on the lookout to expand the AMNH Latin American collections, recognized that a representative fish collection from the region would round out the faunal surveys he now deemed so critical to his broader evolutionary project. So, while research in Latin America had not figured into Eastman's plans prior to his arrival at the Museum, within a few years of his arrival he, too, would head south, ostensibly to pursue research on and collect Brazilian fishes (fossil and extant) for Curator Bashford Dean's compendium and to fill the lacunae in the Museum's existing fish exhibitions.

At first Eastman worked exclusively on Dean's Bibliography of Fishes, but his excellent work quickly merited him an additional appointment as Research Associate in the Department of Paleontology (1915).[2] Already forty-seven years old, Eastman was sensitive to the fact that he was obliged to prove himself if he was going to remain at the AMNH and he rewarded Dean's confidence in him by producing a spate of papers which not only enhanced his, but the Museum's overall reputation in ichthyology.

Eastman's Early History and Start at the AMNH

On the surface, it appeared that Eastman's upward trajectory at the AMNH was geared toward impressing his superiors, as well as guaranteeing the security of his new position. But there may have been other factors driving his ambition. Given the ichthyologist's calm and innocuous scholarly mien, it is surprising to learn that early in his academic career, while he was still an instructor at Harvard, Eastman was involved in an incident so sordid that it certainly could have stopped someone less determined than he would prove to be.

The circumstances of this affair can be summarized as follows: on July 4th, 1900, Charles Eastman shot and killed his brother-in-law, Richard H. Grogan Jr., in the hen yard of their shared Cambridge, Massachusetts property. In more detail, Eastman and Grogan (with their families) lived side-by-side, while the former taught in the Zoology Department of Harvard University. The brothers-in-law had spent the Independence Day holiday together, later dining with their families. Some time after their meal concluded, both men took target practice in the hen yard, where Eastman would claim that his pistol had accidentally discharged. The bullet hit Grogan, who fell to the ground; remarkably, he was able to rise to his feet and angrily charged at Eastman, weapon in hand. The latter grabbed his brother-in-law's gun and fired. As the two men struggled to the ground, Grogan lay mortally wounded and died shortly thereafter.[3]

The charges facing Eastman for killing Grogan were initially dismissed as the result of an horrific accident. No longer under investigation, Eastman left Cambridge and headed west to collect fossils. But further police work, as well as eyewitness accounts of the shooting, cast sufficient doubt on Eastman's characterization of the July 4th event that a grand jury was convened. Evidence presented at that tribunal resulted in Eastman being indicted for murder, and for the remainder of 1900, and into the months leading up to his April, 1901 trial, Eastman languished in a Cambridge jail.[4]

Eastman's trial attracted so much attention nationally that it led The Boston Sunday Post[5] to liken it to the better known, and arguably far more sensationalistic, New England axe murder case involving Lizzie Borden. During Eastman's trial, which lasted from April 22 to May 11, 1901, the ichthyologist continued to portray Grogan's shooting as an unfortunate mishap, a claim which was fiercely challenged by the prosecutor and police. In the end, however, Eastman was acquitted of murder by virtue of his gun having gone off accidentally.[6]

It seems likely that this experience remained with Eastman throughout his life, although it is unclear whether the AMNH scientific staff was familiar with his past when the ichthyologist was brought on board in 1914 (despite the broad play the court proceedings had received in The New York Times[7]; Fig. 8.2). On the other hand, Bashford Dean, Eastman's future supervisor at the AMNH, had certainly been apprised of the chequered details. Even if Dean had not been aware of the trial when it took place in 1901, then he surely learned of the proceedings by 1909 from a letter he received from a Wyoming fossil quarry owner who had previously done business with Eastman. In a rustically written note offering to sell Dean a coveted fossil gar (an early sturgeon), the quarry owner informed him that Eastman (not yet at the AMNH) had purchased a similar fossil from him, and for far more money, at a time when he was under suspicion for "killing Grogan his bro-in Law [sic]."[8] There is, therefore, no possibility that Dean was unfamiliar with Eastman's past, although it is unclear whether he chose to discuss the case with him when the latter was hired at the Museum, or kept mum out of gentlemanly respect. The simple fact is that knowledge of the Eastman trial was readily available to anyone who cared to research it.

Perhaps a more important question is whether the ONI was aware of Eastman's past when they recruited the ichthyologist to conduct espionage in Latin America. Sometime in the Spring of 1917, Eastman agreed to undertake the same type of sleuthing assignment in Latin America as Joe Spinden had done in Central America, although in doing so, he knowingly jeopardized the security of the Museum position which he had worked so assiduously to secure. Nonetheless, Eastman was duly sworn in as ONI Agent 139, after he was likely recruited either by Henry Crampton, with whom he would communicate from the field, or by Charles Sheldon, a "big game hunter and wealthy sportsman who was a friend of every naturalist," including those at the AMNH and who also happened to work for Naval Intelligence.[9]

Was the motivation for Eastman's embrace of espionage based on the supposed glamor of his new assignment, or did it reflect his patriotism and/or disdain for Germany? Was it, perhaps, an attempt to rectify the mistakes of his youth? Or was it likely a combination of all these rationales? Whatever his personal reasons were, when the Museum administrators learned in June of 1917 of Eastman's decision to engage in spy work, they moved quickly to provide him with another transactional *quid pro quo* which paralleled the one already in place for Spinden. Echoing the parameters of the archaeologist's arrangement, Eastman would be allowed to proceed with his government assignment as long as he consented to simultaneously pursue scientific work. In Eastman's case this meant augmenting the Museum's existing collection of fishes with much desired Brazilian specimens (fossil and living). The Museum's understanding of this arrangement was that it reflected parallel objectives. As it turned out, however, Eastman would be hard pressed to fulfill the institution's expectations during his residency in Latin America.

In service to the secrecy of his assignment as well as to circumvent the Museum's ongoing moratorium on overseas fieldwork, an "official" cover story was hastily concocted which offered a rationale to the Museum community for Eastman's travel. Rather than suggesting that Eastman would be conducting research overseas, the Museum higher-ups floated the idea that the ichthyologist would instead function as a kind of cultural ambassador to the scientific institutions of South America. In the public framing of his mission, Eastman would visit natural history museums in Brazil, Paraguay and Uruguay where he would make observations on the state of research in these nations and subsequently publish his findings in the Museum's journal. To solidify that premise, in advance of his trip The American Museum Journal published a news brief which claimed the purported goal of Eastman's trip was to pursue "friendly relations and cooperation with scientific institutions in Central and South

"All the News That's Fit to Print."

The New York Times.

THE WEATHER. Probably showers and cooler; fresh southerly winds.

VOL. L...NO. 16,019. NEW YORK, SUNDAY, MAY 12, 1901.—TWENTY-SIX PAGES. PRICE THREE CENTS.

CHARLES M. SCHWAB ON THE STEEL TRUST

ELECTRIC STRIKE SPREADS.

ENGINE PLUNGED INTO WEST FARMS CREEK

HARD BLOW FOR DESERTERS.

C. R. EASTMAN ACQUITTED

LENA PENSACK'S BODY FOUND.

AUTOPSY ON McCORMICK BOY.

THE ALLIES' METHODS IN CHINA CAMPAIGN

THE DRAINAGE CANAL FIGHT.

PUTNAM'S HILL TROLLEY LINE.

C. R. EASTMAN ACQUITTED

Harvard Instructor Is Found Not Guilty of Murder.

JURY OUT OVER FIVE HOURS

Attorney General Knowlton Denounces the Accused in Strong Terms—The Judge's Long Charge.

CAMBRIDGE, Mass., May 11.—After a long and dramatic trial, Charles R. Eastman, the Harvard instructor, to-night was acquitted on the charge of murdering his brother-in-law, Richard H. Grogan, Jr., while at target practice on July 4 last. The jury, after listening patiently for nearly three weeks to a tremendous mass of testimony, followed by two days of argument, debated five hours and a half before reaching a verdict.

The twelve men filed out of the courtroom at 4 o'clock this afternoon and returned at 9:30 this evening, and when the foreman, looking straight at Eastman, said in a clear voice, so that it was heard

Fig. 8.2 "C. R. Eastman Acquitted". Front page of New York Times of May 12, 1901, with story on Eastman trial, enlarged at right. Downloaded from Times website, not copyright, in the public domain

America;"[10] in other words, he would promote the "scientific Pan Americanism" which President Osborn continued to publicly endorse.

Once again, and with a far more nuanced explanation than the simplistic reason given for Spinden's travel ("temporarily joining up with Morley,") the Museum hierarchy had subverted its own research prohibitions outside the country during wartime. One suspects the rationale provided to the Museum community for Eastman's trip needed to be more sophisticated than that manufactured for Spinden, since at the time of Eastman's departure for Brazil (June, 1917), Spinden had already been absent for two months, far longer than the brief trip Wissler, or his colleagues, had envisioned. Although Spinden's travels do not seem to have been discussed at the Museum, Osborn could not take the risk of Eastman's travel being misinterpreted.

The irony of the creative rationale fashioned for Eastman is that, in retrospect, the ichthyologist actually succeeded, perhaps more than any other AMNH scientist who did "work" in Latin America during the War, in cultivating contacts and arranging exchanges with local museums. Just as the AMNH had announced he would, during his extended stay in Brazil (from June, 1917 until April, 1918) Eastman furthered ties with the Museu Nacional in Rio de Janeiro (arguably the most important natural history institution in South America), while cementing further relations with the Museu Paulista in São Paulo. He also pursued new associations with the geological commissions of the states of Rio de Janeiro and São Paulo,[11] while he simultaneously arranged for future exchanges of collections and scientific papers from these organizations with the AMNH.

Yet, the assignment the U.S. government had given Eastman in Brazil had nothing at all to do with "scientific Pan Americanism," scholarly exchange, or even advancing the Museum's Latin American evolutionary project. Working as designated Agent 139 for the ONI, Eastman's specific task was to forward data on German commercial activities in that country (especially with regard to the sale of manganese) to the appropriate U.S. personnel. These included ONI contacts in Brazil and Curator Crampton at the Council of National Defense, as well as other agencies in Washington. He was also instructed to consider Brazil's reluctance to join the Allied war efforts and to work toward ending that nation's self-proclaimed neutrality. With a view toward accomplishing that goal, Eastman repeatedly questioned his scientific colleagues in Brazil on their views of the European War, comments which he then secretly forwarded to Washington.[12]

President Osborn and curators Dean and Crampton were fully apprised of Eastman's "extracurricular" activities, although his "secret" identity as a spy was not acknowledged among the staff. These three administrators who knew the truth would only raise questions about his mission when Eastman failed to deliver on the promised fish specimens. Bashford Dean, initially, and Osborn, at a later date, would eventually come to the conclusion that the institution had somehow been cheated by Eastman's strict compliance to his ONI assignment. As if to confirm their conclusion *ex post facto*, Eastman's South American scientific findings have never been published.

Eastman "On the Job" in Brazil

Eastman arrived in South America in June of 1917, a little more than two months after the United States declared war against Germany. At the time of his relocation to Rio de Janeiro in June, Brazil was said to be "neutral" with regard to the European War; by the time Eastman departed from the city (April, 1918), Brazil had totally reversed its position, not only pledging solidarity with the Allied cause but even sending a limited number of troops to the European front.[13] Just how integral were Eastman's activities to Brazil's changing stance on the War? The answer to that question lies in an examination of the totality of the activities Eastman undertook while he operated undercover in the South American nation.

When he first arrived in Brazil, it is likely that Eastman carried with him the same orders as his immediate (and ultimately unsuccessful) predecessor had.[14] Those earlier instructions had contained the following admonition "Your mission is first: To investigate the political situation in Brazil with reference to German colonization and activities. Give your confidence to no one. Acknowledge no connection with the ONI to anyone. Make reports only to the persons or addresses specified. Always have in mind that your actions are being observed."

Eastman seems to have followed this protocol to the letter. He was constantly on the move, alternating between Rio de Janeiro and São Paulo as his primary bases of operation. Within these cities, he would frequently change his place of residence, moving from hotels to pensions to clubs,[15] likely in an effort to avoid suspicion. At the same time, his visits to the countryside were as infrequent as his efforts at fieldwork. Ostensibly he was a fossil fish expert working in Brazil for the AMNH, a description which provided him with the requisite "cloak" for intelligence gathering in major urban centers. To facilitate that goal, Eastman assiduously cultivated contacts whom he considered useful, including the U.S. Ambassador to Brazil, Henry Morgan, who coincidentally happened to have been a classmate of Eastman's during his Harvard years. Within the diplomatic community Eastman (who was fluent in French) sought the close counsel of the French minister to Brazil, Paul Claudel.[16]

Whatever details Eastman managed to ferret out from his sources he immediately forwarded to his ONI superiors and/or Henry Crampton. Well before any correspondence is found in the Central Archives between Eastman and Osborn, the ichthyologist had already shared his keen observations on the Brazilian political landscape with Crampton who, at the time, was still at the Council of National Defense. To reach Crampton, Eastman first sent a handwritten note to Bashford Dean who, in turn, contacted Crampton. Eastman's initial concern was disclosing what he considered the greatly underestimated level of German financial involvement in Brazil. "It is possible" he cautioned the invertebrate zoologist in his September 19, 1917 report, "that our people at home do not appreciate the full significance … or the magnitude of alien operations."[17]

One of Eastman's principal assignments was securing a supply of manganese for the Allies, a task which, according to this same September letter, Eastman had already discussed with Crampton before his departure for Brazil in June. Eastman worried that German companies positioned in Brazil before 1914, such as Siemens, were sufficiently invested locally to allow them to obtain special options advantageous to their interests. "Siemens," he wrote to underscore the point, had been "buying up large tracts, paying 20% down, and contracting for payment of the balance on the day that peace terms are signed."[18] While such practices confirm the company's bold confidence in Germany's ultimately successful outcome in the War, their representatives carefully masked their goings-on in Brazil by conducting their business largely through locally owned shell companies. In truth, local Brazilian interests had much to gain from German investment, a fact corroborated in a confidential memo obtained by Eastman from the French minister, Paul Claudel. In his report, the Frenchman expressed a strong desire for the "U.S. and British governments [to] get together and use more forceful means than heretofore for insisting that the Brazilian government put a stop to recent large purchases of (or contracts for) manganese properties by German interests, which are using Brazilian firms as a cloak."[19] By passing this information to Crampton, who presumably would alert the proper authorities, Eastman hoped that Washington would take necessary action.

Interestingly, on October 26, 1917, a little more than a month after Eastman had forwarded Claudel's assessment to Crampton, Brazil unexpectedly switched sides. Now endorsing the Allies, Brazil became a belligerent in the War, and the only nation in South America to send troops to the front (albeit the contribution was limited to several officers and medical staff who assisted in the French campaigns).[20] While the United States was very enthusiastic about having these South Americans onboard (as were the grateful Europeans), not all Brazilians endorsed the decision.

What had convinced the Brazilian government to so dramatically reverse course in its politics? The primary catalyst seems to have been the sinking of several Brazilian merchant vessels by German submarines, an assault which began in 1916 and continued into the next year; the coup de grâce was the torpedoing of a Brazilian ship on October 23, 1917.[21] This continuing German threat to Brazilian-Atlantic shipping signaled trouble for the Brazilian economy which, like most Latin American nations, was disproportionately dependent on foreign trade. Even though the United States purchased more products from Brazil than either Germany or Great Britain, the Germans were not far behind in their total consumption of Brazilian goods. A lot of this trade was in coffee; viewed together, these three nations purchased three-quarters of Brazil's coffee, her major export, before the War.[22] Conversely, Germany's exports to the South American nation (which included industrial goods) had also increased significantly by the outset of World War I. Notwithstanding the severity of German attacks nor the potential for their continuation, Brazil was not in a position to overhaul her economy to suddenly favor domestic manufacture. Germany, therefore, remained a valuable trade partner, and it was obvious to the Brazilian government and the business community that commercial relations with that nation had to be handled with extreme delicacy.

In the meantime, the political class in Brazil remained divided on what position their nation should take toward the European conflict. Some notable politicians, like senior government spokesman Rui Barbosa, criticized and publicly questioned

Brazil's neutrality,[23] but others, like the distinguished senior Brazilian diplomat to Washington, Manuel de Oliveira Lima, were disposed in the opposite direction.[24]

Rhetoric aside, when Brazil finally declared war against Germany, the decision was as much the result of careful behind-the-scenes manipulation (where Eastman likely played a role) as it was the attack on a merchant ship by a torpedo three days earlier. Brazil's decision to join the Allies was widely interpreted as a gesture of solidarity with the U.S.,[25] but Washington was not completely satisfied given the widespread labor unrest throughout the country, and the large number of German-Brazilian immigrant communities whom the United States viewed as potential "fifth columnists".[26] More troubling, as Frederick Luebke has shown in his book Germans in Brazil, was the fact that the Brazilian government allowed large German-owned banks to operate within its borders with impunity. Luebke points out that "Luso-Brazilian merchants continued to patronize German wholesale houses because it was in their private interest to do so."[27] If such traders were embarrassed at the prospect of being seen with their German friends, they had only to use the telephone to conduct transactions and no one would be the wiser. Of further concern to the United States government was the fact that Brazil was culturally indebted to the German/Prussian military tradition, especially with regard to defense. Many in the Brazilian officer corps were considered "Germanized" in terms of their organization, while at least thirty-two high ranking officers between 1906 and 1912 had personally trained with the German army.[28]

In brief, animus toward Germany for interrupting Brazilian shipping continued to be tempered by the sobering reality that Germany remained a necessary trading partner. The reality was that the submarine attacks only momentarily pushed public opinion against the Reich. On the other hand, anti-German paranoia in Brazil was vocal enough (perhaps fed from the inside by agents like Eastman) that the Brazilian Foreign Minister, Lauro Muller, suspected of being sympathetic to the Germans (likely because of his heritage), was forced into resignation.[29]

Historians remain divided on whether a pro-German front was real in Brazil or if the German-speaking communities, located primarily in the southern states, would have supported the Kaiser had Germany prevailed during the War. Recently, Nancy Mitchell had stated unequivocally that "Germans living in Brazil were not a fifth column… Although retaining their German culture, they had become Brazilian citizens and were loyal to Rio, not to Berlin."[30]

Whether the threat existed or not was a moot point in Washington; U.S. war planners believed the specter of pro-German solidarity across the ocean was real enough, as did the contemporaneous generation of U.S. scholars. Consequently, when a scathing report was anonymously submitted to the ONI warning of wavering Brazilian enthusiasm for the War just five months after the Brazilian declaration of hostilities against Germany, its arrival seemed to confirm previously held misgivings. This assessment, dated March 20, 1918, was initially sent to Captain Edward McCauley, Assistant Director of Naval Intelligence in Washington, D.C., but its conclusions were so alarming that McCauley immediately forwarded it on to the State Department.[31]

Even though his name does not appear on it, there is little doubt that this damning report is Eastman's handiwork. To begin with, the author, like Eastman, remained skeptical of the Brazilian "conversion" to the Allied side. Secondly, the account was forwarded to the ONI, precisely as agents like Eastman had been instructed to do when they received vital intelligence. Lastly, the report's references to Dr. Bruno Lobo, the Director of Brazil's National Museum (the Quinta da Boa Vista), with whom the AMNH ichthyologist had established relations, as well as to Paul Claudel, the aforementioned French minister befriended by the AMNH scientist, identify Eastman as the author. In addition, from Captain McCauley this screed made its way to Leland Harrison, the head of Intelligence at the State Department, with a note observing that the contents had been written by an American living in Brazil for the past half year, exactly as Eastman had done.[31]

The March 20 report chides Brazilians for their "almost complete apathy" [my underlining] "toward the momentous issues of the great world-war."[32] The author further advises that "The issues of the war have not struck home to the Brazilians as a people, their passions have not been inflamed, their soul (telle quelle est) [Eastman spoke French, as noted] has not been quickened by any aspect of the war which they regard as vital for themselves, and their only sensitive spot which has been and produces a reaction is their pocketbook."[33]

More evidence that Eastman authored this account appears in the characterization of Dr. Lobo, Director of the Museu Nacional, and Eastman's erstwhile colleague. Lobo is described as being "a highly educated Brazilian" but distinctly naïve in his attitude toward the War. The Director apparently believed, as did many other Brazilians of his status, that the War was "too far off" for his countrymen to be concerned, and that it would likely end in a draw. More worryingly, Lobo actually suggested to the report's writer that England might satisfy Germany by simply allowing her to "take possession of the whole of Russia and do with it as she (Germany) likes."[34] Eastman was apparently so thoroughly exasperated by his colleague's ignorance that he felt it necessary to single him out for censure.

Finally, the anonymous diatribe concludes with a semi-serious remedy for Brazilian apathy for which the author gives credit to Paul Claudel. Brazil should be squeezed "like a nut between the jaws of a nut-cracker," it reads. "One of the jaws is our control of the supply of coal, and the other is placing an embargo on coffee...Brazil will sit up and take notice when Uncle Sam has something to say about war measures."[35] Clearly the report's author was certain that only extreme actions would awaken Brazilians to the reality of the War. For a scientist who not only earned a doctorate in Germany, but spent years studying in that country, Eastman's tone toward the Germans is unapologetically hostile. "Scratch a Hun," he averred, "and you get a barbarian. Here he has a strong foothold. Now is the time to smoke him out. But the Allies must do it."[36] One can only imagine the consternation raised in Washington after Eastman's report circulated through the halls of the State Department.

Eastman Falls Out with the Museum in Favor of Government Service

Disquiet over Brazil's commitment to the Allies continued to color Eastman's approach to ONI activities within his host country, but a very different kind of anxiety was simultaneously reflecting itself among those in the AMNH who understood what Eastman was actually doing in Brazil. This uneasiness had nothing to do with Brazil's shifting political positions but, instead, focused on Eastman for his potential default on his *quid pro quo* with the Museum. Despite what they both thought was a clear understanding of their transactional agreement with the scientist, i.e., that he would be allowed to pursue his twin ONI assignments of convincing Brazil to support the Allies and tracking exports of manganese, in return for collecting numerous specimens on the Museum's behalf, both Osborn and Dean were concerned that Eastman had fallen short on his end. The situation was perplexing: where exactly were the fishes Eastman was supposedly collecting? Had he, hopefully, discovered something exciting which held potential for a new exhibit? Although the scientist periodically relayed hints that he was fulfilling his scientific commitment, Eastman never actually hid the truth from his bosses that the bulk of his time in Brazil was spent engaged in what he euphemistically referred to as "extracurricular activities." If Eastman had indeed let the Museum down, what was to be done about it?

Eastman tried to make the best of what he understood was his deteriorating stance within the Museum. Recognizing that above all Osborn needed to be placated, Eastman offered the Museum president the following excuse on November 26, 1917: "It has not been possible to write at all in detail regarding my work here during the last few months, but am taking advantage of this opportunity to send a personal word by Mr. R. Anderson, a geologist, formerly of Stanford University, who will mention to you in a general way something about the present state of affairs in Brazil with reference to the international situation." "The part taken by our Government in recent developments affecting the international policy of all the South American countries is very considerable," he continued, "and the influence of the U.S.A. has been felt in the southern continent in a way that commands respect and admiration for our diplomatic representatives."[37]

Instead of providing a list of the important fish specimens which he had acquired, Eastman attempted to please his superiors by emphasizing the political importance of his activities. The summary of achievements which he sent to Osborn described the personal friendship Eastman had established with Henry Morgan, the U.S. Ambassador to Brazil[38] and, by implication, his familiarity with others in the U.S. diplomatic corps. With regard to ONI work, Eastman was more circumspect, as he was required to be, and suggested that "My own share in activities has been of a more or less confidential nature, and you will therefore pardon me if I do not refer to them other than in a general way that the results have been more successful than I had reason to anticipate and they [my underlining] desire my service to continue for a certain period longer—a few months possibly as I now have a particular sort of experience and other qualifications which cannot easily be replaced by another."[39]

This was adding insult to injury. Not only had Eastman seemingly neglected his obligation to the Museum, he was now compounding his transgression by asking for an extension of his time in the field of several months. In a coda at the end of his missive, Eastman hastened to add, somewhat disingenuously, "[O]n the scientific side the results are very gratifying, as I hope to be able to report later."[40]

A curious feature of this letter is that Eastman obviously did not believe it necessary to explain to Osborn who the "they" were who required his further service. In yet another, obviously hurriedly-penned note to the Museum president, Eastman utilized code of his own devising to compose a message to his superior. This undated, scrawled and penciled message, written on graph paper instead of letterhead, is certainly one of the most cryptic papers in Eastman's archive. Apparently written solely for Osborn's (sympathetic?) eyes, its hasty construction suggests that Eastman was anxious to get something to his boss quickly. This note is reproduced below, with my interpretation of Eastman's meaning in square brackets and boldface:

> To H.F.O. [**Henry Fairfield Osborn**]
> Morgan [**the Ambassador**] well liked and very fluent
> E's [**Eastman**] personal connection have stood him in good stead
> Br [**Brazil**] good meeting, dispos [**disposed to enter war**???]
> Jes of S.B. ord Dang. Ana auth w't
> In sh d [**should**] be done
> U.Y [**Uruguay?**] arr'd [**arrived**] here C.H infl. [**Chile influenced**] Ar [**Argentina**] coming
> R J [**Rio de Janeiro?**] most import [**important**]
> Ag [A**rgentina?**] get. q. rem
> My own imp [**impression**] dan [**danger**] of spec. [**specific**] int [**interest**] raising? crisis for own end. He confirms
> See Dean [**that is, pass on this information to Bashford Dean**]
> Love,
> Aunt [obviously Eastman][41]

If this "coded" correspondence was meant to follow his November 26th letter, cited above, Eastman was clearly trying to flatter Osborn by subtly including him in the intelligence circle.

That Eastman used Osborn as a conduit to pass on intelligence to others can also be surmised from State Department Counselor Frank Polk's response to a letter he had previously received from Osborn. In answering Osborn, Polk thanked the Museum President for enclosing Dr. Charles Eastman's report on January 29, 1918, a date which was roughly a week after Eastman mailed the lengthy summary of his activities described above to Osborn.[42] Only an excerpt of Eastman's original letter survives in the AMNH Central Archives and is strictly limited to scientific matters. It is tempting to conclude that the earlier pages of this letter, the part that contained references to political matters, strongly hinted to Osborn to alert the State Department.[43]

Notwithstanding all this subterfuge, Eastman's efforts to flatter Osborn by including the Museum President in his espionage efforts ultimately failed, and both Osborn and Dean spent the early months of 1918 feeling shortchanged. Had Eastman recognized how shaky his position had become? Much like Spinden's ploy to cajole Wissler by purchasing the de Armas artifact collection for the AMNH, Eastman attempted to ameliorate the situation by sending a few specimens of fish material to the Museum via a personal envoy, the U.S. consul in Sao Paulo. The consul represented to Eastman that he would make an in-person report to Osborn, a fact which Eastman happily relayed to New York[44]: "The U.S. consul at São Paulo will sail shortly for New York, and I am preparing to send by him a personal letter to you explanatory of my activities, and also several cases of recent and fossil fishes, books, etc."[45] On December 17, 1917 Eastman additionally wrote to Dr. Lucas that he was forwarding four cases of material to New York to be transported by the H.G. Morpurgo company, a shipper who had been recommended by the U.S. consul at São Paulo.[46] A note in the Museum archives, without a date, confirms that one box of books and three of fishes (total value of $300) had arrived, and were logged into the Museum's collections.[47]

It was not until January 23, 1918 that Eastman finally wrote his long-awaited report about scientific matters to Osborn. He summarized his activities as a researcher having spent almost eight months in Brazil in succinct terms: "Very close and cordial relations have been established with the Museu Nacional at Rio and Museu Paulista at Sao Paulo," and "also with the federal and state geological surveys."[48] Curiously, Eastman did not (or decided not to) comment upon the departure of the von Iherings (a father and son team of scientists) from the São Paulo Museum, except to say he learned that they had permanently "retired." This statement totally ignored the fact that Hermann von Ihering left under a cloud of suspicion regarding his German sympathies and that his son, Rudolpho von Ihering had subsequently resigned his own post in protest. Seemingly Eastman's only concern was that now "there is no one to look after the large collection of recent fishes and reptiles."[49] Perhaps the von Iherings were too obviously of German background (the father had been born and trained in Germany) for Eastman's taste. By contrast, AMNH Ornithologist Frank Chapman had written to Osborn only a few days earlier, expressing exactly the opposite opinion, averring that the von Iherings had been shabbily treated.[50] Instead, in his January letter, Eastman voiced enthusiasm for the new Director of the Paulista museum, Dr. Affonso d'Escragnolle de Taunay, a man who had, in his opinion, "proven a good and generous friend."[51]

Eastman continued his report by relaying the news that contact had been established with Dr. J. Pacheco of the São Paulo Geographic and Geological Commission. A day earlier, on January 22, 1918, Pacheco had written to Eastman to thank him in advance for his generous offer to have AMNH scientists study the dinosaurian and reptilian remains from the Brazilian Cretaceous period which had previously been collected by the Commission; Pacheco informed him that these fossils were ready to be shipped to the United States.[52] Yet there is nothing in the

AMNH files to indicate that such a transfer of material ever took place. The few boxes of material which Eastman had previously sent, and which we know did arrive in New York, were summarily dismissed by Director Lucas. Of the thirty specimens of marine fishes included in Eastman's shipment, Lucas griped to Osborn, "I very much doubt if they have any great intrinsic value."[53]

Eastman proposed to his bosses the exchange of these Brazilian-found dinosaurs and other fossil collections for AMNH casts, an idea Osborn rejected out of hand, declaring in a note to Lucas that such an exchange would not be possible given the "depleted condition of our cast collections."[54] Lucas concurred, adding that Pacheco's request for "one or two large-sized models of a restored Triceratops skull is quite out of the question. Aside from the cost of such a cast, none has ever been made."[55] Eastman's efforts to appease the AMNH administration appears to have only made matters worse; the ichthyologist was viewed as having made promises to Brazilian scientists which either could not be fulfilled or were judged unworthy of the effort. Despite his well-meaning and successful contacts with Latin American scientists, and his fulfillment of the goals of "scientific Pan Americanism", Eastman had not improved his standing with his superiors one iota.

Given the lack of evidence of his commitment to scientific collecting, Eastman's request for an extension of his leave of absence presented a new dilemma for the Museum. President Osborn was clearly cognizant of the importance of the ichthyologist's clandestine wartime activities, but that was not what concerned him the most: he could not, or was unwilling to, ignore his own nagging doubts that Eastman had been lax in fulfilling his promise to the AMNH. In the meantime, Eastman had apparently not even bothered to inform his own family that he wanted to stay on in Brazil. On December 18, 1917 (coincidentally the same day that Eastman petitioned Lucas to stay longer in country), Eastman's son, a student at Harvard, innocently wrote to Lucas that he anticipated his father's return shortly, unless "unforeseen possibilities arose."[56]

Osborn and Lucas were perplexed; should they continue to support Eastman's work in Brazil or demand that he return to New York to finish the bibliography of fishes? On March 1st, 1918, Lucas wrote to Bashford Dean, now a Major in the Army Ordnance Headquarters in Washington, to ask him to "take up with the proper parties in Washington, the question of further scientific work in South America at this juncture."[57] This was an oblique reference to the undercover work Eastman had been involved in Brazil during the past year. Eager to reach a conclusion on Eastman, Osborn asked Dean what was the point of the ichthyologist remaining in Brazil, "[N]ow that Brazil is actually in the war"? It seemed to the administrator "that it would be wiser for him to return."[58] Apparently, Brazil's lukewarm response to the War, notwithstanding the fact that they had joined the Allies, didn't interest him in the least.

This was the signal Curator Dean had been waiting for, and he wasted no time taking action. On the same day that Osborn solicited Dean's opinion, Director Lucas (having conferred with Dean presumably by telephone) informed President Osborn that Dean believed that "no more collecting in Brazil is considered necessary…"[59]

In response, without any hesitation, Osborn sent a terse cable to Eastman which demanded that he "Return Museum May first."[60]

Eastman responded with alacrity (and undoubtedly some alarm) that he would return to New York prior to May 1st, having already received similar orders from "headquarters" [clearly a reference to the ONI and not the AMNH]. Considering the pains he had taken to maintain his "cloak", it is surprising that in his response Eastman revealed that he would be returning to the United States onboard a U.S. naval vessel, rather than a commercial one.[61]

On his return voyage, and offshore of Bermuda, Eastman hastily penned a message to Osborn on April 21st, explaining that he had been "ordered to report to Washington as soon as practicable on arrival." The Museum and Osborn would have to wait until Eastman first cleared matters with the ONI for him to return to New York.[62] It was becoming very evident where Eastman felt his priorities lay, and the Museum administrators were growing increasingly frustrated at what they perceived to be unprecedented short shrift on Eastman's part.

Irritation with Eastman only intensified when Director Lucas questioned the scientific value of the fishes which Eastman had shipped back to the Museum.[63] The ichthyologist's contacts with colleagues in Brazil (especially the issue of dinosaur casts) had caused embarrassment[64] and his apparent lack of enthusiasm for his Museum assignment was clearly quite untenable.[65] While it was true that Eastman had been charged with initiating contacts with scientists in Brazil, in the opinion of his superiors, his awkward manner of doing so had left the AMNH looking deficient.

It was Dean who decided to use the predicament Eastman had put himself in to marginalize his position at the Museum; eventually this would lead to his being permanently severed from the institution. Whether it was Eastman's perceived scientific failure, or the expectation that when push came to shove he would favor his government work over that of the Museum, Dean recommended to Director Lucas that Eastman's resumption of the bibliographical work on the fishes be considered "provisional."[66] Seemingly without first consulting Osborn, Lucas reiterated Dean's proposal to Eastman, writing additionally to inform him that not only had his status been altered, but that he would now receive a reduced salary because his future work was considered "provisional in view of all matters concerned."[67] This was a low blow. But, Dean's shenanigans didn't stop there. On May 20, Dean planted the idea in Osborn's mind that Eastman might be expected to continue his ONI work, even though the ichthyologist had expressed his willingness to, once again, "tackle the fossil fish indexing." "If he were again given a task abroad" Dean opined, then all bets were off since "'tis war time and we may have to delay publishing [the bibliography],…if the worst comes to the worse."[68]

There is no evidence in the Museum files of what Eastman thought of Dean's actions but, as if on cue, just a week later Eastman wrote to Osborn that he had been approached by Colonel House with regard to "active war-work." Along with House's son-in-law, Dr. Sidney Mezes, "who is assisting him in some very important work undertaken with the sanction of the President" (a reference not to Osborn but to Woodrow Wilson), the Colonel had made overtures to Eastman to join the secret coterie at the "Inquiry" in uptown Manhattan.[69] According to Eastman, House and Mezes, along with members Drs. Bailey Willis (a distinguished Latin Americanist from California) and Isaiah Bowman of the American Geographical Society, "suggested that I am qualified to render service of material value in connection with this particular board, or bureau, the more precise nature of purpose of which can best be explained to you personally."[70] The work would require of Eastman only a portion of his time he thought, and he assured President Osborn that he would be able to devote a substantial part of his schedule to the continuation of the fish bibliography, especially as the "Inquiry" was located in Manhattan. Clearly, Eastman hoped to strike a sympathetic chord with Osborn in regard to the War because at the end of his letter he pleaded that "I am, as you know, most anxious to engage in some form of active war-work, and the present opportunity seems to be one in which my special qualifications can best be engaged."[71]

It is instructive that Eastman did not feel he had to explain to Osborn who Colonel House or Dr. Mezes were (the latter was already the President of City College of New York). Eastman's letter only obliquely alluded to the initiative which he had been invited to join, because the work of the "Inquiry" was still secret at this point and would only become generally known to the public after the conclusion of the war.[72]

On the other hand, Isaiah Bowman did not hesitate to openly mention the "Inquiry" when he personally contacted Osborn on May 28th, 1918 regarding Eastman's future participation in his organization. That he did so suggests that Bowman considered Osborn of sufficient political stature that he might be included in discussions of national security without question. For that matter, he also shared with Osborn the knowledge that the "Inquiry" was "gathering confidential information for the Department of State." Bowman came right to the point in his letter to Osborn: "Dr. Eastman's experiences in Brazil and his knowledge of languages, besides many other qualities, would seem to make him a very useful man to us, and I wish to inquire if the Museum could loan a substantial part of his time to the "Inquiry" for the purpose of research."[73]

Without so much as saying so, Bowman intimated that the AMNH should be proud that a member of its staff had been selected to participate in this distinguished body. He ended his letter by casually suggesting that the Museum underwrite Eastman's salary, adding that "All of the public institutions and universities to which we have appealed for expert assistants have made most liberal responses by continuing the salaries of members of their staffs in this government work, so that we are required to pay only the travelling expenses of the people engaged upon it."[74] Osborn's response to Bowman was guarded. He informed the distinguished geographer that he would refer the matter to Dean who will, "I am sure…take a patriotic view of the matter."[75]

Later that day Osborn wrote to Eastman with regard to Bowman's proposal, indicating his personal willingness to have Eastman divide his time between government work and the bibliography on fishes. "Dean will take a patriotic view of the matter" he assured him, as well.[76] Perhaps Osborn actually managed to convince himself of that possibility when he sent a draft of his letter to Dean for the latter's approval.

But Osborn could not persuade Dean. Having already reduced Eastman down to "provisional status," the Curator was not about to bend, irrespective of the "Great Cause." It is likely that Dean additionally resented the impression Eastman had made on men like Colonel House, Mezes, Bowman and Willis, and he may have viewed Osborn's willingness to accommodate Eastman with annoyance. Then again, perhaps it was the notion of continuing Eastman's salary, already knocked down to "provisional", which angered Dean. It is unclear whether Henry Crampton influenced Dean in any way, but the zoologist certainly would have resented Eastman for being offered a position on the "Inquiry" for which he, himself, had been rejected. In any event, Dean's response was decidedly negative.

That Dean and Osborn had spoken by phone later in the afternoon is evident because in a postscript[77] to his response to Eastman, the President informed the ichthyologist that since writing earlier in the day he had been in communication with Dean. "We agree that your work is important and should continue if the Government is prepared to pay for it. [However], I do not feel that the Museum should do more than it has done already." Osborn reminded Eastman that the Government had already paid for part of his expenses in Brazil and probably could continue to do so.[78]

The ongoing exchanges between Eastman and Osborn on the one hand, and Dean, Lucas and Osborn, on the other, provide the evidence of how the Museum president was subtly persuaded that the AMNH should now totally eliminate Eastman from its staff. On June 29, 1918 Eastman informed Osborn that in conjunction with his work for the American Geographical Society, as well as for the U.S. Geological Survey, he might have to relocate temporarily (for the summer) to Washington, D.C. where yet another important position awaited him at the War Trade Board, an entity created to "control the foreign commerce of the United States during the war."[79] Somewhat naïvely, Eastman explained that he had asked Lucas and Dean for permission to participate and that "no objection has been raised by either of them to allowing war-work to take precedence temporarily of research in the bibliographical and ichthyologic field." But, as he still was planning to continue his work for the Museum ("I do not anticipate any very serious interruption"), Eastman also asked President Osborn to continue his salary which had now been reduced, reflecting his "provisional" status, to $200 per month.[80] Osborn's response to Eastman indicated that he was on board with the arrangement, although he thought "there should be some definite understanding as to the division of time." He closed his response with the heartfelt sign-off "best wishes for this patriotic service."[81]

The same day in which he wrote to Eastman, Osborn transmitted a copy of his response, along with a copy of Eastman's original request of June 29th to Dean. He further implored Dean (in consultation with Lucas) to advise him as to "any action which you both recommend."[82] What Osborn perhaps did not know was that Director Lucas had already received a solicitation for a reference on Eastman from the War Trade Board[83] to which he had curtly responded that "so far as I know Doctor Eastman is thoroughly reliable; he is undeniably patriotic."[84]

Dean, however, saw this as a *carpe diem* moment; he could now move to rid himself of this thorn in his side once and for all. In an acrimonious tone, Dean informed Osborn on July 12 that, in consultation with Lucas, he was recommending against Eastman's request to continue at the Museum for the following four reasons: (1) Eastman had not fulfilled the position for which he had been initially hired, e.g., specifically for bibliographical work. "Dr. Eastman has given us a very inadequate quid pro quo…from about fourteen months ago…he has turned in results which I think might have been gathered within one or two months;" (2) his leave of absence in South America produced no results for the bibliographic project; (3) he had not only been paid by the Museum, full pay, but in addition had received funding from the Government for his services in South America; and (4) "We believe that the Museum has done its duty by the Government in respect to Dr. Eastman during the past year and we are convinced that if the Government now required the services of the Doctor, it is in a position itself to pay his stipend."[85] Dean closed his letter to the President with a postscript explaining that as he had already informed Eastman of his conclusions and that it would not be necessary for Osborn to follow with his own letter.[86] He hoped the matter was now settled.

Undoubtedly shaken by Dean's negative reaction, but still hopeful, Eastman wrote again to Osborn on July 13, beseeching him to allow him to continue the special research which he had been doing for the Museum, while he pursued his work half-time for the government. He even offered to accept a further reduction in pay to $100 per month for his scientific work, without which he thought he would not be able to cover the cost of his travel to and from Washington as well as his living expenses there.[87]

Osborn took his time forwarding a response to Eastman, writing several drafts of his letter[88] in which he agreed to further reduce Eastman's salary to see if the new arrangement might work. But, after receiving Osborn's draft response to Eastman, Dean was adamant. "E [Eastman] should either 'cut bait or fish'. If he remains in Washington as he suggests he may have the best will in the world to divide his time but he can't. It is not in the air!…August might see either on or off the payroll??".[89]

When Osborn finalized his response to Eastman he gently paraphrased Dean's sentiments. "I know that it is very difficult to put one's mind on scientific work at the present time" he wrote, "because one's emotions are so constantly drawn upon; I experience this difficulty myself…Hence I am afraid that the arrangement you suggest will not work out satisfactorily—your mind should be free for the Government work. So please do not hesitate to make an adjustment at Washington which will enable you to remain there and give it your undivided attention." In a masterful final stroke, Osborn concluded, "I know that Curator Dean will be very sorry to have you give up the index [of fishes] and so will I who appreciate most highly the thought and enterprise which you have given this great undertaking."[90]

With no other course of action available, Eastman sent in his formal letter of resignation on July 31, 1918 and was immediately taken off the payroll[91]; pleading his cause, he was able to recoup his July salary and the principal on his pension.[92] After petitioning Osborn yet again, he was additionally granted a continuance on his non-paid status as a Research Associate in Paleontology.[93] In August of 1918,

Eastman, unburdened by the demands of the AMNH, was duly appointed to the War Trade Board, Intelligence Division,[94] in Washington, D.C. which, according to him, "was hot after Hun firms in Brazil, and my experience there during last year is proving very helpful."[95] Eastman had worn "two hats" for the course of the war; in the end, his intelligence objectives clearly won out over his research.

In a sense, the AMNH had made the choice easy for Eastman by simply removing him from the payroll. He could now continue his government assignments without the nagging guilt of ignoring his scientific work. The irony was that Eastman's career in Washington was actually quite short-lived. During the month of August, or in early September, Eastman contracted the "Spanish influenza," which had already reached world-wide pandemic status. Perhaps he was already feeling the effects of the disease when he reported in his last letter to Osborn that the conditions under which he was working which were "exceedingly trying and impose a severe tax on one's energy."[96]

At the end of September, still experiencing the effects of his bout with the disease, Eastman was sent back to New York to recuperate. He was admitted to a naval hospital in Long Beach, Long Island, even though his wife lived close by in the town of Lynbrook. The hospital Eastman was sent to had formerly been a grand hotel in Long Beach (long a playground for wealthy New Yorkers) but had been taken over by the government for the rest and relaxation of returning soldiers and sailors; as an ONI agent Eastman, apparently, qualified to be admitted. Unfortunately for ONI Agent 139, on the night of September 27, 1918, shortly after his arrival at the hospital, Eastman decided to take a post-prandial walk along the boardwalk. In the darkness, and unfamiliar with his surroundings, he dropped off the end of the walkway and subsequently drowned.[97]

According to the obituaries published after his body was recovered, the mysterious circumstances of his death were not questioned at the time. Yet, it remains a curious fact that Eastman did not return to his home in nearby Lynbrook to recuperate, but instead entered a government facility in Long Beach. Was his wife afraid of the virus, or was he tracking someone, or something, at the naval hospital/hotel? Had he suffered a psychotic break brought about by the stress of his current position, or by the severance of his connection to the Museum and committed suicide? Or was he murdered?

Unfortunately, there are no answers to these questions. Considering his caustic dismissal from the AMNH, it is curious that following his death the Museum hastily arranged for a party of scientists to attend his funeral in Lynbrook[98] and sent flowers to his widow.[99] Only a few days later Dean made an offer to Professor Gudger, head of the Department of Biology, Normal College in Greensboro, North Carolina to finish the fish bibliography for which Eastman had originally been engaged.[100]

There is little doubt that Eastman had achieved the intelligence objectives in Brazil outlined for him by the ONI, and that he was able to maintain his cover seemingly without incident. His appointment to the Intelligence Division of the War Trade Board attested to the importance and breadth of his knowledge of South American economics which he had gained from that posting. In retrospect, it would seem that Eastman had few misgivings about crossing the liminal boundaries which separate science from politics, even though his actions ultimately resulted in his dismissal from the Museum. Yet, one is left wondering if Eastman would have been satisfied to return to a career in science after the war ended, had his life not been cut short? It is easy to imagine that like George Cherrie (as will be seen in the next chapter) and Henry Crampton, Eastman's patriotic service to the government might easily have extended beyond the conclusion of the War.

Ironically, despite their annoyance with him, Eastman's successful pursuit of contacts with Brazilian scholars and scientists was exactly what the AMNH administration had advertised he would achieve during his South American mission. As far as his Brazilian colleagues were concerned, Eastman had demonstrated that there was one "Pan American" scientific community which was capable of engaging in broad intellectual exchange.

Yet Eastman received no kudos for any of these efforts at the AMNH. Tragically, while his career at the American Museum had once seemed destined for greatness, Eastman disappeared from the ranks of its distinguished scientists, almost without a trace.

Notes

1. "Obituary of Charles R. Eastman," New York Times, September 30, 1918. See also Bashford Dean, "Charles Rochester Eastman," in Science, N.S., Vol. XLIX., No. 1258, February 7, 1919, pp. 139–141 and "Obituary of Charles R. Eastman," Cambridge Tribune, Vol. XLI, No. 32, 5 October 1918.
2. Capsule biography from Vertical File, Library, AMNH, compiled by Robin Smith, n.d.
3. "The victim with his dying breath is said to have accused Eastman of murdering him." From the Press Democrat, Vol. XXVII, No. 139, 23 April 1901. See also "Trial of C.R. Eastman," The New York Times, April 27, 1901.
4. "Eastman Case From Shooting to Verdict," Boston Sunday Post, 12 May, 1901.
5. *Ibid.*
6. *Ibid.*
7. "C.R. Eastman Acquitted: Harvard instructor is found not guilty of murder," The New York Times, May 12, 1901.
8. R. L. Craig to Bashford Dean, Kemmerer, Wyoming, December 1, 1909, Vertebrate Paleontology Archive, AMNH.
9. This description of Sheldon is provided by Roy Chapman Andrews in his memoir, Under a Lucky Star: A Lifetime of Adventure, New York, Viking Press, 1943. Reprinted by Speath Press, 2007, see p. 147. Sheldon, who told Andrews, "I am in Naval Intelligence," apparently was responsible for recruiting the Museum scientist for the ONI.
10. From the American Museum Journal, 1917, p. 501.
11. In a letter to Frederic Lucas, Rio de Janeiro, December 18, 1917, CA, 1069, Eastman described the manner in which each of these state geological commissions maintained separate museums which, according to him, were more than willing to place material at his disposal. Unfortunately at the time the material was too bulky to send to New York.
12. Eastman's report on the attitudes of Brazilians toward the War is contained in a communication sent by Captain McCauley, Assistant Director of Naval Intelligence to Office of the Counselor, State Department on April 25, 1918. It is located in NARA Record Group 59, in the series "Letters received by Leland Harrison", declassified September 15, 1982. Its contents are discussed later on. Harrison was by this point the head of the Intelligence sector of the State Department, known as the Bureau of Secret Intelligence.

See Samuel D. Kleinman, "State's Spies: The Bureau of Secret Intelligence and the Development of State Department Bureaucracy in the First World War," Honors Thesis, History Department, Georgetown University, 9 May 2016.

13. José Maria Bello, A History of Modern Brazil, 1889–1964, trans. James L. Taylor, Stanford, 1966, p. 233.
14. Jeffery M. Dowart, The Office of Naval Intelligence… p. 103. These orders, to Agent 6, are dated March 2, 1917. Eastman's predecessor was Lcdr. Edward Breck, who spoke excellent German and posed as a wealthy German professor, but was quickly outed as a fraud. Eastman, on the other hand, operated as the scientist which he was. For a discussion of Breck, see Eric Setzekorn, "Intelligence in Defense of the Homeland: The Office of Naval Intelligence in World War I: Diverse Threats, Divergent Responses," in Studies in Intelligence, Vol. 61, No. 2 (Extracts, June 2017), pp. 50–51.
15. Between late November, 1917 and the end of December that year, Eastman moved from the Hotel Albion in Rio to the Hotel Splendid in that city, to the Anglo-American Club in São Paulo (mid-December) and back to the Splendid by the 18th. This arc of his travel may be traced in the letterheads of the various locales from which Eastman posted letters to the AMNH. See CA, 1069.
16. Eastman comments on Morgan in his letter to Osborn, Rio de Janeiro, November 26, 1917, CA, 1069; and on Claudel to Crampton, September 19, 1917, CUSMC, Crampton, Box 1.
17. Eastman to Crampton, September 19, 1917, CUSMC, Crampton, Box 1.
18. *Ibid.*
19. *Ibid.*
20. Bello, A History of Modern Brazil…, p. 233.
21. These attacks began in the Spring of 1916 and continued into 1917. See Martin, Latin America…, pp. 47–55. See also E. Bradford Burns, A History of Brazil (New York, Columbia University Press, 1970), pp. 255–256.
22. Burns, A History…, p. 217.
23. Bello, A History of Modern Brazil…, p. 233. Rui Barbosa spoke against Brazilian neutrality while addressing the Law Faculty in Buenos Aires.
24. On Oliveira Lima's Germanophile stance see Schulze, Frederik, "German Immigrants (Brazil)" in Ute Daniel, Peter Gatrell, Oliver Janz, Heather Jones, Alan Kramer and Bill Nasson, 1914–1918-online. International Encyclopedia of the First World War, Freie Universität Berlin, Berlin, 2014.
25. Martin, Latin America…, p. 70. The rest of the Allied nations also congratulated Brazil.
26. At least 200,000 Germans had moved to Brazil between the 1820s to the 1930s. But the total number of pro-German residents in Brazil was unknown, as was their political allegiance. See Frederik Schulze, "German Immigrants (Brazil)"… for a discussion of what was perceived as a "German Peril." See also Chap. 11 on Rollo Beck for his photographs of Brazilian labor demonstrations.
27. Frederick C. Luebke, Germans in Brazil: A Comparative History of Cultural Conflict During World War I (Baton Rouge, Louisiana State University Press, 1987), p. 195.
28. See Thomas E. Skidmore, Brazil: Five Centuries of Change (New York, Oxford University Press, 1999), p. 94. Frank McCann, "Brazilian Foreign Relations in the Twentieth Century," in Wayne A. Selcher, Brazil in the International System: The Rise of a Middle Power (Boulder, Westview Press, 1981), p. 5 says Brazil's officer corps was Germanized: "its troops used German drills to train with their Mauser rifles, while at Vila Militar outside Rio de Janeiro, the very kitchens were German in style and equipment."
29. Muller was actually forced into retirement. See Bello, A History…, p. 233.
30. Nancy Mitchell, The Danger of Dreams: German and American Imperialism in Latin America, 1895–1914 (Chapel Hill, The University of North Carolina Press, 1999), p. 151.
31. Report submitted by Captain McCauley, USN, Assistant Director of Naval Intelligence to Leland Harrison, Office of the Counselor, State Department, in NARA, Record Group 59, Series "Letters received by Leland Harrison, Pro-German Attitudes, 1916–1918."
32. *Ibid.*, p. 1.

33. *Ibid.*
34. *Ibid.* p. 6.
35. *Ibid.*, pp. 6–7.
36. *Ibid.*, p. 16.
37. Eastman to Osborn, Rio de Janeiro, November 26, 1917. CA, AMNH, File 1069.
38. *Ibid.* Eastman wrote that "I am personally familiar with the excellent results obtained by our ambassador to Brazil, Mr. Morgan, friend and former classmate of mine at Harvard, and would rank him second to none among all our representatives in South America."
39. *Ibid.*
40. *Ibid.*
41. Eastman to HFO, n.d., CA, AMNH, File 1069.
42. Frank Polk, Counselor, Department of State to Osborn, Washington, January 29, 1918. CA, AMNH, File 1069.
43. "Excerpt from letter of Dr. Charles Eastman to President Osborn" January 23, 1918. CA, AMNH, File 1069.
44. Eastman to Osborn, São Paulo, Dec. 10, 1917. CA, AMNH File 1069.
45. Eastman to Osborn, São Paulo, Dec. 11, 1917. CA, AMNH File 1069.
46. Eastman to Lucas, Rio de Janeiro, Dec. 17, 1917. CA, AMNH File 1069.
47. The note (n.d.) reads "Making entry today: 1 Box Books, 3 Boxes Fishes. Value $300." A note penciled in the margin says Eastman.
48. "Excerpt from letter of Dr. Charles Eastman to President Osborn" January 23, 1918. CA, AMNH, File 1069.
49. *Ibid.*
50. Chapman to Osborn, Ormond, New York, January 20, 1918. CA, AMNH, File 1004.
51. "Excerpt from letter of Dr. Charles Eastman to President Osborn" January 23, 1918. CA, AMNH, File 1069.
52. Dr. Joviano Pacheco, Comissão Geographica e Geologica do Estado de S. Paulo, to Eastman, São Paulo, January 22, 1918. CA, AMNH, File 1069.
53. Lucas to Osborn, New York, April 10, 1918. "Re: Brazilian Fishes." CA, AMNH. File 1069.
54. Osborn to Lucas, New York, April 3, 1918. CA, AMNH, File 1069.
55. Lucas to Osborn, New York, April 10, 1918. CA, AMNH, File 1069.
56. Alvan Clark Eastman to Lucas, Cambridge, Dec. 18, 1917. CA, AMNH File 1069.
57. Lucas to Dean, New York, March 1, 1918. CA, AMNH File 1069.
58. Osborn to Dean, New York, March 9, 1918. CA, AMNH File 1069.
59. Lucas to Osborn, New York, March 9, 1918. CA, AMNH File 1069.
60. Osborn to Eastman, New York, March 11, 1918. CA, AMNH File 1069.
61. Eastman to Osborn, Rio de Janeiro, March 19. 1918. CA, AMNH File 1069.
62. Eastman to Osborn, offshore Bermuda, April 21, 1918, CA, file 1069.
63. Lucas to Osborn, New York, April 10, 1918. "Re: Brazilian Fishes." CA, AMNH. File 1069.
64. Osborn to Lucas, New York, April 3, 1918. CA, AMNH, file 1069.
65. Lucas to Osborn, New York, "Re: Brazilian fishes", April 10, 1918. CA, AMNH, File 1069.
66. Dean to Lucas, May 6, 1918. CA, AMNH, file 1069.
67. Lucas to Eastman, May 15, 1918. CA, AMNH, file 1069.
68. Dean to Osborn, May 20, 1918. CA, AMNH, file 1069.
69. Eastman to Osborn, New York, May 27, 1918. CA, AMNH, file 1069.
70. *Ibid.*
71. *Ibid.*
72. The work of the "Inquiry" was only made public in December of 1918. Cf. "Inquiry of the American Geographical Society for the Information of the Peace Commissioners," Science, December 13, 1918, Vol. 48, Issue 1250, pp. 590–592. For a detailed discussion of the organization see Gelfand, The Inquiry….
73. Bowman to Osborn, New York, May 28, 1918. CA, AMNH, file 1069.
74. *Ibid.*
75. Osborn to Bowman, New York, May 31, 1918. CA, AMNH, file 1069.

76. Osborn to Eastman, New York, May 31, 1918. CA, AMNH, file 1069.
77. Postscript to above.
78. *Ibid.*
79. Eastman to Osborn, New York, June 29, 1918. CA, AMNH, file 1069. The activities of the War Trade Board, created in October, 1917 were first publicly discussed in Report of the WAR TRADE BOARD (Washington, Government Printing Office, 1920).
80. Eastman to Osborn, New York, June 29, 1918. CA, AMNH, File 1069.
81. Osborn to Eastman, New York, July 5, 1918. CA, AMNH, file 1069.
82. Osborn to Dean, New York, July 5, 1918. CA, AMNH, File 1069.
83. C.A. Cotterill, War Trade Board, Washington, D.C., to Director Lucas, June 27, 1918. CA, AMNH, file 1069.
84. Lucas to Cotterill, New York, June 28, 1918. CA, AMNH, file 1069.
85. Dean to Osborn, New York, July 12, 1918. CA, AMNH, file 1069.
86. Postscript to above.
87. Eastman to Dean, New York, July 13, 1918. CA, AMNH, file 1069.
88. Osborn to Eastman, several drafts. New York, July 17, 1918, CA, AMNH, file 1069.
89. Dean to Osborn, Riverdale, New York, July 20, 1918. CA, AMNH, file 1069.
90. Osborn to Eastman, revised letter probably sent on July 20, 1918 after receiving Dean's comment. The actual letter does not appear in the files.
91. Eastman to Osborn, New York, July 31, 1918. CA, AMNH, file 1069.
92. Eastman to Lucas, Washington, August 28, 1918. CA, AMNH, file 1069.
93. Osborn to Eastman, New York, August 6, 1918. CA, AMNH, file 1069.
94. Eastman revealed his position to Osborn in a letter written from Washington, August 31, 1918. CA, AMNH, file 1069. The Report of the WAR TRADE BOARD… p. 454 notes Eastman's participation in that division.
95. Eastman to Lucas, Washington, August 28, 1918. CA, AMNH, file 1069.
96. Eastman to Osborn, Washington, August 31, 1918. CA, AMNH, file 1069.
97. "Obituary of Charles R. Eastman," New York Times, September 30, 1918.
98. A note dated October 2, 1918, in CA, AMNH, file 1069 alerted colleagues at the Museum to the funeral service the next day at the Episcopal Church in Lynbrook, Long Island. The train departed from Penn Station at 2:06 pm and arrived at 2:46. The note suggests taking a return train at 3:30.
99. Dean to Osborn, New York, October 8, 1918. CA, AMNH, file 1069.
100. *Ibid.*

Chapter 9
The Two Amigos: Frank M. Chapman and George Cherrie, ONI Agent 245

Among the best known scientists from the American Museum of Natural History who chose to work in Latin America before World War I were Frank M. Chapman and George Cherrie (Fig. 9.1). The latter had already worked in Latin America prior to being hired by the Museum and had served as curator of birds at the National Museum of Costa Rica, in San José. He continued research as an independent scientist through the 1890s,[1] before he was hired to work in New York, at the AMNH. By contrast, Chapman, who was about the same age as Cherrie, and who had worked exclusively for the AMNH for the first half of his career, remained mostly interested in North American bird populations before turning his attention southward. In his autobiography, Chapman observed that "South America has played so large a part in my life that it is difficult for me to realize that I was in my forty-eighth year when I first visited that continent. Colombia was my landfall [1911] and I now look upon my expedition to that country as an adventure of my boyhood."[2]

Chapman switched his focus at the start of the 1910s, linking up with Cherrie in the pursuit of specimens for Osborn's grand Latin American project. Leo Miller, their AMNH colleague and more than twenty years their junior, joined their group soon after, and the threesome was tasked with collecting mammals (large and small) as well as avian specimens. (Chapman's work remained largely limited to acquiring bird exemplars). The intrepid trio scoured most of South America in search of material to both confirm Osborn's long championed evolutionary theories and to contribute to the Museum's growing collections of Latin American flora and fauna. In the course of their extensive ventures, they traveled the length and breadth of Colombia, Peru, Chile, Argentina and Paraguay; in the case of Miller and Cherrie, these travels also included Brazil, where they distinguished themselves by accompanying former President Theodore Roosevelt during his aforementioned precarious trip down the River of Doubt (Fig. 9.2).

The adventures of the threesome were recounted to enthusiastic readers in the pages of the American Museum Journal. Leo Miller additionally wrote a hefty volume detailing his expeditions, which was published in 1918 and contained more than

R. M. Delson, *Scientists or Spies?*, The Latin American Studies Book Series,
https://doi.org/10.1007/978-3-032-18117-6_9

Fig. 9.1 Left, Frank Chapman, 1910, AMNH Asset ID 131487; right, George Cherrie, date unknown, AMNH Asset ID 101591

Fig. 9.2 Left to right: George Cherrie, Theodore Roosevelt, Leo Miller, 1913, Brazil; AMNH Asset ID 236924

seventy illustrations and maps. Intriguingly entitled In the Wilds of South America,[3] it read much like a turn of the century boy's adventure tale. Miller used his talents as a storyteller to summarize his "over 150,000 miles trip"[4] to the regions of "Colombia-March, 1911, to September, 1912; Venezuela-November, 1912 to June, 1913; British Guiana [present day Guyana]- July to October 1913; the Roosevelt-Rondon South

American Expedition, mostly in Brazil but covering a part of Paraguay, with stops in Uruguay and Argentina-October, 1913-to June, 1914; Bolivia-May 1915, to January, 1916, touching at Panama, Ecuador and Peru en route; and finally, Argentina-January to September, 1916."[5] Chapman and Cherrie each wrote separate memoirs, although these were not published until the 1930s.[6]

In the period prior to the United States entrance into World War I, Chapman, Cherrie and Miller remained committed to collecting in South America. But when the United States declared war against Germany in April, 1917 Leo Miller chose to enlist in the armed forces rather than continuing his scientific studies. Had he not chosen to enter the military, it was likely that the AMNH would have granted the young scientist permission to pursue research in South America after the outbreak of war, if for no other reason than to provide him with a cover for some important government assignment. However, unlike Chapman and Cherrie, Miller was still young enough to qualify for the draft, and sometime soon after April, 1917, he joined the United States Army and was rapidly promoted to the rank of Lieutenant.[7] The Museum Letter of July 6, 1918, an in-house organ which kept the Museum staff abreast of the wartime activities of its employees, reported that Lieut. Leo E. Miller had been married on April 2nd of that year and that the couple was now living in Dallas, Texas, where the Lieutenant was engaged "as an observer in the Aviation Corps."[8] One wonders whether his ornithological training had prepared him for his assignment? If so, it would fall into the category of "assignment by association," similar to the one Crampton received when he evaluated typologies of flying machines and war materiel for the Council of National Defense based on his ability to produce evidence-based conclusions through classification.

Although neither Chapman nor Cherrie enlisted in the armed forces (both men were too old), each would make valuable observations about the political status quo in Latin America which went far beyond matters related to Museum work. Chapman's direct participation in the nation's war effort began with his stint at the Red Cross in Washington, D.C., where he was on assignment from the summer of 1917 until the Spring of 1919. Based on his prior experience in Latin America, the Red Cross returned Chapman to the region as a "Special Commissioner" in late 1918. Ostensibly on a goodwill tour to thank those who had contributed to the Red Cross, his duties now included reporting on local political (especially labor) conditions, as the War drew to a close.

Cherrie volunteered for intelligence work with the ONI, just as Crampton had done, and served as Agent 245 of the intelligence service, an affiliation which, according to his descendants, he particularly relished.[9] He likely received an early appointment as Crampton had (late 1916?), making him one of the earliest recruits not just from the Museum, but from academe in general. Cherrie completed the bulk of his surveillance a few months before Spinden and Eastman joined the service, although he continued reporting to the ONI into the 1920s (see Chap. 11).

Both Chapman and Cherrie (Miller was not part of this venture) returned to South America in mid-1916 to collect specimens in Colombia, Argentina, Brazil and Paraguay. While the specific details of Cherrie's ONI Latin American assignment of 1916–7 are not available, his expedition plans for the Museum required him to traverse regions of Latin America which were of deep concern to Washington, especially those countries which had remained stubbornly neutral in their political stance on the War. While planning for their future South American trip from his Brooklyn home, Cherrie informed Chapman (the expedition's leader)[10] that he was requesting special equipment which he now considered imperative for his research, including a "moving picture camera." It is unclear whether that specific ask was fulfilled, but Cherrie did take extensive photographic equipment with him when he left New York in May of 1916[11] and produced hundreds of photographs. He would not return to New York until February of the following year, having used his reconnaissance through Argentina, Paraguay and Brazil to successfully alert local populations to the threat of German interference in the region.

Cherrie and Chapman began their 1916 expedition to South America in Ecuador. The two men collected together in the Andean highlands, especially in the upper Páramo zone, before Cherrie left Chapman in July to make his way south to Argentina. His final destination was Brazil, where he hoped to collect specimens in the swampy *pantanales* of the state of Mato Grosso. For public consumption he was "officially" continuing the explorations which had been initiated earlier during the Roosevelt-Rondon Expedition. Relieved of the responsibilities he had endured while accompanying the intrepid ex-President down the River of Doubt, Cherrie could now, theoretically, "investigate closely the life histories of the remarkable birds and mammals of the region"[12] which Roosevelt had previously "explored." In pursuing this fauna, Cherrie spent two months at Puerto Pinasco on the Paraguay River, after which time he proceeded upriver to the Brazilian settlement of Descalvados (near Cáceres), in the alluvial plains of the *pantanales,* where for an additional three months he collected water birds. According to the reports he submitted, he managed to amass an extraordinary amount of material in a relatively short period of time, resulting in a haul of three thousand skins of birds, four hundred mammal skins and some fifteen hundred photographs.[13] By any measure, this was a spectacular achievement.

But it would appear that collecting was not all Cherrie was doing while he was in South America; in this regard, his diary provides singular clues into his likely ONI assignment, without specifically revealing its purpose. His entries from August, 1916 note that before setting out to collect in Brazil, Cherrie spent almost an entire month in Buenos Aires, Argentina, where he was able to recuperate from the debilitating sinus attacks which had plagued him in the "high (and cold) altitudes of Peru and Bolivia."[14] To amuse himself while he rested in the Argentine capital, Cherrie spent several days exploring the sights of the city, including a Sunday stroll through the well-respected Zoological Garden, a site he found astonishingly devoid of visitors. "Not much like the Bronx on a Sunday," his diary notes, referring to the popular New York attraction, the Bronx Zoological Park (aka The Bronx Zoo). Nevertheless, there were interesting South American wildlife exhibits which intrigued him, and he spent his visit profitably photographing some unusual specimens to add to his and the Museum's collection.

The day after his zoo adventure, however, Cherrie took some unusual steps for an individual who presumably was merely prepping for a collection trip into the Brazilian interior. He called upon the Brazilian Minister to Argentina, to whom he presented the credentials which he had obtained from the U.S. military attaché stationed in Buenos Aires, Colonel David L. Brainard. Armed with an official introduction, Cherrie learned from the Brazilian Minister that conditions were quite unstable in the state of Mato Grosso, his next destination. Unfortunately for him, an attempt to quell a local revolt saw Brazilian federal troops taking up positions in the very region where Cherrie planned to collect. This was surely not a very auspicious beginning for his own explorations and Cherrie's diary glumly allowed that "[My] activities may be nipped in the bud."[15]

In retrospect, it was odd that Cherrie received his credentials from Colonel Brainard, the U.S. military attaché in Buenos Aires, instead of the local U.S. counsel. Perhaps Cherrie took that route because as an authorized ONI agent he was expected to maintain contacts with U.S. military officials. On the other hand, as confirmed by Spinden's actions in Central America, or Eastman's in Brazil, there do not seem to have been regulations prohibiting ONI personnel from communicating with, and soliciting the advice of political appointees. What is intriguing about Colonel Brainard in particular is that he was quite famous in his own right by the time Cherrie dealt with him. For one thing, he was a survivor of the Army Expedition (The Lady Franklin Bay Expedition) of 1881, which had established a U.S. scientific station in the Arctic. He followed this experience with a tour of duty in the Philippines during the Spanish-American War, after which he was transferred to Buenos Aires to assume the post of military attaché and where he served as advisor to Cherrie and, as it turned out, to Chapman, as well.[16] Moreover, both men continued to report to and receive counsel from Colonel Brainard throughout the course of the U.S. involvement in the War, even after the officer returned stateside.

Cherrie went ahead with his Brazilian plans despite the threat of local interference. In the days leading up to his departure for Brazil, and following his meetings with both the Brazilian Minister and Colonel Brainard, Cherrie took some time to attend an International Cattle Show in Buenos Aires. Ostensibly his visit was to familiarize himself with the breeds of Aberdeen, Angus, Holstein and Hereford cattle which had been brought from Europe to Argentina decades before, and which now flourished in the *pampas*.[17] But, why should Cherrie have been interested in cattle other than the fact, perhaps, that he was born in Iowa? Certainly the AMNH had no particular interest in beef cattle, and bovine studies were not part of Cherrie's research plans.

A less obvious explanation of Cherrie's attendance at the cattle show is that the experience allowed him to gather information which was clearly outside the parameters of his work for the Museum, but which had everything to do with the politics of the moment. Was it mere coincidence that when Cherrie arrived in Buenos Aires in August of 1916, and presumably received his ONI assignment, the Germans had just embarked on a new, and insidious, program of biological warfare geared to spreading ganders and anthrax virus to Argentine livestock and destroying that industry? Were Cherrie's orders to either disclose or stymie this plot?

Argentina had long supplied Great Britain with beef exports, a commerce which constituted a vital portion of the nation's economy. The trade began in the late nineteenth century, and continued unabated well into the second decade of the twentieth century, notwithstanding the outbreak of War and German interference (or threatened interference) with Atlantic shipping lanes. Indeed, the impact of hostilities on the long-enduring partnership between Argentina and England had been relatively limited and the fact that beef continued to be shipped to Britain where it was supplied to British troops on a regular basis can largely be attributed to Argentina's neutral political stance in the War. But by 1916, two years into hostilities, the Germans rationalized that rather than allow the trade to continue there might be greater political advantage for them in disrupting the Argentine beef supply entirely, a goal which could be achieved through biological weaponry. Once the Germans recognized that if successful, this disruption might prove disastrous for the Allies they devised a scheme to undermine the Argentine cattle industry, and cause irreparable harm to the existing supply chain.

The means via which the Germans chose to achieve their aims were simple enough; they would attempt to embed live disease cultures into tiny ampoules of glass which were, in turn, hidden in sugar cubes. Such cubes could then be fed to livestock and, as the cattle gnawed on their treats, their teeth would crack open the vials, exposing the animals to unfiltered infection.[18]

Successful implementation of this scheme would certainly have wreaked havoc in the Allied food chain to Europe. Just a year earlier, a German spy network in the United States had successfully employed a similar technique, and had inoculated mules slated for delivery to British troops with disease. The results devastated the continuing supply of these animals to Europe.[19] (As noted earlier, Spinden had tried unsuccessfully to hire mules while he was in Venezuela in 1915, a situation which he attributed to the desperate Allied requests for that nation to forward mules to the European front.)[20]

Had Cherrie been asked by the ONI to keep his eye on the vast cattle ranches of the Southern Cone, which coincidentally were located in the same area where he planned to collect specimens for the AMNH? Was that the explanation for the time he took to acquaint himself with the prominent breeds of Argentine cattle? Could it have been his mission to spread the word to local ranchers about the deadly German anthrax/ganders plan? Such conjecture seems more than likely as Cherrie spent the remainder of his trip traveling from one cattle ranch to another after he left Buenos Aires, and before he returned to New York.

As it turned out, one of Cherrie's first stops after leaving Buenos Aires was Puerto Pinasco, on the Paraguay River. It was in this part of the Chaco region (halfway between Asunción in Paraguay and Corumbá in Brazil), that the "New York and Paraguay Concern", an enormous cattle estancia of some 90 square leagues of land, had its headquarters. Because of the "abundance of marsh land, *llanos* and thick forest", Cherrie wrote to Chapman, he would be "compelled" to stay on the *estancia* ranch for a month or so. In the meantime, if Chapman needed to reach him, he was to write to him in charge of their mutual contact, Col. Brainard.[21]

Further evidence that Cherrie's interest in cattle was not coincidental is provided by the fact that wherever Cherrie ventured in the Chaco of Mato Grosso he conspicuously sought out the companionship of cattlemen, several of whom were Americans. At the end of October, for example, Cherrie met Mr. Ramsey, the Texan manager of the Brazil Land, Cattle and Packing Company, at the office of an American dentist in Corumbá (where Cherrie was having a half dozen cavities filled).[22] The dentist, in turn, introduced Cherrie to Mr. Reis, the owner of yet another estancia to which Cherrie quickly got himself invited to "study the water birds and make some pictures."[23] It seems highly plausible that Cherrie was spreading the warning about potential German anthrax/gander poisonings while he simultaneously went about his announced purpose of collecting and photographing the local fauna. It is also clear from his diary that he was in constant contact with Col. Brainard, who continued to act as his advisor.[24]

In November of 1916, Cherrie wrote to Chapman that he now found himself in Descalvados, Mato Grosso, Brazil a short distance outside the town of Cáceres. Located in the approximate geographic center of the *pantanales* swampland, and about 600 miles from the town of Corumbá, Descalvados was the headquarters of the Anglo-American Brazil Land, Cattle and Packing Company. The collection of buildings which constituted Descalvados was largely geared to processing meat via the *charque* (or drying) method, which produced jerky as an end product. A traditional food of the region, *charque* was also proving to be a viable and practical source of meat for the Allied troops, and one which was considerably easier to transport than either refrigerated beef, or beef on the hoof.

But the situation in this region was anything but calm. Not only were German efforts to sabotage the chief export of the area worrisome but, as noted above, Cherrie was equally preoccupied with ongoing regional strife. While he was still in the Paraguayan Chaco, an area which borders on Brazil, Cherrie warned Chapman that there was a "revolution brewing in Matto [sic] Grosso" and that Corumbá, the major Brazilian port on the Paraguayan River was under "marshal [sic] law."[25]

It is likely that Cherrie was referring to the ongoing conflict known as the "*Contestado*" (a prolonged contest over ownership of the territory between the Brazilian states of Paraná and Santa Catarina), which had spilled over into the state of Mato Grosso. This territorial struggle pitted hapless backland settlers, followers of a charismatic and messianic cult figure known as João Maria, aka "the Monk," against the authority of the Brazilian Federal Government. Like similar contemporaneous Brazilian visionaries, João Maria "put himself ahead of both the government and the church."[26] Considered a "savior" by poor uneducated backlanders, conversely, the Brazilian government viewed João Maria as a criminal. This struggle between them had begun well before Cherrie reached Paraguay, as the Brazilian Minister had already informed him, and it lasted into late 1916.[27] Cherrie encountered these "revolutionaries" as well as the federal troops on the cattle ranches he had been scoping. According to the account he sent to Chapman, "The government troops 'confiscate' the cattle of the revolutionists [sic];

while the revolutionists just help themselves to the cattle belonging to the federals.[28] In either case, Cherrie concluded that "plenty of cattle are being sacrificed."[29] Obviously a continuation of this conflict posed as serious a threat to the Allied troop food supply chain in Europe as did the prospect of German biological warfare.

Despite the various distractions which faced him, Cherrie managed to observe and to collect an enormous number of specimens which he then photographed for the Museum's benefit. Among the animals which caught his attention were rhea, flamingoes, "paroquettes" [sic] and cowbirds.[30] During his travels, he had made contact with Dr. Friburg, a local scientist in charge of a small museum and zoo (the Jardín Zoológico y Botánico) in Asunción, Paraguay. Writing to Chapman on September 3, Cherrie reckoned that he would be able to convince Dr. Friburg to make exchanges of previously unstudied skins with the AMNH.[31] Yet, by the end of the month, following a return trip to Asunción, and after reassessing the German threat to the region, Cherrie had drastically changed his opinion of his Latin American colleague. Eschewing Dr. Friburg's generous offer to exchange material with the Museum, Cherrie could not bring himself to deal with the man, ranting in a letter to Chapman that "Friburg is a rabid [his underlining] German!".[32]

In November, 1916, Chapman warned Cherrie that when his work concluded he should book passage on a Brazilian, rather than a British ship, for his return trip to New York. Although he expected that Cherrie knew the situation, the prospect of German submarines, which now appeared in the Western Hemisphere with some frequency, made it likely that any British ships which carried beef for the British army would become targets. In the meantime, Chapman (already stateside) wrote that he would be meeting with Colonel Brainard in New York and would relay Cherrie's "reported progress" to him.[33]

Prior to arranging for his passage home Cherrie had to overcome several obstacles, including limited train service, before he could make his way out of Mato Grosso. The distance between Cherrie's outpost in Campo Grande and the railroad station in Porto Esperança, where he needed to make a connection to the coast, would take at least fifteen hours to cover; yet another full day of travel was required before reaching São Paulo. Once in the city, Cherrie would be obliged to board one more train, to Rio de Janeiro, where his next orders awaited him. As he wrote to Chapman, "Until I get there [Rio de Janeiro] I will not know if I am to make a longer stay in the country."[34] Unfortunately for Cherrie, he remained stuck in Descalvados until early January, when the rainy season had arrived in earnest, hampering his collecting. Despite his disappointment, he had still managed to secure specimens of several species which had not been observed during the earlier Roosevelt expedition.[35]

Cherrie's diary reports that he did not arrive in São Paulo until January 17, 1917, whereupon he immediately caught a train to Rio de Janeiro.[36] He was relieved to find no instructions from either Col. Brainard or the Museum in his mail at the U.S. Embassy, a finding which Cherrie interpreted to mean he was to "start home at once"[37] as both his assignments (for the AMNH and the ONI) were now complete. In the time that remained before his voyage, Cherrie took the opportunity to explore the city. What he experienced in the tropical capital of Brazil precipitated a spate of caustic observations in his diary; taken together, Cherrie's frequent comments on the city and its "Carioca" residents can scarcely be considered charitable. "Rio seems to me" he confided to his journal, "to have more than her share of the flotsam and jetsam of civilization. In the streets everywhere are seen the wrecks of human driftwood."[38] He persevered through the city, even though he was stricken by a brief bout of fever and still managed to fit in a visit to the National Museum (the Quinta da Boa Vista), the most distinguished natural science institution in Brazil. Unlike Eastman, who would visit the same museum a half year later, and Chapman, who had personally recommended the current director of the museum (Dr. Manoel Ribeiro) to him,[39] Cherrie found little to his liking. "The department of Vertebrate zoology gave me the shivers" he reported in his journal. "There is scarce a decent looking bird or mammal in the entire collection. Most are poor scarecrows [his underlining]. No effort is made to make the exhibits interesting. No information is given as to life habits or food…it is a storehouse of mummified curios."[40]

Cherrie departed the city and after an uneventful passage on the S.S. *Rio de Janeiro*, arrived in New York during the middle of February.[41] His experiences in South America, however, do raise some questions: first, why did an ONI agent report to an army colonel, rather than the U.S. naval attaché in Argentina for the duration of his assignment? Did this mean that there was no reliable U.S. naval attaché (or U.S. consul) in Argentina with whom he could make contact in mid-1916?[42] Secondly, were the efforts to avoid contamination of the Argentine beef supply successful? The answer is that with regard to the overall German biological warfare in Argentina, "the efficacy of the[ir] 1916 effort is unknown."[43] Yet Cherrie had certainly been strategically placed in the axis of the targeted Argentine/Brazilian/Paraguayan cattle industry to have quietly warned ranchers of this danger, and to have made a real difference. Interestingly, a year and a half later, in early 1918, the Germans once again attempted an inoculation scheme, only this time it was aimed at Argentine horses. Far more successful on this try, Argentine horse exports to France and Italy essentially ceased as a result.[44]

If his involvement in thwarting the first German campaign to poison livestock in Argentina and the Southern Cone (or any other ONI business he might have been engaged in, for that matter) necessarily had to remain secret, Cherrie could openly take pride in amassing an impressive number of specimens and photographs during his travels. Always a consummate collector, his 1916–7 trip to South America yielded exemplars of such unusual mammals as the Giant Anteaters and big Swamp Deer, which he documented with numerous photographs. From this haul of specimens, Cherrie considered those of *Rhea*, Wood Ibis and Woodhewers especially noteworthy.[45]

What Cherrie did not do during this time in the Southern Cone was to advance the needle of "scientific Pan Americanism." More to the point, his observations of the few colleagues and/or collections he encountered was either hostile or downright condescending. Frank M. Chapman's experience in the region would not be altogether different.

Frank Chapman, Ornithologist and Red Cross Commissioner

Despite Cherrie's extensive travels in the southern hemisphere, his compatriot at the Museum, ornithologist Frank M. Chapman, was considered the Museum's point man on all things South American. Unlike Cherrie, Chapman had not even visited the area before 1911, and his prior Latin America experience had been limited to intermittent trips to Cuba in 1896 and 1897, where Chapman found himself in the midst of a Cuban colonial war for independence against the Spanish. Both of his visits occurred before the sinking of the U.S.S. *Maine* (February 15, 1898), an event which resulted in the direct involvement of the U.S. in the Cuban war, which subsequently became known as the Spanish-American War. Nonetheless, Chapman did make some keen remarks about the political situation in that unhappy island, drawing attention to the youth of the Spanish troops who served on the island and to the smallpox epidemic which was decimating their ranks with as much efficiency as Cuban bullets.[46]

Chapman also made a few brief expeditions to Mexico (Yucatan and Vera Cruz) to collect small mammals and birds, as well as trips to Trinidad, in the British West Indies.[47] But collecting this material was apparently incidental to his early career; when he began working full-time for the AMNH Department of Ornithology in 1888, he earned his reputation by mounting exhibitions of birdlife of New York City and its surrounding suburbs rather than on Latin American ornithology.[48] His attention later turned to creating a broad-themed gallery of North American birds; this would be housed in a new exhibition space boasting painted backgrounds, against which prepared bird specimens could be displayed as if they were being seen in their native habitat.[49] While not everyone was impressed with Chapman's departure from the static way in which specimens had previously been showcased, others recognized that his pioneering efforts paved the way for future exhibitions which could more accurately depict animals in their native environments for the public.[50]

Seemingly, all of Chapman's previous experiences were but a prologue to his coming explorations in South America. Perhaps the most important influence on the direction of his future research were the observations of bird life he had made earlier at Mt. Orizaba, Mexico. At this site, Chapman had been able to demonstrate that the bird population was not homogeneous, but stratified, a feature which was the result of distinct climate zones. He wondered if the same sorts of patterns might characterize the bird populations of the Andes which, if they did, could help explicate the course of zonal evolution in that area. This intriguing question spurred Chapman to leave the familiarity of North America in 1911 and undertake an expedition to the mountainous areas of Colombia and Ecuador where he and his Museum colleagues would eventually amass an impressive haul of thousands of birds. As he was able to confirm, these specimens did indeed differ substantially from each other depending on the widely differing zones in which they had been collected. The zones he collected in ranged from the humid Tropical zone, to the sub-Tropical zone, followed by the Temperate zone and, lastly, to the Páramo (or Puna, in Peru) zone found at a height of 15,000 feet; these disparate areas could not be more different from each other. Chapman's quest to understand the zonal evolution of South American birds began in 1911, and he took more than six years to complete his findings. Ultimately his conclusions were published in a monumental 726 page study, Distribution of Bird-life in Colombia, an achievement which appeared in the Bulletin of the American Museum of Natural History (1918)[51] and earned him the prestigious Daniel Giraud Elliot Medal from the National Academy of Sciences.[52]

To a great extent, Chapman's work in South America was supported by private subsidies from outside the Museum, which included the wealthy benefactors Malcolm MacKay, Daniel E. Pomeroy and George B. Case.[53] In addition to receiving financial support from these individuals, Chapman also obtained $1000 from the American Geographical Society, funding which allowed him to travel to Lake Titicaca in Peru for ethnographic observation. This was a grant that had been awarded as a result of an auspicious meeting Chapman had with Isaiah Bowman from the AGS, in 1915.[54] Assured of sufficient support, Chapman was able to explore the breadth and length of the entirety of the Andes Mountains extending through the nations of Colombia, Ecuador, Peru, Bolivia and Chile. Within this span, Chapman would collect birds which inhabited areas ranging from the lowland, humid Tropical Zone all the way to specimens from the highland Páramo. The result was a superlative avian collection which provided sufficient data not only to distinguish between bird zones (his initial goal), but also to analyze bird distribution across these same regions.[55]

The six years Chapman spent in collecting and analyzing these data allowed him to cultivate valuable contacts with local scientists which he (and the Museum) would come to rely on for future projects in Latin America. Additionally, many of the high level U.S. personnel befriended by the ornithologist during this time would prove be of utmost importance to the Museum after the United States became a combatant in World War I. Among these was Leland Harrison, whom Chapman encountered in 1912 when the former served as the U.S. Chargé d'Affaires in Bogotá. As the local representative for Washington, Harrison was made personally responsible for facilitating Chapman's (and also Leo Miller's) expeditions, a task which he accomplished with considerable skill. Not only were the expeditions officially approved by the heretofore reluctant Colombian government, but Harrison was also able to secure permits for Chapman and Miller to carry firearms and ammunition into the field.[56] A few years after serving in South America, Harrison returned to Washington, D.C. where he was appointed Head of the Bureau of Secret Intelligence of the State Department, which would later evolve into the Federal Bureau of Investigation.[57] In his new role, Harrison's jurisdiction extended beyond the United States to Latin America, where he maintained close communication with several ONI agents.[58] While not a designated ONI agent himself,[59] Chapman renewed his contact with Leland Harrison when he began his 1916 South American expedition, and it is more than likely that it was Harrison who put both Chapman and Cherrie in touch with Col. Brainard in Argentina.

From the Museum's perspective, Chapman's careful work in Colombia represented an inflection point for future cooperation between the AMNH and Bogotá and, by extension, with the rest of South America. The Museum was fully cognizant of Chapman's diplomatic successes with the often hostile Colombian Government, acknowledging his skill by appointing the ornithologist to head the AMNH delegation to the Second Pan American Scientific Congress in 1915. As discussed earlier, Chapman accepted Osborn's invitation to attend the meeting without hesitation,[60] seemingly as anxious to promote the Museum's research agenda in South America as was his boss.[61]

In regard to this objective, Osborn wrote to Director Lucas, "I think it very important, in connection with our forthcoming increased exploration in South America, that we should establish most friendly relations with the South American Republics."[62] It was clear that the Museum president viewed the upcoming Second Pan American Scientific Congress as an excellent opportunity to "make them [their Latin American colleagues] better acquainted with the progress of scientific work,"[63] although in expressing this opinion it would seem that he was inferring that the learning curve for Latin American scientists was well behind that of North Americans. One wonders if Osborn had any idea that the fear of being perceived as backward in science had served as one of the original rationales for excluding U.S. colleagues from the first three Latin American scientific congresses.

When the Washington meeting concluded, however, Chapman came away disappointed in the results. Not even trying to hide his dissatisfaction from Osborn, he complained that "[t]he program of the Congress was evidently designed for the furtherance of the Monroe Doctrine and Pan-Americanism, rather than in the interest of science," and "that there was no section in which either a biologist or physical geographer could find a place. Nevertheless, the numerous receptions and entertainments which were given, offered many opportunities for meeting of delegates from the south, and I feel that I have furthered our interests in South America through the connections which were made on these occasions."[64] It is likely that Chapman also resented the fact that the papers he was slated to present on Colombia, as well as his outline for the Museum's future plans in Latin America, were not scheduled within the framework of the Congress, but instead relegated to the much smaller audience of the Congress of the Americanists, which had been meeting concurrently in the nation's capital.[65]

Notwithstanding his personal feelings about the outcome of the Congress, Chapman could not deny that his participation at the event paid off handsomely for his further research in South America. Within a month of the Congress's conclusion in January, 1916, Chapman was already planning a return expedition south in which he would be accompanied by George Cherrie.[66] Writing to Director Lucas about his upcoming trip, Chapman requested that the AMNH solicit a letter of introduction for him from Secretary of State Lansing, with whom he had become familiar through the Washington meeting. Chapman correctly interpreted that Lansing's real interest in the Congress lay in "the promotion of Pan-Americanism." He therefore suggested that Lucas "make it clear to him that the interests of Pan-Americanism would be advanced by my journey. I, of course, plan to visit all the museums on my route, with the object of promoting mutually satisfactory relations between them and us."[67]

Pleased by Chapman's understanding of the Museum's growing interest in Latin America, Osborn took it upon himself to send a personal letter to Lansing, referencing the fact that Chapman had represented the Museum at the Congress and underscoring the urgency of the ornithologist's plan to "visit the various institutions of learning and museums on his route, in order to promote mutually satisfactory relations between them and us."[68] Three days later, Osborn took another chance and requested free rail transport to Mt. Chimborazo (Ecuador's highest mountain) for the entire Chapman party from the New York office of the Guayaquil and Quito Railroad. By way of explanation, he revealed that the Museum was planning to exhibit the specimens Chapman collected in that locale in "what is called a Habitat Group of the life of the higher altitudes, or Páramo, of Chimborazo." He hoped the pitch would pique the interests of the rail men, explaining that "[S]uch a group is, in effect, a panorama, in which a painted background approximately twenty-five feet in length would adequately show Chimborazo and the surrounding country; while the foreground would reproduce in detail the vegetation and bird life of the region." He assured the railroad board that such exhibitions garnered "wide popular interest,"[69] implying that providing railroad travel to the area could potentially attract tourists. Following this logic, a contribution to Chapman's budget would set the wheels in motion for such a future outcome.

Both Lansing and the railroad reacted positively to Osborn's requests: the Secretary of State indicated that he would write on Chapman's behalf and promised to provide letters of introduction to all U.S. diplomatic and consular officers in South America,[70] while the president of the railroad company alerted his general manager to render to Chapman, "such assistance and…courtesies as may be possible."[71] Osborn also contacted Domício da Gama, the Brazilian ambassador to the United States, and solicited his help in securing free passage for the ornithologist on the Lloyd Brasileiro steamship line.[72]

Consequently, when Chapman sailed for Ecuador on May 6th, accompanied by his wife and their son,[73] the trip was the result of considerable diplomatic and commercial intervention. Their voyage took them via the new Panama Canal, where Chapman noted in his diary that he was "thrilled by the trim orderliness of the locks and the machinelike precision with which our ship was handled."[74] From there the party sailed to the port city of Guayaquil on the Ecuadorian coast where, after meeting up with George Cherrie, they disembarked and made their way to the capital of Quito. Chapman and Cherrie began at once to make preparations to ascend the mountains and conduct explorations at an altitude of 11,000 feet, a height which represented the upper limits of where cultivation was possible; from this level, which would then become the staging area, the party could progress even higher into the Páramo.[75]

After collecting in the highlands, the group returned to Guayaquil where they boarded another vessel for the Peruvian port of Callao. For the next 23 days, the intrepid pair traversed substantial Peruvian territory, arriving in the Andean city of Cuzco on July, 24th.[76] At this point the two men split up (as had been prearranged) and went their separate ways, with Cherrie traveling to Buenos Aires (as noted earlier in this chapter) and thence to Corumbá (Brazil) to continue the "collecting agenda" of the Roosevelt-Rondon Expedition, while Chapman, his wife and the rest of their group made their way to Lake Titicaca to witness the annual feast of Copacabana with its combined aspects of indigenous ritual and Catholicism.[77] This decidedly ethnographic portion of his trip was financed by the aforementioned grant from the National Geographic Society, which was interested in promoting cultural awareness of Latin America.[78]

From Peru, the Chapman party took a train to Arica in Chile, a journey that descended some 15,000 feet to the Pacific Ocean. When they reached the shore, they boarded a ship bound for Valparaíso.[79] In between those countries, Chapman made a brief stop in La Paz, Bolivia, where he made "a most favorable arrangement with the government museum for the first of birds in return for their identification."[80] In keeping with his promise to visit "all" the science museums in South America,[81] Chapman similarly visited the Philippi collections of birds in the National Museum of Santiago, Chile, after which he wrote a long, newsy letter to Sherwood proclaiming that "the trade relation part of my journey is therefore panning out most satisfactorily."[82] More importantly, however, Chapman felt his experience had provided him with "a personal knowledge of the 'lay of the land' in this part of South America," giving him "a new grasp of the whole situation in the direction of our further work here, and [which] will greatly expedite the ends we have in view."[83] Spirits obviously high, Chapman cheerfully observed "Fords here every ten inches–make the biggest noise of any car in town–and a Santiago Chauffeur loves his horn."[84]

From Santiago, Chapman boarded another train to journey across the Andes, this time moving eastward toward Mendoza, Argentina, where he would spend time examining the excellent collections of birds at the Carlos Reed Museum. As proof of Chapman's diplomatic skills, Professor Reed, Director of the eponymous museum and a professor at the local university, willingly agreed to exchange specimens with the AMNH.[85] While in Mendoza, Chapman also enjoyed a brief reunion with AMNH colleagues Leo Miller and Howarth Boyle, who arrived in Argentina after collecting together in Bolivia.[86]

From the provincial city of Mendoza, Chapman made his way to the Argentine capital of Buenos Aires, where he met with Dr. Roberto Dabbene, Curator of Birds of the National Museum. Here again, as had been the case with Reed, Chapman was able to arrange for an exchange of "much-needed material from the Argentine."[87] Dabbene also requested that his distinguished visitor address the Argentine National Society of Natural Sciences and the Ornithological Society of La Plata, an honor which Chapman readily accepted.[88]

According to his diary, these efforts were warmly received[89]: a pamphlet which memorialized his address to the La Plata Ornithological Society contained Dr. Dabbene's remarks, highlighting the ornithologist's relationship to Roosevelt, as well as Miller's and Cherrie's assistance during the River of Doubt expedition.[90] Despite his halting Spanish, the AMNH ornithologist had, apparently, successfully conveyed the Museum's interest in South American biology, as well his belief that protection of the Argentine avian patrimony should be a primary goal for the nation.[91] Chapman had also emphasized the desire of the AMNH administration to pursue contacts with colleagues in the hemisphere,[92] for which a receptive La Plata Ornithological Society graciously bestowed a corresponding membership on him, by way of reciprocation.[93] So far, Chapman's 1916 trip had not only gone according to plan but had met with noticeable success. Interestingly, while he visited Buenos Aires, Chapman was in regular communication with Colonel Brainard, as was Cherrie, although the latter had already left the Argentine capital bound for the cattle ranches near Corumbá.

In contrast to Cherrie's experiences in Paraguay and Brazil, the ongoing European War had not seriously intruded on Chapman's expedition. But that situation was about to change as he continued his expedition into Brazil. He was informed, with "extreme regret,"[94] that Dr. Herman von Ihering, Director and Founder of the Museum at São Paulo, Brazil had been forced from his post apparently because of his German background. Personal feelings about his former colleague notwithstanding, Chapman still decided to make a visit to the Museum on his journey northward through Brazil and spent "a most profitable day at the museum." The AMNH received the promise of a future exchange of several hundred specimens from southeastern Brazil.[95] Evidently, the antipathy Chapman had felt (and later voiced, see below) at the von Ihering firing had been conveniently set aside in the interest of Museum business.

From São Paulo, Chapman proceeded to Rio de Janeiro where he visited with Ambassador Morgan and learned for the first time that while he was in South America, German submarines had been attacking and sinking ships off the eastern coast of the United States.[96] This knowledge caused him considerable consternation and worry that the English ship on which he had already booked passage home would be a potential target, especially as its cargo was beef destined for the Allied troops in Europe. Displaying considerable caution, Chapman decided to take that steamer only as far as Barbados, its next destination,[97] where he imagined he would be able to book another ship to New York. While on the Caribbean island, he awaited clarification on the situation from the Museum, an excruciatingly slow process given that cables were already being held up by censors despite the fact that the U.S. had not yet joined the War effort. To add insult to injury, U.S. flagged vessels, which had seemed a safer bet, were suddenly far scarcer in the Atlantic than earlier in the year. Following a short stay in Barbados, where he took the opportunity to observe some birds unknown to him, Chapman had no choice but to return home on the next British steamer which came to the island. Unbeknownst to him at the time of his booking was the fact that the new ship was also sailing with a full cargo of beef destined for the troops, despite the ship captain's a priori assurances to the Germans that he was not engaged in transporting British food supplies. Chapman reckoned this made the captain a "marked man,"[98] and for the entire trip home Chapman and his wife were riddled with anxiety.[99] Luckily, they arrived home in November of 1916 without incident, several months prior to Cherrie,[100] having experienced none of the latter's encounters with hostile Germans or South American troops. In short, thus far the War had not seriously impacted the trajectory of Chapman's career, and it is fair to say that Chapman's interest in politics remained minimal ("detached" was his own assessment), even after the April, 1917 declaration of war against Germany. An ability to insulate himself against wartime involvement, however, would ultimately prove impossible to sustain.

In late May, 1917, Chapman's good friend and sponsor, George B. Case, pressed upon him the urgent need for workers at Red Cross headquarters in Washington. Although he initially wavered on deciding to join this effort, Chapman found it impossible to ignore the further entreaties of Henry P. Davison, a New York banker and, more importantly, treasurer of the AMNH, who also happened to be President Wilson's appointed Wartime Director of the Red Cross. Chapman subsequently "allowed" himself to be persuaded to join this service and was rewarded with a trip to the capital in Davison's plush private rail car.[101]

For the next two years, and with Osborn's approval, Chapman volunteered for Red Cross work, leaving his Museum research behind. During his first 12 months in Washington, he served as the editor of the organization's in-house publications and occupied himself with writing a rather dry historical overview of the organization. Like his peers, Chapman would often interrupt his work for a leisurely lunch at the Cosmos Club, "the unofficial headquarters of the new army of scientists for the first time called to the assistance of their country at war."[102]

Despite his duties and absence from the Museum, Chapman continued to be viewed by the AMNH as the linchpin to Osborn's broad project on Latin American evolution. In January of 1918, President Osborn took the opportunity of sending "Major" Chapman[103] (presumably an Army rank, since the Red Cross had been placed under the aegis of the War Department), a highly complimentary encomium written about him by the eminent ichthyologist C.H. Eigenmann. In the latter's estimation "Chapman is the one man in North America whose visits to South America are entirely idealistic, whose endeavors are not looked upon with more or less suspicion. He ought by all means to go ahead with his South American work. I hated to see him quit for even so unselfish and highly patriotic self-denying work for the Red Cross as he is doing. However this war is not going to last forever and I hope to see Chapman enlisted for S.A. again. I envy him his facile Spanish and understanding of the South American…".[104]

Enclosing Eigenmann's letter with his own, and addressing his staff member as "My Dear Doctor Chapman," Osborn praised his curator's "diplomatic services in South America." He begged him to write a popular article on the museums of South America for the Museum Journal, suggesting that such a piece "would be extremely welcome in the South" [America], and could be combined with discussions of Eastman's museum visits (although, as seen in the previous chapter, the ichthyologist never formally wrote up his experiences). The degree of intimacy of Osborn's salutation (quite unusual for him, and usually reserved for close friends) was equally odd in that Chapman did not have a formal doctorate; it was obviously meant as a flattering gesture.[105]

Responding to his boss, Chapman returned the favor by addressing his leader as "My dear President Osborn."[106] He modestly protested that Eigenmann had overrated him. But, he added, "I do believe in the importance–indeed necessity, of personal contact with our colleagues in South America as a means of allaying their jealousy and securing their cooperation."[107] To illustrate what he meant, Chapman reported that his own publication on Colombia had "set a standard for faunal work in South America,"[108] and had earned Col. Roosevelt's praise,[109] among others. And, just in case Osborn had missed the point, he modestly added that "The Colombia minister appeared to be impressed with…the [my] Colombian Memoir, saying…that courtesies such as ours would go far toward annulling the effects of our action in regard to Panama."[110]

In regard to Osborn's request for an article on South American museums, however, Chapman was unexpectedly, and decidedly, obtuse. Such a piece, he adamantly asserted, would be impossible to write. "In writing one I should experience much the same difficulty that, as a rule, I encountered in South America in expressing to the Curators concerned my opinion of their collections."[111] In stark contrast to the enthusiastic reports he had previously sent from the museums of La Paz, Santiago, Mendoza, Buenos Aires and São Paulo,[112] Chapman now chided these very institutions, claiming to have been totally unimpressed with what he had seen. "The Santiago authorities, for example, have shown a criminal disregard of the first principles of labeling in destroying all the original labels on Phillipi's types—replacing them with printed tags! Those in power in São Paulo have treated von Ihering outrageously, while in Montevideo and Rio the Museums—so far as birds and mammals are concerned—are doing either no work at all or work that should be left undone."[113] "It is," he mused, "much easier to slide over these places with the tongue than with the pen."[114]

What had caused Chapman to completely reverse his positive opinion of South America's natural history museums, and to now view the same institutions with such obvious hostility? Where was the vaunted "idealism" which Eigenman thought he had perceived in Chapman's Latin American work, and why was it that Chapman now considered the very same South American colleagues whom he had previously described as gracious hosts to be "jealous" and generally suspicious individuals? While it is hard to know for certain what lay behind Chapman's change of heart, the reluctance of the South American countries with which he was most familiar (including Argentina, Chile and Brazil, initially) to join the Allied cause had likely colored his response. Of course it was also true that during his earlier travels, because he had hoped to borrow specimens from these museums, being diplomatic was obviously a prerequisite. Perhaps once he was installed in the wartime climate of Washington he reasoned that he could, at last, give vent to his true feelings. Contrasted with Eastman's very positive attitude toward some of the same Latin American museums, this new attitude of Chapman's was not only damning but it flew in the face of Osborn's hopes of expanding "scientific Pan Americanism."

While there is no memorialization in the Museum's archives of Osborn's response to Chapman, it is likely that his unexpectedly negative attitude influenced the Museum president's decision to recall the ornithologist at once from Washington. Suddenly threatened, and concerned for the security of both of his positions, Chapman turned to George Case, his boss at the Red Cross, to enlist his help in dealing with Osborn. Case wrote to Osborn only two days later, arguing that "Dr. Chapman has been of the greatest possible use to us from the time he came here and we have laid out to him now a plan which has splendid possibilities... It should be borne in mind that when a man has gone through the period of war organization of the Red Cross, as has Dr. Chapman, he has absorbed the spirit of the policies...with this in mind we are going to ask Dr. Chapman...to undertake a trip to South America."[115] Clearly, Chapman's extensive travels in South America and familiarity with countries in the southern half of the continent made him an excellent candidate to meet with and personally thank expatriate American communities for their generous contributions to the Red Cross. In light of ongoing worker unrest and changing geopolitical alignments,[116] we can imagine that Chapman would also have been well-placed to report on the political climate in the nations he was to visit.

Chapman followed up with an additional letter to Osborn on the same day (April 3, 1918), in which he attempted to rationalize his continued stay at the Red Cross by explaining that he had been asked to become a Red Cross Commissioner, a prestigious position which, in his words, was "a great opportunity to do something for the Red Cross, even more for the Country, and not a little for the Museum." Although he could "not go into details," Chapman appealed to Osborn's empathy for "scientific Pan Americanism" by referencing "the wide possibilities this mission offers for establishing sympathetic relations between us and our Southern neighbors."[117] Was he being disingenuous? He forwarded these remarks to the chief Museum administrator seemingly as if he had not severely criticized these very same neighbors in his earlier correspondence and had not disavowed the prospect of writing Osbon's much desired report on the status of these very museums. Now he did not hesitate to assure Osborn that the State Department would confer a semi-official status upon him and provide the coaching for his future assignment which would prepare him to visit several of the American Red Cross chapters in Latin America. Without explaining how it might be possible, Chapman said he anticipated that he would emerge from this trip somehow better fit "to pursue my scientific researches."[118]

It is unclear if Chapman's appeal made sense to Osborn, but the administrator neither wished to appear unpatriotic nor to relinquish any possible benefit which the Museum might secure by having Chapman return to Latin America. Perhaps more importantly, he was also reluctant to irritate Treasurer Davison. Nonetheless, before he rendered a final decision, he handed both Case's and Chapman's letters to Director Lucas to solicit his advice. Lucas viewed this request no differently from the proposals of those other Museum scientists who had joined the ONI. He therefore proposed the same *quid pro quo* that Spinden, Eastman and all the other scientists drawn to government work had previously agreed to, i.e., that the Red Cross could send the ornithologist to South America with the express understanding that he would be simultaneously representing the Museum. It would be, in Lucas' newly minted phrasing of this transactional behavior, a "combination expedition."[119] Osborn pounced on this by now familiar formula and forewarned Chapman, who had seemingly lost track of this fact, that "our future scientific relations with South America are of great political importance."[120]

With Osborn on board with this two-way arrangement, in June of 1918, several months before his role as Commissioner was set to begin, Chapman secured the services of Englishman Harry Watkins to collect specimens for the AMNH during the time when he would be otherwise occupied by his Red Cross mission. Watkins had already been resident in Peru for some fifteen years and had previously worked with Chapman in the Urubamba Valley during his 1916 expedition. Chapman wrote that he felt confident in recommending this locally knowledgeable collector to represent the Museum, further assuring Osborn that "…this excellent man [who] can effectively carry forward our South American work at a time when we are helpless so far as sending museum expeditions is concerned,"[121] an oblique reference to the moratorium. Once again, Osborn would be able to maintain the pace and scope of collecting in Latin America despite the Museum's wartime fieldwork prohibition. This was not an offer he was likely to refuse.

But reading between the lines of Chapman's June 3 letter, it should have been obvious to Osborn that his "best man" in South America would have little time to pursue his scientific work. Chapman tantalizingly held out the hope to Osborn that he would meet up with Watkins for a month of collecting in Peru[122] after his Red Cross work had concluded. This meeting never materialized; realistically, by this point in the War, Osborn should have understood that Chapman's scientific work would take a backseat to his broader mission, as it had for the other Museum scientists who were on loan to the government. When the ornithologist confirmed in that same letter that he had scheduled appointments with the State Department "to consider certain details of my mission",[123] his admission should have raised red flags, although Osborn had little choice except to acquiesce to his new assignment.[124] Was it, therefore, a clever ruse used to support Chapman's cover (as he had done for Spinden, Eastman, Cherrie and Crampton), or naïveté, when Osborn had the Museum's Acting Director petition the Collector of Customs to allow three trunks containing Chapman's camp and collecting equipment to be forwarded to Lima, Peru?[125] Perhaps he thought, if nothing else, Watkins would be able to make good use of the equipment.

These trunks were duly dispatched to Peru during the summer but, as events transpired, Chapman would not leave Washington for his South American mission until October of 1918. If Osborn was miffed by this delay in the ornithologist's plan, there is nothing to indicate it in the Museum archives. In his letter to Osborn dated June 3, 1918, Chapman suggested that he would explain everything personally to the Museum president when he returned to New York from Washington at the end of the month; it was clear that he was hoping to mollify his boss.[126] His autobiography indicates that he used the brief period before his departure from Washington to finish several publication projects for the Red Cross and to improve his Spanish language skills which, as noted earlier during his formal speech to the La Plata Ornithological Society, were decidedly lacking, notwithstanding Eigenmann's glowing assessment. Amusingly, both Chapman and his wife were tutored nightly in Spanish by Augustin Iturbide, great grandson of Emperor Augustin I, the ruler who had briefly led Mexico after it declared independence from Spain in 1821. The great grandson, banned from his own country on orders of Porfírio Díaz, apparently supported himself by tutoring students while he lived in exile in Washington.[127] Chapman also used some of these interim weeks to visit former President Theodore Roosevelt at his home at Sagamore Hill, on Long Island. There the two men, who knew each other from the River of Doubt expedition and had corresponded on occasion, discussed the scientist's current Red Cross mission.[128]

Chapman was finally able to depart Washington for Havana, Cuba, and arrived there on October 4, 1918; his wife followed two weeks later.[129] In contrast with his earlier travels, this time Chapman's voyage was unusually guarded. Instead of proceeding by ship from New York to Florida, where he could have easily made a connection to Cuba, he traveled overland, presumably still as worried about German submarines at sea as he had been in 1916.[130] As evidence of the extraordinary clout that his position as a Red Cross Commissioner carried, on his way south Chapman stopped at the Marine Corps Training Camp at Parris Island, where he was granted the unheard-of privilege of visiting his newly enlisted eighteen year old son.[131] From there Chapman traveled by rail, at night, to Key West[132]; he likewise sailed from the Florida Keys to Havana, Cuba under the cover of darkness.[133] We can speculate on why he felt it necessary to travel this way, but certainly his personal safety was a concern.

During his stay in Havana he not only visited with American and Cuban Red Cross personnel but gained access to the highest ranking officials, including the American Minister and, surprisingly, the President of Cuba, Mario García Menocal.[134] From Havana he made his way to the erstwhile colony of the Isle of Pines, where he visited with Americans who had established themselves there as lumbermen and citrus estate owners.[135]

Returning to Havana, Chapman met up with his wife, and the two set sail for Panama on October 26, 1918, on board a Spanish steamer. Upon reaching the Canal Zone, Chapman took time from his schedule to celebrate the fifteenth anniversary of Panama's independence from Colombia in a public theater, from the safety of the American Minister's box. In retrospect, Chapman's participation in such an event on November 3 seems incredibly impolitic, especially given his earlier boast to Osborn that his Colombian bird memoir had single-handedly eased the lingering tension between that country and the United States over the latter's role in the breakaway of its Panamanian province.[136]

Chapman and his wife remained in the Canal Zone for several more days after the event and were still there when peace was declared on November 11, 1918.[137] Curiously, instead of having his mission cancelled, as one might have imagined with the cessation of hostilities, Chapman continued on to South America. He would later observe ominously in his autobiography, "We have since discovered that the Armistice did not end the war."[138]

Was Chapman prescient or, more likely, had he been briefed by the State Department about Washington's continuing political concerns for South America? It was likely the latter, because it was becoming apparent that "German intelligence activity in the Western Hemisphere shifted from North America to the Southern Cone as 1918 wore on."[139] Even with the end of the War in sight, espionage was still being hotly pursued in South America.

To make matters even more worrisome for Washington, by July, 1918, the German spymaster in Peru had made contact with representatives of Japan, which had been cooperating with the Allies for several years.[140] Japanese agents were eager to "buy into this network [of German spies],"[141] probably related to commercial conflict with the United States, but which presaged the closer contact of World War II. Apparently, the idea of penetrating a still lucrative South America via an already existing surveillance nexus of resident sleuths was an irresistible proposition; it certainly made for decidedly strange bedfellows. As the war was concluding, a carefully cultivated friendship between spies of Germany and Japan blossomed in Peru, Chile and Argentina. It is probably no coincidence that Chapman was specifically ordered to visit all three of these nations. These visits would afford him ample opportunity to make observations and to report back to Washington if he witnessed anything of concern, which may have been his underlying task. In effect, the position of an unassuming Red Cross Commissioner was an even better cover than that of an ornithologist collecting for the Museum!

The threat of a newly-coalesced German-Japanese axis of spies was matched by the specter of worker unrest and insurrection, spurred on by Bolshevik and (anarcho-) socialist recruiters. Threatened United States interests in the Southern Cone, including those at copper mines and ports where European emissaries actively sought to connect local laborers and intellectuals to global political upheavals, had to be addressed.

Building on European worker actions, strikes spread to South America during 1918 and 1919.[142] The Peruvian port of Mollendo saw a shipyard strike in March, 1918, following almost two years of demands for higher pay and shorter workdays.[143] In Callao, the port of Lima, the maritime union was unable to follow suit, but Chilean port workers in Valparaiso made some headway.[144] In January, 1919, several uprisings took place in Argentina, as mining unions were drawn into the struggle (and the German-Japanese axis of instigators).[145] As he was traveling in the region, Chapman could easily have been asked to report on worker sentiment there.

As a consequence, after leaving Panama in November, 1918, and over the course of the next few weeks, Chapman and his wife visited many mining and port communities along the Pacific coast of South America, as well as in interior cities. Their journey southward from the Canal Zone took them to the ports of Callao and Mollendo, Peru, and Valparaíso, Chile, as well as the inland cities of Arequipa, Peru and La Paz, Bolivia. Everywhere the Chapmans traveled in the Southern cone, they exhibited extreme caution about their personal security. Accordingly, they traveled courtesy of American-owned railways in those nations, which provided them with the private cars of the owners, assuring them the ability to reach their destinations in anonymity. Once again, they crossed Lake Titicaca, as they had in 1916, but, in contrast to their previous passage, this time they sailed on a small boat which catered to indigenous passengers, rather than take the more commonly used tourist vessels which might have exposed them to scrutiny. In late December, via the Antofagasta Railway, the Chapmans arrived at the American copper mining settlements at Chuquicamata and Sewell, Chile.[146] For public consumption, Chapman's mission was to visit the nearly seven hundred American staff who lived in these communities to thank them for their generous Red Cross contributions. But, at the same time, he also could have reported back on the propensity for the local (indigenous) workers in these American enterprises to rebel; observing employee-management relations may have been one purpose in visiting these mines. In his notes, Chapman commented upon the fondness of the workers for drink which, coupled with the "dangers of the mine workings and of snow avalanches," had led to Sewell becoming a "dry town."[147] The copper mined at these locales was important not just to the American mine owners, but to broader U.S. interests during the War; worker uprisings were simply not going to be tolerated. At one point, Chapman reflected that "The monthly reports of the welfare service, both here [Sewell] and Chuquicamata, are filled with carefully recorded data that should be of much value to sociologists." As an example, he drew attention to "a Sewell report, showing the decrease in the number of minor camp disturbances, following the introduction of a mechanical piano."[148]

It wasn't just the miners who raised concern. Recent historiography of this period points out that "[P]ort and maritime workers' unions in Peru and Chile leaned heavily toward radical politics in the first decades of the twentieth century…[and] through labor organizing, [they] found their way to a politics of anarcho-syndicalism."[149] It was no coincidence that German agents operated in such centers after the war had officially ended,[150] nor was it likely by happenstance that Chapman spent a considerable portion of his trip making observations in Valparaiso, as well as other Chilean and Peruvian ports, before attempting to reach what was perhaps the epicenter of such labor upheaval, the city of Buenos Aires.

Unfortunately for Chapman, a general strike at the Buenos Aires port in January, 1919 resulted in his postponing a planned visit to that city. Starting out weeks before January from Santiago de Chile, he had attempted to reach the Argentine capital via a wildly circuitous route across the Andes. This initially involved crossing the mountains via mule, then switching to motor boat and finally to a wagon as he made his way eastward to the Argentine Lake District and to the town of Bariloche. Eventually, when he reached the resort coastal town of Mar del Plata, (which he refers to in his account as "The Argentine Newport)", he was invited to address a crowd at a public fund raiser being given for the benefit of a local hospital. According to Chapman's Autobiography, the event was marred by the behavior of attendees from the upper classes who turned their backs on him while he talked, an indication of their disdain for the Allies and/or Americans."[151] It is odd, given his assignment, that in his official report on his activities to the Red Cross, Chapman failed entirely to mention this incident.[152]

Chapman finally was able to reach Buenos Aires in early February, 1919, where his arrival followed a week in which hundreds of striking workers in the city were killed by government forces (an event commemorated in Argentine history as "La Semana Trágica)."[153] Following a brief stay in the Argentine capital, he quickly proceeded to Montevideo, Uruguay where he met up with Red Cross officials. He had expected to go north to São Paulo and Rio de Janeiro, but the still unsettled labor conditions in Buenos Aires, along with the uncertainty of booking a ship from Rio to New York (and, perhaps, memories of his frightening 1916 voyage), caused him to change his mind. Instead of proceeding northward, he hurriedly retreated across the Andes to Valparaíso. There he and his wife boarded a ship bound for the United States only to find when they reached Panama, that they had been unexpectedly summoned by the Red Cross to proceed across the ocean to Paris. There Chapman was to "report in detail on the situation in South America"[154] at the Peace Conference at Versailles.

There is no correspondence, or reference, in either his official report to the Red Cross or his Autobiography, of what the AMNH thought of Chapman's new assignment. That the ornithologist had completely reneged on his promise to undertake a collecting expedition after his travels was clear enough. But Chapman was now a person of political consequence who was far better informed to report on the political upheavals of post-War South America to the Peace Conference at Versailles than perhaps anyone else in the United States. No wonder that Percy Alvin Martin, the distinguished U.S. historian whose definitive study on Latin America and the War was published in 1925,[155] requested that Chapman supply him with material for his book. But curiously, rather than respond directly to the historian, Chapman simply turned his letter over to the Red Cross, which sent Martin a copy of Chapman's bland official report.[156]

While Frank M. Chapman was never an agent of the ONI, there can be no doubt that while he represented the American Red Cross in the closing days of the War, and in post-bellum South America, his observations on the political scene would have been as important to the United States Government as if he had been a bona fide agent. The Red Cross did engage in political activity during the War. In the summer of 1917, several scholars, including C.E.A. Winslow from the Public Health Department at the AMNH, went on a significant, but ultimately unsuccessful, mission to Russia in an attempt to reach out to the Kerensky government faction and/or the Bolsheviks.[157] Chapman, by contrast, was at least in a position to relay an accurate portrait of the state of labor relations in the ports and mines of Southern Cone countries in the months directly following the conclusion of the War.

However, Chapman's political activities did little to nothing to advance the cause of "scientific Pan Americanism," or the profile of the AMNH in Latin America during and immediately after the War. It is also fair to conclude that during his tenure at the Red Cross, his contributions to South American ornithology remained at a standstill. Perhaps more disturbing were the wavering feelings he expressed toward his Latin American scientific colleagues. One wonders which of his expressed sentiments truly represented his thoughts on the scientists he encountered during his travels? Was he really "Mr. South America," as his colleague Eigenmann's letter intimates? Had he ever positively viewed the state of science in the Americas, as his 1916 letters indicate, with their pleasant commentary on local museums and scientists, or did he actually believe that the state of science in these areas was as poor as his wartime refusal to write Osborn's desired essay suggests?

In an interesting footnote to Chapman's view of Latin America, research by a new generation of Latin American scholars has singled out the ornithologist for his demeaning attitude toward his scientific colleagues in the hemisphere.[158] We might consider their use of the term "imperialist" to describe Chapman's attitude a bit harsh, but the hostility the ornithologist voiced toward his Latin American counterparts apparently has not escaped the notice of historians, and perhaps not even his colleagues at the time.

How should F.M. Chapman's service during the war be characterized? Were his patriotic activities as Red Cross Commissioner the only ones he undertook? If he had also served his nation by providing intelligence on diplomatic hotspots to Washington and those gathered at the Peace Conference at Versailles, he could have taken legitimate pride in both open and clandestine endeavors. Surprisingly, in contrast to the proud admission of intelligence work by Joe Spinden,[159] Chapman never mentioned any such effort in his autobiography. It is possible that he was hesitant because he was more senior than Spinden and more concerned about his reputation, especially following Boaz' 1919 attack on intelligence-gathering colleagues.

It is true that we have no direct evidence that Chapman engaged in intelligence gathering for the U.S. government. Yet his relationship with Leland Harrison, Head of the Bureau of Secret Intelligence of the State Department; his subsequent diplomatic work for the Red Cross; and his meeting with state department representatives before his 1918 expedition combine to suggest that he may have been tasked with reporting on conditions in the countries he visited. Even if he did not, his meetings with scientists and museum staff in Latin America prepared him to act as a scientific diplomat for U.S. interests in the region (although his negative views about his counterparts did not support such activities). In fact, as opposed to his museum colleagues Cherrie and Spinden, Chapman set aside his scientific work for several years to carry out the political tasks for which he had volunteered. In sum, it is reasonable to conclude that George Cherrie and probably Frank Chapman played significant roles in securing intelligence from Latin America during the critical wartime period and beyond.

Notes

1. See "George K. Cherrie" in Tom Taylor and Michael Taylor, Aves: A Survey of the Literature of Neotropical Ornithology (Baton Rouge, Louisiana State University Libraries, 2011).
2. Frank M. Chapman, Autobiography of a Bird-Lover (New York, Appleton-Century, 1933), p. 226.
3. Leo E, Miller, In the Wild of South America: Six Years of Exploration in Colombia, Venezuela, British Guiana, Peru, Bolivia, Argentina, Paraguay, and Brazil (New York, Charles Scribner's Sons, 1918).
4. *Ibid*. An excerpt from the cover.
5. *Ibid*, p. VIII.
6. George Cherrie's memoir was Dark Trails: Adventures of a Naturalist (New York, G.P. Putnam's Sons, 1930), while Chapman's autobiography (see above) was published in 1933.

7. "The Following Assistants in the American Museum of Natural History have enlisted in the army…Harold E. Anthony…and Leo. E. Miller." Science, NS, Vol. XLVI, No. 1177, p. 58.
8. Museum Letter, July 6, 1918. CA, AMNH, file 1069 ("Military or Special Service)."
9. I [R. Delson] had the pleasure of meeting with Cherrie's descendants while they were visiting the Library of the American Museum of Natural History some years ago. I was happy to confirm for them that, yes indeed, "Grandpa" had been a spy for the government. Cherrie is listed as Agent 245 in the "Office of Naval Intelligence Designations," Appendix 4, in Harris and Sadler, The Archaeologist…, pp. 370–380.
10. Cherrie to Chapman, Brooklyn, New York, February 6, 1916. Cherrie correspondence, Department of Ornithology Archives, (hereafter DOA), AMNH.
11. From the notes on Cherrie in the Vertical File, AMNH library, compiled by Robin Smith.
12. *Ibid.*
13. *Ibid.*
14. Cherrie Diary, Peru-Ecuador, Matto [sic] Grosso. DOA, AMNH. Entry of Sunday, August 13, 1916.
15. *Ibid.*
16. Glenn M. Stein, "General David L. Brainard, U.S. Army: Last Survivor of the United States' Lady Franklin Bay Expedition (1881–84)", downloaded from Academia.edu, p. 20. Originally published in The Polar Times, July 2008/January 2009.
17. See Cherrie Diary, Peru-Ecuador, Matto [sic] Grosso. DOA, AMNH. entry of Tuesday, August 15, 1916.
18. See Bisher, The Intelligence War…p. 81.
19. *Ibid.*, pp. 80–81. This German sabotage of animals slated for the War continued well into 1918. See *Ibid*, p. 199. Cherrie's actions may well have stalled German biological warfare successes for a year and a half.
20. *Ibid.*, p. 121.
21. Cherrie to Chapman, Puerto Pinasco, Paraguayan Chaco, Rio Paraguay. September 3, 1916. Roosevelt Expedition of 1916, Correspondence of George Cherrie with Frank Chapman, DOA, AMNH.
22. Cherrie Diary, Peru-Ecuador, Matto [sic] Grosso. DOA, AMNH, entry of October 26, 1916.
23. *Ibid.*
24. *Ibid.*
25. Cherrie to Chapman, Puerto Pinasco, Paraguayan Chaco, Rio Paraguay. September 3, 1916. Roosevelt Expedition of 1916, Correspondence of George Cherrie with Frank Chapman, DOA AMNH.
26. José Maria Bello, A History of Modern Brazil…, p. 224.
27. *Ibid.*, p. 232.
28. Cherrie to Chapman, Descalvados, Rio Paraguay, Matto [sic] Grosso Brazil, November 16, 1916. DOA, AMNH.
29. *Ibid.*
30. Cherrie to Chapman, Puerto Pinasco, Alto Paraguay, September 26, 1916. DOA, AMNH.
31. Cherrie to Chapman, Puerto Pinasco, Paraguayan Chaco, Rio Paraguay. September 3, 1916. DOA, AMNH.
32. Cherrie to Chapman, Puerto Pinasco, Alto Paraguay, September 26, 1916. DOA, AMNH.
33. Chapman to Cherrie, New York, November 18, 1916. DOA, AMNH.
34. Cherrie to Chapman, Descalvados, Matto [sic] Grosso, Brazil, November 20, 1916. DOA, AMNII.
35. Cherrie to Chapman, Descalvados, Brazil Land, Cattle and Packing Company, December 6, 1916, DOA, AMNH.
36. Cherrie Diary, Peru-Ecuador, Matto [sic] Grosso DOA, AMNH, entries of January 17, and 18 1917.
37. Cherrie Diary, Peru-Ecuador, Matto [sic] Grosso DOA, AMNH, entry of January 19, 1917.
38. Cherrie Diary, Peru-Ecuador, Matto [sic] Grosso DOA, AMNH, entry of January 20, 1917.
39. Chapman to Cherrie, New York, November 18, 1916. DOA, AMNH.

40. Cherrie Diary, Peru-Ecuador, Matto [sic] Grosso DOA, AMNH, entry of January 21, 1917. Perhaps because it was a Sunday, Cherrie apparently did not get to meet Dr. Ribeiro.
41. Cherrie Diary, Peru-Ecuador, Matto [sic] Grosso DOA, AMNH, entry of February 18, 1917.
42. See Captain Wyman H. Packard, A Century of U.S. Naval Intelligence (Washington, Department of the Navy, 1996), p. 8.
43. Bisher, The Intelligence War…, p. 199.
44. *Ibid*.
45. Cherrie to Chapman, New York, February 21, 1917. DOA, AMNH. Cherrie was being very modest. According to the published "Notes" section of the American Museum Journal, Vol. 17 (1917), p. 151, Cherrie sent home "three thousand bird skins and four hundred skins of mammals, together with fifteen hundred photographs and a wealth of valuable data."
46. See Chapman's remarks on Cuba reprinted in Suzann L. Goldberg, Michael J. Novacek, and Giraldo Alayón, The History of Scientific Relations between Cuba and the American Museum of Natural History (AMNH), 160 Years of Collecting and Collaborating (1857–2017) (New York, American Museum of Natural History Special Publication, 2017), p. 25.
47. William King Gregory, Biographical Memoir of Frank Michler Chapman, 1864–1945 (Washington, National Academy of Sciences, 1948), p. 119.
48. *Ibid*.
49. *Ibid*., p. 121.
50. *Ibid*.
51. Chapman's entire oeuvre is listed in Bibliography of Frank M. Chapman, DOA, AMNH.
52. Gregory, Biographical Memoir…, pp. 123–125.
53. *Ibid*., p. 124.
54. Bowman and Chapman first met at Yale University in late September, 1914, Bowman to Chapman, September 2, 1914, DOA, AMNH. By the following year, Chapman was able to report to Osborn that the relationship between him and Bowman was good, Chapman to Osborn, New York, September 8, 1915, DOA. The promised monies arrived in the summer of 1916: Gilbert Grosvenor, Director and Editor, National Geographic Society to Henry F. Davison, Treasurer, AMNH, Washington, D.C., July 29, 1916. CA, file 370.
55. Gregory, Biographical Memoir…p. 124.
56. Alvey A. Adee, Acting Secretary of State to Leland Harrison, Esquire, American Chargé d'Affaires, Bogotá. October 24, 1912, NARA, RG 59, box 29/1, folder 2.
57. See Kleinman, "State's Spies…", p.53. Harrison became the Head of the BSI in April, 1916.
58. *Ibid*., p. 60 discusses ONI agents reporting to Harrison.
59. *Ibid*.
60. Osborn to Chapman, June 28, 1915. CA, AMNH File 632.
61. Chapman to Osborn, June 30, 1915. CA, AMNH File 632.
62. Osborn to Lucas, July 7, 1915. CA, AMNH File 632.
63. *Ibid*.
64. Chapman to Osborn, January 7, 1916. CA, AMNH File 632.
65. *Ibid*.
66. Cherrie to Chapman, Brooklyn, New York, February 6, 1916. Cherrie correspondence, DOA.
67. Chapman to Lucas, April 3, 1916. CA, AMNH File 370.
68. Osborn to Secretary of State, Robert Lansing, April 4, 1916. CA, AMNH File 370.
69. Osborn to Board of Guayaquil and Quito Railroad, New York, April 7, 1916. CA, File 370.
70. Secretary of State Robert Lansing to Osborn, Washington, April 10, 1916. CA, File 370.
71. E.H. Norton, President Guayaquil and Quito Railway Company to Osborn, New York, n.d. CA, File 370.
72. Osborn to Domicio da Gama, Ambassador, Brazilian Embassy, Washington, D.C., May 4, 1916. CA, File 370.
73. Chapman, Autobiography…, p. 249.
74. *Ibid*.
75. *Ibid*., p. 252.
76. *Ibid*., p. 266.

77. *Ibid.*
78. As late as July 29, 1916 the National Geographic Society had not forwarded the second half of the promised stipend ($500 each half) to Chapman to complete his Peruvian expedition. Sherwood to Gilbert H. Grosvenor, Director and Editor, National Geographic Society, Washington, July 25, 1916. CA, File 370. Grosvenor to Sherwood, July 29, 1916. CA, File 370.
79. Chapman, Autobiography…p. 268.
80. Chapman to Sherwood, August 20, 1916, Santiago de Chile. CA, File 370.
81. Cf. note 68.
82. Chapman to Sherwood, August 20, 1916, Santiago de Chile. CA, File 370.
83. *Ibid.*
84. *Ibid.*
85. Chapman, Autobiography…, p. 270.
86. *Ibid.*
87. *Ibid.*
88. *Ibid.* Unfortunately for the ornithologist, it was only at the last moment that Chapman learned he was expected to deliver this lecture in Spanish, a realization which sent him into an immediate panic. To avoid potential embarrassment to himself and the Museum, Chapman made a stealthy appeal to the branch manager of City Bank in Buenos Aires to have a staff member translate his remarks into Spanish. He subsequently gave his talk, reading from this cribbed paper and not entirely sure of what he was saying to the audience. Chapman delivered his talk on September 11, 1916, with what he hoped would be "proper emphasis on its more important passages." *Ibid.*, p. 271.
89. *Ibid.*, p. 271.
90. Remarks of Dr. Roberto Dabbene in "Recepción del Dr. Frank M. Chapman dada por la sociedades Argentina de Ciencias Naturales y Ornitológica del Plata, el 11 de septiembre de 1916," Physis (Revista de la Sociedad Argentina de Ciencias Naturales), t. II, pp. 439–448.
91. Chapman, Autobiography…,, p. 6.
92. *Ibid.*, p. 9.
93. *Ibid.*, p. 12.
94. *Ibid.*, p. 271.
95. *Ibid.*
96. *Ibid.*, p. 273.
97. *Ibid.* On October 23, 1916 the Museum cabled Chapman to report that the State Department would not advise on what line to take and, moreover, the Censor had held [up] the first cable. CA, File 370.
98. *Ibid.*, pp. 273–274.
99. *Ibid.*, p. 274.
100. *Ibid.*
101. *Ibid.*
102. Frank M. Chapman, One Year with the American Red Cross: June, 1917–June,1918. A Record of How the American People Came to the Support of Their Country at War. Washington, D.C., I.S. No. 2, July, 1918. His remarks on the Cosmos Club are published in Chapman, Autobiography…, p. 278. Crampton also frequented the Cosmos Club while he was stationed in Washington.
103. The Museum Letter of July 6, 1918 refers to Chapman as "Major Frank M. Chapman." CA, File 1069 ("Military or Special Service").
104. C.H. Eigenmann to Osborn, Pittsburgh, January 12, 1918. CA, File 1069 ("Military or Special Service of the Museum Force").Osborn to Chapman, January 16, 1918. CA, File 1069.
105. *Ibid.* It is possible that Chapman received an honorary doctoral degree, but not from Columbia University. On March 19, 1918 Osborn had written to John Pine to ask his help in securing a degree from Columbia for Chapman. HFO papers, AMNH, Box 5, Folder 1, "Frank Chapman, correspondence." The Columbia archives' list of honorary degree recipients does not include Chapman.

106. Chapman to Osborn, January 20, 1918, CA, File 1069.
107. Chapman, Autobiography…, p. 1.
108. *Ibid.*
109. *Ibid.*
110. *Ibid.*, p. 3.
111. *Ibid.*, p. 2.
112. Chapman to Sherwood, Santiago de Chile, August 20, 1916, CA, File 370.
113. Chapman to Osborn, January 20, 1918, CA, File 1069.
114. Chapman, Autobiography…, p. 2.
115. George B. Case to Osborn, Washington, D.C., April 3, 1918. CA, File 1069.
116. Bisher, The Intelligence War…, pp. 249–255 discusses German-Japanese contacts and worker strikes in 1918–1919; see also Joshua Savala, "Ports of Transnational Labor Organizing: Anarchism along the Peruvian-Chilean Littoral, 1916–1928," Hispanic American Historical Review, Vol. 99, no. 3, (August, 2019), pp. 505 et seq. These activities are considered in more detail below. There is no definitive evidence of a sub rosa reporting assignment for Chapman, but it seems reasonable to suggest that there was more to his work than meets the eye.
117. Chapman to Osborn, Washington, D.C., April 3, 1918.
118. *Ibid.*
119. Director Lucas to Osborn, New York City, April 5, 1918. CA, file 1069.
120. Osborn to Chapman, New York City, April 5, 1918. CA, file 1069.
121. Chapman to Osborn, June 3, 1918. CA, file 370.
122. Chapman, Autobiography…, p. 2.
123. *Ibid.*, p. 3.
124. Osborn to Chapman, June 18, 1918. CA, file 370.
125. Acting Director to Collector of Customs, New York. July 23, 1918, CA, file 1069.
126. Chapman to Osborn, June 3, 1918. CA, file 370.
127. Chapman, Autobiography…, p. 279.
128. *Ibid.* p. 289. Chapman discusses his visit to Roosevelt in the context of learning of the former President's death in January of 1919.
129. *Ibid.* p. 280.
130. Frank M. Chapman, "Notes from a Traveler in the Tropics," Bird-Lore, Vol. XX, No. 6, November–December, 1918, pp. 393–397.
131. Chapman, Autobiography…, p. 280.
132. *Ibid.*
133. Chapman, "Notes…", p. 394.
134. *Ibid.*, p. 395.
135. Chapman, Autobiography…, p. 280.
136. *Ibid.* For a history of the American colony on the Isle of Pines, Cuba, cf. Michael E. Neagle, America's Forgotten Colony: Cuba's Isle of Pines (Cambridge, Cambridge University Press, 2016).
137. Chapman, Autobiography…, pp. 283–284. Cf. note 114 for his comments on Colombia.
138. *Ibid.*, p. 284.
139. Bisher, The Intelligence War…, p. 236. The Japanese-German arrangements in the Southern Cone were confirmed to Washington by Agent 148 of the ONI (not connected to the Museum) on April 29, 1919. NARA, Military Intelligence Division, RG 165: Records of the War Department General and Special Staffs, 1860–1952, Security Classified Correspondence and Reports, 1917–1914, file number 10987–290/81, "Potential Enemies to the United States in Southern South America. (Forecast for Future)."
140. Bisher, The Intelligence War…, p. 248.
141. *Ibid.*, p. 250.
142. *Ibid.*, p. 251.
143. Savala, "Ports of Transnational…", pp. 505–508.
144. *Ibid.*, pp. 508–512.

145. Bisher, The Intelligence War..., pp. 251–259.
146. Chapman, Autobiography..., pp. 284–288.
147. *Ibid.*, p. 291.
148. *Ibid.*, p. 292.
149. Savala, "Ports of Transnational...", p. 503.
150. For example, Hans von Riedel, Germany's head of intelligence in Peru (who had opened the link to Japan), operated out of Valparaiso but was available to quickly get to Mollendo or Callao (Lima), etc. if needed. Bisher, The Intelligence War..., pp. 248–249.
151. Chapman, Autobiography..., p. 294.
152. I am indebted to the Hoover Institution, Stanford University, for providing me with a copy of Chapman's official report to the Red Cross, "The American Red Cross in Latin America." C. 1920.
153. See Bisher, The Intelligence War..., pp. 251–255.
154. Chapman, Autobiography...pp. 295–296.
155. Martin, Latin America....
156. Letter of Emmet W. While, Manager, Insular and Foreign Division, American Red Cross, to Professor P. A. Martin, June 22, 1920. This letter is archived with Chapman's official report at the Hoover Institution.
157. For a discussion of the very political Red Cross mission to Russia see Neil V. Salzman, Reform and revolution: the life and times of Raymond Robbins (Kent, Ohio, Kent State University Press, 1991).
158. Maria Margaret Lopes and Irina Podgorny, "The Shaping of Latin American Museums of Natural History, 1850–1990." Osiris, 2001, Vol. 15, No. 1, pp. 108–118.
159. See Chap. 7, text and note 72.

Chapter 10
Rollo H. Beck: Aka the "Birdman"

There is little doubt that AMNH staffers Chapman, Spinden, Cherrie, Crampton and Eastman volunteered for, and undertook singularly important roles for the U.S. government during World War I. But there were several other American Museum scientists who found themselves similarly involved in gathering intelligence for the U.S. government under differing circumstances. A brief examination of their experiences is warranted to complete the current discussion of the elision between science and politics.

It was noted earlier that prior to the U.S. entrance into the War, and while he was collecting for the Museum in Cuba, American Museum mammalogist Harold Anthony (Fig. 10.1, left) received what were likely coded messages regarding the presence of German activity on the island.[1] His explicit account of the ongoing political upheaval and the role of the revolutionaries in that nation in the first months of 1917 was enthusiastically received by AMNH President Osborn and, somewhat surprisingly, forwarded to local newspapers for publication.[2] But unlike his other colleagues who volunteered for work with the ONI (or Red Cross, or War Trade Board), Anthony, along with Leo Miller, enlisted in the Army and served stateside during the course of the War. Yet, as the next chapter will reveal, Anthony maintained his connection to the military well past the time of his service and submitted observations on the political situation in Ecuador to military headquarters in Washington in 1920, two years after the conclusion of hostilities in Europe.

Barnum Brown (Fig. 10.1, right) joined the AMNH Vertebrate Paleontology Department in the 1890s[3] and first worked in Patagonia, but later shifted his research to Mexico and Cuba.[4] His firm knowledge of Spanish and familiarity with broader Latin America made Brown a likely candidate for recruitment for Latin American undercover intelligence, and he was indeed approached by the ONI for the same type of spying assignment that Spinden had taken. But the paths taken by these two AMNH scientists differed greatly: unlike the archaeologist who faithfully spied for the ONI in Central America until the Spring of 1919, Brown was recruited

R. M. Delson, *Scientists or Spies?*, The Latin American Studies Book Series,
https://doi.org/10.1007/978-3-032-18117-6_10

Fig. 10.1 Left, Harold Anthony, 1917, AMNH Asset ID 36609; right, Barnum Brown, 1919, AMNH Asset ID 37243

but seems to have only briefly joined Naval Intelligence and worked as Agent 233 in Mazatlán, Mexico under the *nom de guerre*, Warren Palmer Brown. Crampton, as noted earlier, blamed Osborn for "letting the cat out of the bag," when he revealed in a public setting that Brown had been recruited to do intelligence work. Obviously, the ONI had little use for an outed agent, and in Crampton's opinion that was why Brown was not recognized as a *bona fide* ONI agent.[5]

Brown's account of the matter suggests that it was his own decision to refuse the Naval intelligence position which had been offered to him.[6] Perhaps his sleuthing experience in Mexico had not been to his liking or, alternatively, he was told to keep his assignment secret, even in his personal notes. Whatever the reason was, unlike his colleagues Spinden and Eastman, Brown spent most of the critical wartime year of 1917 in residence at the Museum, a situation that would end abruptly in October of that year when he was enticed by an offer from the Anglo-American Oil Company, working in tandem with the U.S. Treasury Department, to prospect for petroleum in Cuba.

A paleontologist by training, not a geologist, Brown was first obliged to hone his petroleum surveying skills by participating in an "internship" with a San Antonio, Texas oil company.[7] After this stint, Brown was fully equipped to assess the geology and oil producing potential of the Cuban districts he was assigned to: Matanzas and Pinar del Rio. These efforts, as his daily notes indicate, were subsidized by the Anglo-American Oil Company (affiliated in the United States with Standard Oil), rather than the AMNH.[8] As many as fifteen other companies were simultaneously engaged in drilling for oil in Cuba when Brown arrived there, but none had been successful, primarily due to a lack of proper equipment and geological knowledge.[9] It was obvious that with oil resources an increasingly vital component to winning the War (in retrospect, as valuable a commodity as manganese and nitrates), a well-informed paleontologist like Brown, who understood geology and stratigraphy, and who was capable of evaluating the fossil record for potential evidence of oil deposits, would constitute a major asset to any petroleum company.

Based on his observations in Cuba, however, Brown was forced to report to Anglo-American Oil that there was actually limited petroleum potential in that nation. Although disappointed that their quest did not pan out, the oil company was apparently so impressed by Brown's thoroughness and professionalism that they subsequently requested he join their ongoing geological surveys in Abyssinia (Ethiopia).[10] Brown was willing to do so, but first he had to square things away with the Museum. AMNH records note that on July 1, 1919, with the War over, the paleontologist temporarily severed his ties to the Museum, resigning his position as Associate Curator of Fossil Reptiles. This allowed him to enter into an arrangement whereby he could work for Anglo-American full-time, unencumbered by Museum demands on his time. Given President's Osborn's transactional insistence that those of his staff who acted as government agents devote at least some of their time in the field to Museum work, notwithstanding the pressure of their Washington assignments, Brown's resignation makes perfect sense. In any event, on August 1, 1921, when his African commission was completed, Brown petitioned to be allowed to resume his AMNH position.[11] In a preemptive effort to guarantee his successful return to the Museum, Brown's Anglo-American contract had stipulated that he be allowed to secure specimens for the AMNH during his employment hiatus.[12] He also felt it necessary to forewarn Dr. W. D. Matthew, the Curator of the Department of Vertebrate Paleontology, that "the Cuban and Abyssinian work is at present not for general discussion so please just tell Professor Osborn of my plans."[13]

At least two other scientists from the American Museum staff worked in intelligence gathering capacities during World War I, although neither of them on Latin American assignments. The first, Curator Charles-Edward Amory Winslow of the Public Health Department, the initiator of the highly acclaimed Food Conservation exhibition of Spring, 1917, asked for and received permission from President Osborn to join a Red Cross mission to Russia that summer. The purpose of the trip was to assess and report on that country's deteriorating health and nutritional conditions, in the early stages of what became the Russian Revolution. As noted earlier, this mission was unsuccessful, and its participants were grateful just to return home unscathed.[14]

The other scientist, Roy Chapman Andrews (Fig. 10.2), became ONI Agent 241 on June 10th, 1918, the only AMNH staffer recruited by the service to operate

Fig. 10.2 Roy Chapman Andrews: left, 1920; right, in field gear, Mongolia, 1925; AMNH Asset IDs 104936 and 258467, respectively

outside of Latin America.[15] According to his ONI file card, this well-known mammalogist was posted to China, where he was to be referred to at all times by his code name, Mr. Reynolds, rather than his real surname. It would appear that only he and Barnum Brown received code names, although the others may have used aliases while in the field. Andrews departed from New York for Shanghai on June 29, 1918, ostensibly to conduct research for the Museum. His records indicate, however, that in his faux identity as Mr. Reynolds, Andrews simultaneously served U.S. interests by acting as a direct conduit for intelligence gathering in China. He maintained this dual identity throughout 1919, notwithstanding his redeployment from Shanghai to Mongolia, where presumably the ONI wanted to keep an eye on the ongoing conflict in that region between the advancing Russian Bolsheviks and the remaining Czarists. Curiously, it was Frank Chapman[16] who recommended Andrews for his ONI service rather than Henry Crampton, although the zoologist, as seen earlier, continued to be associated with the intelligence service after the War and had been planning a return trip to the Pacific at roughly the same time that Andrews was assigned to Mongolia.[17]

Rollo Beck: Collector of Birds… And Perhaps Intelligence

The final member of the cadre of scientists who were involved in the *sub rosa* politics of World War I was Rollo H. Beck, the peripatetic bird collector referenced at the beginning of this book. Beck would work for the AMNH in South America in the years leading up to World War I, and in the Caribbean before and during the U.S. entrance into the fray. In the process he amassed vast collections of oceanic as well as land birds, while he simultaneously made his own keen observations on the political scene. Unlike the other AMNH scientist/agents whose intelligence work for the ONI, Red Cross or the War Trade Board is documentable, Beck's covert activities must be deduced from his notes and letters rather than from official government records. Beck, aka the "Birdman", accompanied by his long suffering wife Ida Menzies Beck (Fig. 10.3), left behind a diary (frequently incomplete and often with days curiously missing) in his own hand, a prolific trove of personal letters, literally thousands of skinned birds which he shot and cleaned (often with Ida's help) and a large number of superb photographs taken of the various regions in which he worked during the course of his Latin American explorations. His iconographic oeuvre contained few touristic images, but neither was it limited to recording biological specimens or views of collection sites. Instead, Beck's work contained many unexpected and highly charged political images, including those of the parades of military forces and military installations of various Latin American nations. He similarly seemed fascinated with street demonstrations which he documented in Brazil and the Caribbean. In capturing these views, Beck's camera provided insight into the political moods and machinations of nations considered strategically important to the U.S. government before and during the War, and whose political views toward the European conflict were deemed equivocal.

It is worth noting at the outset that although the regions where Beck had been hired to collect oceanic birds in South America consisted primarily of coastal zones and remote offshore islands along the Pacific coast, in contrast the Caribbean portion of his collecting was devoted to acquiring land birds living at the highest elevations. It may not be coincidental that in all of the locales where Beck collected, from Corral on the Pacific coast of Chile, to Río Gallegos in Argentina, all the way north to the mountainous regions of the Dominican Republic and Haiti, there were pockets of German agents, often using the "covers" of business owners or bankers, who regularly reported back to the German Homeland. Moreover, in the parts of the Caribbean where the intrepid Beck would spend weeks collecting, concentrations of revolutionaries frequently menaced local officials and represented a real threat to local law enforcement, as well as to U.S. Marine Corps troops which were currently occupying Hispañola (Haiti and the Dominican Republic). As a seemingly innocent bird collector, Beck was able to explore and transmit observations as well as photographs of the territories in which Germans and/or rebels were known to be active in the Caribbean, to either consular or military authorities. In short, Beck's photographic oeuvre, which is filed in the AMNH iconographic collection and which contains images that are frequently

Fig. 10.3 Rollo H. and Ida M. Beck, 1916 (passport photo); AMNH Asset ID 236773

accompanied by terse and/or ironic descriptors, reveals far more than is perceived at first glance. When Beck's Latin American expedition concluded in August, 1917, he had spent more time cumulatively south of the border than all of the other AMNH scientists previously discussed, making his experience all the more valuable to the Museum and the U.S. government alike.

Beck also differed from the other scientists considered thus far because, unlike them, he was not a full-time employee of the AMNH, but worked on a contract basis, underwritten by private interests. In that capacity, he spent the better part of five years in Latin America collecting material which would eventually become part of the permanent ornithological collection of the Museum. His lengthy journey to the Southern Hemisphere began in late 1912, and for the next three and a half years he collected along the southernmost coasts of South America. In the Spring of 1916, Beck and his wife made their way northward to St. Thomas, in the Danish Virgin Islands, where his West Indian explorations would begin. These lasted a further year until the summer of 1917, after which time the couple finally returned to New York.

Back in 1912, Beck became the head of a private expedition to South America financed by Dr. Leonard Sanford, a noted surgeon and amateur ornithologist from New Haven, Connecticut, with input from his colleague, Frederick Brewster, a wealthy New Haven businessman with ties to the New Haven Railroad. Beck and Sanford had first become acquainted a few years prior to their 1912 South American collaboration, when the California-born Beck had already earned a reputation as a consummate collector for the California Academy of Sciences. Sanford offered Beck an opportunity to work in Alaska to which the collector readily agreed, and by 1911, he had successfully completed most of his sponsor's goals for the Alaskan expedition.[18] But it is clear from their correspondence that Sanford wanted Beck to continue working for him, and to prolong his stay at Little Kiska Island, in the Alaskan Rat Islands.[19] It is also obvious from this same letter that Sanford intended Beck to undertake another expedition when the Alaskan work concluded, although neither the date nor exact locale of the future project was specified. Cryptically, and perhaps significantly, in a post-script, Sanford warned the collector that he should "say nothing to any one of the Washington plans."[20] It is impossible to know what he was referring to. Did he have in mind something connected with the Smithsonian in Washington, D.C., or with the state of Washington? Or, was he perhaps, implying that Beck would be working in the future in some as yet unnamed capacity for the United States government?

As it turned out, Dr. Sanford had determined that after Alaska, Beck should turn his attention to collecting pelagic birds of South America. Unfortunately for the project, in November of 1912, Sanford unexpectedly fell ill before the two men could finalize their plans. In need of an urgent operation, Sanford shifted responsibility for Beck's South American trip to Frank Chapman at the AMNH, a logical choice as Chapman was, by then, arguably the most respected ornithologist on the East Coast. Additionally, as referenced earlier, in 1911 Chapman had recently shifted gears and had begun to collect in the southern hemisphere, thus becoming familiar with the territory.

Sanford also solicited additional funding for the venture from businessman Frederick Brewster, ensuring that (notwithstanding Chapman's temporary supervision) this continued to be a private, rather than Museum-sponsored, undertaking. Chapman was instructed to advise Beck that "Dr. Sanford, with whom [you] will now be associated with his friend Mr. Brewster, of New Haven, desires to have you sail for South America at the earliest possible moment, under a two-year contract."[21] Beck's payment would be $150.00 per month, plus expenses; wherever possible, Beck was to hire local hunters, carry two or three 12 gauge guns, plus a large supply of ammunition "to have you increase your own efficiency [in shooting down specimens] to the utmost possible limit."[22] With regard to the material collected, all of it was to be forwarded in boxes to the Museum marked "Department of Birds, Sanford Collection, American Museum of Natural History, New York City,"[23] where it would be made available to scholars as a study, not exhibition, collection.

On December 5, 1912, Brewster wrote to Beck directly in order to straighten out what he considered to be some confusion in the directions the collector had received from Chapman. For one thing, he insisted that the ornithologist obtain a telegraphic code (either the Lieber or A.B.C. code were his suggestions) since "cable costs are high to South America and your trip is going to be a long one, we might save some money."[24] There is nothing in any of his records to suggest that Beck immediately followed through with this instruction, although later correspondence indicates that in March of 1916, using the proprietary AMNH telegraphic code "Museology," Sanford relayed the following cable to Beck via intermediaries in Punta Arenas, Chile: "Incorderai Beck Incritique." A note accompanying that message explains that "Incorderai" meant "Shipment of—has not yet arrived", while "Incritique" meant provide data on when the shipment would be made.[25] This cryptic note suggests that Sanford had been given access to the Museum's telegraphic code, if he needed to use it. It is unclear what Beck was shipping, although presumably the cargo was birds and/or photographs.

Little time was wasted in putting Sanford's South American plan into motion. Ida and Rollo Beck (the twosome constituted the whole of the so-called "Brewster-Sanford Expedition") set sail from San Francisco on December 4, 1912 aboard the S.S. *Pennsylvania*, bound for the port of Balboa, Panama. They reached that city by the end of the month, changed vessels and proceeded southward to the coast of Peru, where Beck had hoped to shoot and collect some of the oceanic birds responsible for leaving encrustations of guano droppings on the rocks along the Pacific coast. For decades, this avian by-product had been collected and turned into commercial fertilizer, and its sale abroad provided substantial income for investors as well as for the Peruvian government. Local officials, however, were leery of letting Beck shoot these pecuniary birds and viewed his unusual permit request as suspect.[26] Faced with the bleak prospect of failure at the outset of his expedition, Beck had no choice but to change his plans. Instead of remaining on the coast, he followed a suggestion which Chapman had made earlier[27] and made his way inland to Lake Junín.

Up to this point the American Museum of Natural History had not been directly involved in Beck's expedition. By previous agreement, whatever material was collected during the expedition would be forwarded to Museum headquarters in New York, where it would be labeled the "Sanford Collection" and presumably remain private, available to others only for limited study purposes. But in order to secure the permits needed for Beck to engage in the desired coastal collecting, Brewster was forced to rely on the Museum's growing access to the authority of the State Department. "In as much as it was through the Museum that the State Department requested your permit, it is practically necessary that the Museum have a direct interest in this trip of ours," Brewster informed Beck. "I have therefore agreed and so written Doctor Sanford, that we should be glad to furnish the Museum with habitat groups."[28] Without question, this decision changed the entire nature of Beck's mission: the AMNH could now exert its influence at any point in the expedition. Among other requests, the Museum wanted Beck to obtain prime specimens suitable for mounting in Chapman's habitat displays, rather than limiting the scope of his collecting to material destined for private study.[29]

As an end goal to the collaboration with Sanford, both President Osborn and Director Lucas anticipated that Beck would complete a written analysis of the birds he had collected at the expedition's conclusion. In that assumption, however, they were sadly mistaken. Unbeknownst to them, Beck was no scholar; quite unlike the other members of the Museum's highly educated scientific staff, he was certainly not given to publishing reports. While he was clearly capable of supplying the Museum with cleaned-up versions of his notes, when it came to written output, as Chapman tersely communicated to Osborn, Beck "is far better fitted to serve us in the field than in the study."[30] Perhaps that was because Chapman understood that Beck had left school before completing the eighth grade.[31] Sanford, likely also aware of Beck's shortcomings, naturally concurred with Chapman, informing Osborn that "there is no possible chance of his [Beck] working on a narrative volume, certainly at present or in the near future. This work is simply distasteful to him and I do not believe that he can be persuaded to attempt it."[32]

Beck's lack of education apparently never surfaced as an issue for the Museum, although his letters and official communications contained numerous spelling errors. Overall, nothing in his notes or written oeuvre could be considered remotely erudite. It is not surprising, therefore, that the responsibility for interpreting Beck's South American collection was eventually assigned to a true scholar, Robert Cushman Murphy, who wrote and published what is now considered the classic, two-volume work on Oceanic Birds of South America in 1936.[33] Murphy's opinion of Beck was lukewarm, at best. He viewed Beck's work as "an aggregation that includes very nearly all species of sea birds recorded from South America," which made him, according to Murphy, an "experienced and peculiarly efficient collector."[34] This was certainly damning with faint praise.

Back in Peru, Beck waited for the State Department's intervention to yield results, spending his time collecting specimens in the Andean highlands. It was not until the beginning of April, 1913, already more than three months after the official start of the Brewster-Sanford expedition, that Beck returned to Lima, having finally received his permit to collect coastal specimens by means of diplomatic intervention. Beck would be allowed to proceed southward along the Peruvian coast, past the port of Callao (Lima), to legally initiate his South American oceanic bird collection.

For the remainder of 1913, and into the middle of the next year, Beck traversed the southern Pacific coast, which eventually landed him on the Chilean coast, as well as the Peruvian. Beginning in Antofagasta, in the north, he made his way to the southernmost Chilean port of Punta Arenas where he arrived in late June, 1914. Along the way his reconnaissance took him through numerous ports and islands, including Talcahuano, Coronel, Corral, and Juan Fernandez. While his stated goal was to amass specimens of and photograph as many oceanic birds as possible, his broader activities suggest that other influences may have been at work. Was it mere coincidence, or an indication of pre-planning that almost all of the areas Beck visited, or circumnavigated, were in some way connected with German interests in the region and/or the production of nitrates, both of which had been of great concern to the United States government since the turn of the century?[35] Was he providing information on Chileans of German background to Washington at the same time that he proceeded with the Sanford project?

Such conclusions seem likely when we place Beck's collecting activities in the broader historical context. Chile had actually been on Washington's radar since the aforementioned "*Baltimore* Crisis" of 1891, but it was during the time of Beck's voyage southward that Chile became truly problematic for the United States government. It was fairly obvious that despite her professed political neutrality in regard to the brewing world war, Chile's leanings were strongly influenced by the nation's longstanding connections to Germany and its military. Moreover there was no indication that this long-standing relationship would terminate because of hostilities in Europe. One early indicator of the nation's predilections was the decision to appoint Emilio Körner, a German citizen, to the post of Commander of the entire Chilean Army, a position which he held until 1910. This gave Körner ample time to impose a Prussian tradition of militarization on his adopted South American troops (Fig. 10.4).[36] Additionally, as in Brazil, substantial numbers of German settlers had emigrated to the South American nation in the nineteenth century, and German-Chilean communities had emerged along the southern Chilean coast. Even into the early twentieth century, these groups continued to maintain a strong sense of identity with the German Motherland, notwithstanding the fact that they held Chilean citizenship by virtue of their birth. In parallel, there was also considerable German capital tied up in the mining and production of Chilean nitrates and copper, materiel vital to the manufacture of weaponry. All told, there was substantial evidence of "fervent support for Germany in the Army, in German commercial establishments in Santiago and Valparaíso, [and] in the German colony in the south."[37]

But that was not the entire extent of German involvement in Chile. In November, 1914, the German government tried to mount a relief expedition which was slated to sail from Chile to Southwest Africa (Namibia) in an effort to defend the German holdings in Africa against the British. A force composed of some five hundred Chilean/German volunteers vowed to fight for the Motherland, inspired by Germany's unexpected naval victory against the British in the Pacific Battle of Coronel which had been waged at the beginning of the month (see below). But a subsequent British win in the Battle of the Falklands (December, 1914) made the possibility of a German fleet crossing the ocean without a British counterattack highly unlikely; as a result, plans for any transoceanic expedition were scrapped. The July, 1915, surrender of the Germans to the English in Southwest Africa finally rendered the broader scheme moot, but the potential for mobilization of pro-German sympathizers in Chile had alarmed Washington politicians, notwithstanding the fact that their own direct participation in the European struggle was still several years away.[38]

Washington's concerns were further exacerbated by the increasing presence of German intelligence officials, known collectively as the *Etappendienst*, who integrated themselves into communities up and down the Chilean coast. By drawing upon those German officers who had previous naval and commercial maritime backgrounds and who were already resident in the major Chilean cities, the German government was able to create viable intelligence networks.[39] As these networks evolved, officers from German commercial shipping lines would routinely travel under the cover of darkness to the ports of Corral, Coronel, Talcahuano and Valparaíso to collect telegrams of interest to Berlin and relay the contents (via wireless) to German warships anchored offshore. There were also

Fig. 10.4 Chilean soldiers, Punta Arenas, 1915; top, group photo by Rollo H. Beck. Image copyright California Academy of Sciences Archive, Rollo H. and Ida M. Beck collection, Box 43, with permission; bottom, lined up outside barracks, AMNH Asset ID 237546

secret German wireless stations (see Fig. 10.12, in the Cuban mountains, for an example) located on the coast of Ecuador and Peru.[40]

What is most interesting is that if we compare the locales in which Beck spent months amassing bird specimens before and during the early years of World War I with the places of interest to the German intelligence network and, by extension, to the United States, there is an extraordinary degree of overlap. This was true not only for the ports located on the coast of Chile but also for Chilean offshore islands in the

Pacific Ocean. From late 1913, for example, Beck spent several months in the Juan Fernandez Islands (perhaps best known as the setting used by author Daniel Defoe for his Robinson Crusoe tale), where hilly terrain and easily camouflaged anchorages had made this region a convenient hiding place for German agents. Was it likely that Beck would have spent all of December, 1913 into mid-February, 1914 in the Juan Fernandez Islands, with its limited population, and be unaware of German interests or its large number of German-Chilean inhabitants? In a letter written to his mother from this locale, Beck explained that he had been living with a local fisherman, and that he had hired a boat to do further exploration.[41] While it is clear that Beck was legitimately collecting birds, might he not also have been reporting on German activity to Washington, using his Museum-linked mission as a cover while he skirted the rugged coastline of this island group?

From his photographic oeuvre it is clear that Beck also spent time in the port city of Valparaíso, another setting where German agents could readily be found. It was the fact that Beck was in Valparaíso which led Museum officials to request that he photograph the 1913 goodwill publicity tour of former president Theodore Roosevelt, which he undertook before the start of his Brazilian River of Doubt adventure. Beck could hardly refuse such an entreaty, especially given Osborn's willingness to use the Museum's connections in the State Department to facilitate the Brewster-Sanford Expedition.

As requested, Beck dutifully followed the ex-President, managing to immortalize Roosevelt's sour mood as he made his way through the crowds gathered to see him in Valparaíso. As noted earlier (see Fig. 3.3), Roosevelt's surly countenance in these images likely reflected the less than stellar reception he had endured earlier in the capital city of Santiago.[42] It is unclear whether Beck forwarded these images to New York immediately or made them available at a later date.

Following this assignment, the Becks continued southward to the southern city of Punta Arenas, where they arrived just prior to the event universally acknowledged as the catalyst for World War I: the June 28th, 1914 assassination of Austrian Archduke Franz Ferdinand and his wife in Sarajevo. Although Beck would collect and photograph numerous oceanic birds (and their eggs) in Punta Arenas in the coming weeks, his choice of this city (the self-proclaimed "southernmost town in the world") for his base of operations is intriguing; Punta Arenas had rapidly become "a thriving center of intelligence and secret logistics early in the war."[43] Given this fact, Beck's photographic essay of the city taken in February, 1915 raises serious suspicions about the full nature of his assignment, as his work consists largely of images of Chilean soldiers drilling in the streets of that city, their commanders and their barracks, hardly views which evoke the casual tourist, much less a bird collector.[44] Was Beck sending signals to someone about the strength of Chile's military in the Far South? In this context it is especially intriguing that in a letter home, Ida informed her mother that "Dr. Sanford thinks Rollo's photographs are wonderful if you please and is having the most interesting ones enlarged."[45] She does not explain for whom and why. It is likely that she did not know, herself.

The couple stayed in Punta Arenas until early August, 1914 at which point they temporarily left their base and sailed northward on a commercial vessel to Buenos Aires, Montevideo and Mar del Plata. In the Argentine capital, Ida and Rollo were able to collect their mail and receive further instructions on how the expedition was to proceed. It was during their absence from Punta Arenas that the first real naval battle of World War I was fought to a deadly finish off the coast of Coronel, Chile. In what was a decided rout, the German East Asiatic Squadron surprised and defeated an unsuspecting British Navy fleet on November 1, 1914, which resulted in a staggering loss of over 1600 lives for the British forces.[46]

The Becks were still in the Argentine capital when this battle took place. Then, without any explanation offered in his diary or letters, the couple suddenly left the city and returned straight away to Punta Arenas, in the Magellan Straits. There is just one mention of the Battle of Coronel in Beck's letters home to his family, and it is contained in a brief note Rollo penned to "Ruth" (his sister?) on November 19, 1914 while he was on board the R.M.S. *Oriana*. Writing from a location some ninety-three miles north of the Falklands, Beck told her that "[T]raveling by water and English boats is possible to develop a little excitement especially if the German fleet shows up here which had a victorious battle off Chile [Coronel] few days since according to German reports. The Japanese boats are reported everywhere about however by the time we leave here suppose the Germans will be disposed of if these Britishers aboard can be believed."[47]

The Becks settled back into the Punta Arenas region where, from the end of 1914 until April 1, 1915, Rollo dedicated himself to exploring the Cape Horn region. The twosome were already in the city on December 8th, 1914, when the British launched a retaliatory strike against the Kaiser's navy, ambushing the German squadron off their own Port Stanley in the Falkland Islands. In contrast to the Coronel disaster, in this instance the British were able to prevail; only one German ship, the SMS *Dresden*, managed to escape the carnage, and it sailed posthaste to a hiding spot in the waters off Punta Arenas.[48]

When the search began for the missing German vessel, Beck was ideally situated in Punta Arenas to be on the lookout. British intelligence searched in vain for the *Dresden* in the waterways around the city and ultimately failed to find the vessel. In the meantime, pro-German sympathizers (i.e., German-Chilean colonists) somehow managed to provision the ship and bring food and fuel to its crew, as it limped cautiously from one hiding place to another. Was Beck part of the hunt? He certainly was in the right place at the right time. His mission of collecting and photographing birds in the same Magellan Straits in which the *Dresden* had taken shelter certainly would have allowed him to search "under the radar." Beck spent about two months, from the end of November, 1914 to the end of January, 1915, working north and west in the Cape Horn area, during which time he left Ida by herself in the city.

By the time Beck returned to Punta Arenas, the *Dresden* had somehow managed to avoid detection and escaped. Sailing stealthily northwards along the Chilean coast, the warship eventually reached the Quintupeu Fjord, near Puerto Montt, where it once again sheltered from sight. There, too, local German-Chilean settlers secretly brought supplies to the ship, allowing her to emerge, fully repaired, on February 14th, 1915.

Although she did manage to destroy one British ship after she had left safe harbor, the *Dresden* finally was cornered by English forces in the Juan Fernandez Islands, where the German commander of the vessel hastily scuttled the ship in the face of overwhelming opposition and his reluctance to surrender under such humiliating circumstances.[49]

Beck continued collecting in the Punta Arenas area until just after the *Dresden* sank at the end of March, when he moved on to Ushuaia and Tierra del Fuego in search of specimens. Once again, Ida was left in Punta Arenas, where she stayed through the harsh and lonely austral winter of 1915. She complained to "Ina" that it would be stupid making a fuss about her isolation because "he [Rollo] is coming back."[50] She also indicated that she had received several letters from her husband sent as he was leaving the community of Río Gallegos, Argentina, coincidentally another well-known German colonist settlement.[51]

By the Spring, in October, 1915, Beck reunited with Ida, and the couple left Punta Arenas for good, bound for Port Stanley, in the Falkland Islands. Later that month, Ida again wrote to "Ina" that she and Rollo had taken time to explore a nearby penguin rookery. After that, Rollo collected by himself for a month while Ida spent her time in a local hotel. Without explaining why, her letter suggested that it would not have been possible for her to accompany him.[52] Later, Ida complained that she was worried Rollo might not be back by the Christmas holiday. "We never know what we are going to do or how long we are going to stay in a place…Rollo is [currently] out on a small sailing cutter collecting some of the outer islands…there was a young man who was to have brought me birds to work on while Rollo was away but he hasn't brot [sic] any so far."[53] Was that because Rollo was otherwise occupied? At the end of January, Ida happily reported to her mother that her husband had indeed made it back for Christmas, not because of the holiday but as "something had to be done to the boat."[54] She also wrote that both of them were getting "fed up" with the life they were living. Their sole entertainment was listening to the sermons of a local Baptist preacher, an Englishman who otherwise worked at a wireless station in the Falklands.[55] Of course, as seen earlier in the case of Spinden's assignments, wireless stations during the War frequently intercepted German messages or, alternatively, transmitted them. There is no indication in either Rollo's or Ida's notes which, or if, the preacher's did.

In mid-February, 1916, the Becks finally left the southernmost portion of South America and made their way northward to Montevideo en route to further collecting in Brazil. In an apologetic letter, Rollo attempted to explain to his mother why he had not written to her from the Falklands: "I did not like the attitude of the tuppering [sic] officials in the place so wrote no letters and received no mail…I was unable to get to certain islands I wished to visit as the only available boat was owned and sailed by a naturalized German who became a Britisher only 2 years before the war started and he might have been a spy or I might have been one for all the officials knew. My letters might have been forged or might have been secured from some one else seemed to be the idea entertained by the Government the way I sized up their action."[56]

Until he accessed his mail in Montevideo, Beck was unsure of his future plans. When he finally received his orders, he learned he was to proceed northward to Brazil where, as his next letter to his mother revealed, he planned to collect birds on the Ilha da (Island of) Trinidade, several hundred miles off the Brazilian coast. He lamented that the war had hindered his collecting in the Falklands but that, nonetheless, he had been able to secure a "good bunch of photographs of 3 different kinds of penguins and other interesting birds but the especial desiderata of the trip were pretty scarce…the war played hob with the expedition."[57]

As soon as he reached Bahia, Brazil, Beck attempted and failed in his effort to visit the island; local officials were reluctant to grant such a privilege, a situation which echoed his earlier troubles in Peru. Sanford and Brewster quickly alerted Osborn to Beck's predicament[58]; and a flurry of letters ensued from the AMNH to relevant political contacts in an attempt to grease the wheels. Osborn even solicited the aid of Domício da Gama, the Brazilian Ambassador to the United States, apparently exerting sufficient pressure on him to have Beck granted priority status for his visit.[59] Shortly after their exchange, Da Gama triumphantly informed Osborn that a permit for Beck's travel to the Ilha da Trinidade had been issued by the Brazilian government on May 18th, 1916.[60] A few days later an absolutely astonished American Consul in Bahia wrote to Museum administrators suggesting that the permission that had been granted to Beck was rare indeed: "Similar permission has been refused to many Brazilians and it is only due to the Ambassador's interest and the high character of the American Museum that an exception had been made in the case of Mr. Beck."[61]

As it turned out, the Brazilians had very good reason for denying Beck, or anyone else, access to the Ilha da Trinidade. In September of 1914, the Germans had secured a harbor on the island from which their intelligence service thought it safe to outfit and sometimes convert large, formerly civilian, ships into naval vessels. One such ship was the *Cap Trafalgar* which was in the process of being provisioned with coal when a British ship spotted her. A naval battle between the Germans and the British resulted in the sinking of the converted vessel.[62] Nonetheless, hidden German anchorages were still a possibility on the island, which is located hundreds of miles from the mainland and was poorly patrolled by Brazilian interests.

Beck was offered the courtesy of passage to the island with a group of Brazilian naval officers but he declined, instead deciding to engage his own boat.[63] Unfortunately for him, this vessel was badly in need of repairs. Beck's journal entry of June 27, 1916 reflected his growing frustration upon learning that not only were an additional three months required to render that boat seaworthy, but that the captain of another boat whom he had subsequently hired was now refusing to take him on the grounds of predicted bad weather. As a result, the increasingly irate Beck decided to skip Trinidade altogether and to leave Brazil in favor of traveling northward to the West Indies, an idea Sanford had proposed earlier.[64] This substitute venue now became the official expedition plan; the Becks would leave Bahia, and South America for good, and on June 30, 1916 and travel to the Greater Antilles to collect bird specimens.

While he waited for Sanford's formal approval of the change in plan, Beck used the interim period to shoot and collect some rather ordinary Brazilian land birds, which he found near the city of Bahia, especially at the mouth of the São Francisco River. Of greater interest for this discussion is the additional time he spent photographing the city itself, again with his unusual emphasis on political events rather than touristic sites. Here too, as he had done in Punta Arenas, Beck photographed large parades of soldiers on several days during March and June, 1916 (Fig. 10.5). He also captured views of crowds gathered to listen to what he labeled "patriotic speech"[65] and "oratory." Most tourists would have taken images of the magnificent gilded baroque buildings at the city's center but Beck chose otherwise. Why? One explanation is that at this point in the course of the War, Brazil had still not joined the Allies, and predicting which way the political headwinds of this strategically and economically important nation would blow was increasingly difficult for Washington. Did Beck's images capture sympathy for the Germans or opposition? And to whom was he communicating this knowledge?

It is also unclear what Rollo and Ida Beck did with their time between June 30, when they finally left Bahia, and August 10, 1916 when the couple landed on the island of St. Thomas, the administrative center of the Danish Virgin Islands. The almost two month long interval is left blank in Beck's notes, with the exception of a brief mention of his stay in Barbados. While on that island, Beck convinced a local man to swim offshore and capture four Audubon Shearwaters in their nest for his collection. But other than that occasion, little Museum work was done for a month and a half while, most likely, the Becks took a well deserved rest.[66] When they arrived in the Danish Virgin Islands via steamer, they knew that the final months of the Sanford-Brewster expedition would be spent in the Caribbean region and a year later they were, as planned, back in New York City.

Whether fortuitously or by design is unclear, but the Becks found themselves in the Danish Virgin Islands at a very propitious moment for the political future of that colony. The Danish office in Copenhagen which administered the colonies of St. Thomas, St. Croix and St. John had been trying since the 1890s to entice the United States to purchase their territories, a request which up to August, 1916, had been met with lukewarm interest at best. However, just as the Becks arrived in St. Thomas, in August 1916, the United States government had a noticeable change of heart and began negotiating in earnest for the transfer of the Virgin Islands to U.S. control. Of the three islands under Danish administration, St. Thomas held the greatest interest for the United States, not only because it possesses an excellent harbor, but because it also presented strategic possibilities for the location of coaling stations from which to provision ocean-going vessels, including those en route to the Panama Canal. The Government of Denmark strongly favored these negotiations, and when a plebiscite was held in that country later in the year (December, 1916), it revealed that at least 64% of the metropolitan population also supported the proposed sale of the colony to the United States for 25 million dollars.[67]

But what of the Virgin Islanders, themselves? How would they react to the sale of their homeland? Would they resist the United States takeover and complicate the transfer of ownership? One way the United States could gauge the willingness of the people to accept the change of ownership would be to observe their reaction to official government pronouncements. Newly arrived in St. Thomas, Rollo Beck

Fig. 10.5 Above, military procession in the street of Bahia, Brazil, March or June, 1916, AMNH Asset ID 237267; below, crowd listening to “patriotic speech”, Bahia, Brazil, March or June, 1916, AMNH Asset ID 237268

Fig. 10.6 Crowd in St. Thomas listening to discussion of U.S. purchase of Danish Virgin Islands, August, 1916, AMNH Asset ID 236992

began almost immediately to photograph a series of outdoor meetings in which details of the sale of the colony were explained to local inhabitants. Beck's images of these meetings portray a polite gathering of onlookers listening attentively to the description of the pending transfer,[68] and there is nothing menacing in these images (Fig. 10.6). Nevertheless, a few sturdy Danish naval personnel, in uniform, are seen standing watch at the outskirts of the crowd, just in case of a disturbance. In addition, Beck snapped views of the harbor in St. Thomas, which was also of strategic interest to Washington. Perhaps mindful that someone might wonder why he was taking photographs of this anchorage, he had Ida pose (as a decoy?) in the foreground of his composition, where she stood fetchingly holding a sun umbrella; the photo also included a milk boy, presumably to show local interest (Fig. 10.7). Photographed from the hills high above St. Thomas, the image clearly conveyed the parameters of the harbor and its entrance, as well as the surrounding islands; the next image in sequence focused on the harbor itself.[69]

The sale of the Virgin Islands was finalized in January 1917, when the United States and Denmark signed a formal ratification treaty. But while the official proclamation announcing the cessation of Danish rule and the simultaneous onset of United States leadership took place on January 25, 1917, the arrival of the first U.S. Governor to the islands was postponed until April 7, 1917, contemporaneous with the U.S. Declaration of War against Germany. Although Beck had left St. Thomas in September, 1916, and was in the Dominican Republic in late March of 1917, quite unexpectedly he left Hispañola to return to St. Thomas just in time to capture images of the new governor's disembarkation and the official political transfer.[70] Was this coincidence, or more likely, had Beck been assigned (by Washington? by Sanford?) to capture the event? Another curious feature of Beck's Virgin Island experience is that several of the photographs he took of St. Thomas were reproduced in one of the earliest historical accounts of the territory in English. Written by the former U.S. Vice-Consul to the Virgin Islands, Luther K.

Fig. 10.7 Above, Ida Beck and "milk-boy" in hills above St. Thomas harbor, Danish Virgin Islands, August, 1916, AMNH Asset ID 101889; below, closer view of ships in harbor, AMNH Asset ID 101890

Zabriskie,[71] the book credits Beck for his work but does not indicate whether Beck was freelancing or providing such images *gratis*. Had he solicited additional work without the knowledge of either Sanford or the AMNH to make ends meet, or did the parties involved simply not care?

While still in St. Thomas on his initial visit in August of 1916, Beck hunted and collected a few birds when he was not otherwise engaged in photographing political events. Several miles offshore he found terns, paroquets [sic] and other small birds.[72] But as none of these species was of real interest to Sanford, the Becks decided not to linger in St. Thomas but, instead, to move westward to San Juan, Puerto Rico. Once in the city, Rollo deposited Ida in a hotel and then continued on to the Dominican Republic, which occupies the eastern portion of the neighboring island of Hispañola. This portion of Beck's expedition coincided roughly with the onset of a multi-year "peace-keeping" occupation of the Dominican Republic by U.S. Marines. The action taken by the United States government proved quite controversial on the island, and many of the prominent Dominican political elite were loudly voicing their opposition to U.S. occupation just as Beck arrived in their country. Beck was determined that such sentiments would not deter him as he traveled by steamer from Puerto Rico to the Dominican capital city of Santo Domingo.

However, when he finally reached the capital on the afternoon of September 23, 1916, what startled him was not the local birds, but the large presence of U.S. Marines who had already set up a base of operations in the city.[73] Under what were clearly martial law conditions imposed by the "peace-keeping" occupation, Beck immediately and correctly assessed the situation, recognizing that he would now be obliged to obtain permission from the local American Vice Consul before proceeding anywhere in the country. If nothing else, his prior experiences in Peru and Brazil had certainly taught him the necessity of clearing arrangements with local officialdom a priori.

But the Consul in Santo Domingo was anything but encouraging. "I beg to inform you" the official replied to Beck's formal request to begin collecting in the countryside, "that the principal towns of this country are occupied by American marines…and that I do not consider it safe for you to travel in territory not under actual military control of our forces."[74] Thwarted, but undeterred, Beck requested that instead he be allowed transit to Haiti (the country which occupies the western portion of the island of Hispañola), where United States forces had also recently been installed. This request was similarly denied, obliging Beck to now remain in Santo Domingo. In an earnest attempt to fulfill his obligation to Sanford, as well as the AMNH, he traveled from the city into the surrounding countryside to make a few preliminary observations. Not trusting the stability of the region, he was always armed on his daily sojourns; his notes suggest that "the natives did not seem unfriendly tho [sic] the same looked sideways at gun which was in case till [I was] out of town, revolver loaded in handy pocket."[75] His daily trek took him a few miles along river banks into the mangrove swamps where ordinary herons, quail and doves roosted. When he finally managed to hire a local boat and sail upriver from Santo Domingo, he found nothing of great interest until he encountered and shot a white wing dove roosting near its nested eggs.[76]

By late October, after hearing shots in the vicinity where he had been collecting, Beck decided that Santo Domingo was no longer safe. He therefore decided to leave the capital city for the town of Sánchez, on the Samaná peninsula, where he hoped to find more stable conditions, notwithstanding the rumors that "some bandits are reported in the hills."[77] Unfortunately for Beck, the local birds he found in Sánchez were equally disappointing, differing little from those he had just shot in the city of Santo Domingo; he would need permission to penetrate further into the interior of the country if a worthwhile collection was to be made. Such plans depended on political circumstances and, of necessity, Beck would have to wait for a "slightly cooler feeling in this country."[78]

For the month of November 1916, Beck was obliged to stay around Sánchez, where he photographed U.S. Marines and captured images of possible German interests in the region. In surprisingly candid photos (which suggest that he had be able to secure an extraordinary level of acceptance by the U.S. military), Beck captured views of Marines landing ashore (Fig. 10.8) and in formation, the cars used by the quartermaster to supply the troops, and Marines enjoying what Beck called "shore leave" (Fig. 10.9).[79] In other images Beck portrayed the Marines at their barracks, sailors standing in front of "Ma and Pa Hagen's Hotel Alemán" [German Hotel] (Fig. 10.10), and various street scenes of Sánchez.[80]

Perhaps the most notable of his images was one he captured of a Marine (a white man) with his arms around local girls (see Fig. 1 in Preface), whom Beck labels in another view of the same grouping, "Black beauties."[81] It is hard to know what he was saying with the offhand title he gave to the composition. Was it his way of commenting on the blurring of the color lines in the Dominican Republic, which ran counter to the attitudes held toward interracial relationships in the United States at this time? Was it an assessment of the morality of the occupying forces? It is hard to know exactly what he meant by capturing this image but the very fact that Beck memorialized it speaks volumes.

Fig. 10.8 U.S. Marines landing at Sánchez, Dominican Republic, November, 1916; AMNH Asset ID 236964

Fig. 10.9 U.S. Marines on shore leave, in Sánchez, Dominican Republic, November, 1916; AMNH Asset ID 236967

Fig. 10.10 U.S. Navy sailors outside "Hotel Alemán", Sánchez, Dominican Republic, November, 1916; AMNH Asset ID 236970

While Rollo was exploring Sánchez, Ida waited anxiously in San Juan. Despite being safely ensconced in Puerto Rico, a U.S. territory since 1898, she complained to her mother in a letter that she "wanted to spend Xmas with him [Rollo] regardless of birds."[82] This was about as testy as she got. Unbeknownst to her, at roughly the same time that she was writing her letter, Rollo had left Sánchez to travel inland into the agricultural communities of La Vega, an area also safeguarded by U.S. Marines.[83] In his journal Beck wrote that he had been "Up in the hills all day tho [sic] marines think it rather dangerous…last week revolutionists tore up some rails." He added "natives I'd pass on trails were not hostile appearing tho [sic] I was dressed more like a native than an American."[84] Bravado notwithstanding, Beck needed to convince Major Arthur Marix, the officer in charge of the local Marine encampment, to grant him the required shooting permit to enable him to make collection in La Vega worthwhile.

With ample warning from the Marines, and perhaps with an agreement to pass on the coordinates of any rebel activity he might observe, Beck ultimately received permission from the officer in charge to explore the La Vega countryside. It is clear from the photos Beck took of Marines while he was in this region that he not only had contact with the troops, but enjoyed intimate access to their everyday routines and barracks.[85] In retrospect, it would seem unlikely that the USMC would allow such unfiltered access to the troops if Beck's actions had not been sanctioned at the highest levels. Clearly Beck managed to maximize his time in the region by shooting and skinning substantial numbers of local birds. He stayed in this region until December 11, 1916, when he once again returned to Sánchez where he waited for orders to arrive from New York.

During the time Beck explored La Vega, Chapman had just returned to New York after spending months in South America, following his harrowing voyage from Brazil via Barbados. Apparently under time constraints, Chapman did not respond to Beck's urgent request for further instructions until December 16, 1916. When he finally got around to answering Beck, Chapman told him that both he and Sanford concurred that if the situation proved too dangerous, Beck should alter his plans and relocate to Santiago de Cuba, after a possible stop at Navassa Island. On the other hand, Chapman also subtly intimated that nothing held as much promise for collecting birds as the mountains of the Dominican Republic, if Beck could somehow manage to get there.[86]

Simultaneously, Sanford began a campaign at the AMNH to have Beck conduct a future expedition to the South Pacific, but this time with the Museum's express backing from the onset of the project. Such plans could be formalized only after the collector returned to New York, but President Osborn signaled his transactional commitment by agreeing to "help such an [future] expedition, in return for collections."[87]

Beck was not privy to any of these plans as he was still in the Dominican countryside. Finished with the La Vega area, Beck departed from Sánchez on December 19th and returned on an overnight steamer to the capital city of Santo Domingo where, to his great surprise, he was met at the quay by a very anxious Ida who had left San Juan a few days earlier.[88] On the same day that the couple reunited, Beck also paid a visit to the officer in charge not only of the capital city, but of the broader U.S. occupation of the country, Brigadier General Joseph H. Pendleton. In a subsequent letter to Sanford, Beck relayed Pendleton's assessment that "they [the military] wouldn't send less than 50 marines on a trip through the interior, 10 would be worse than none at all." But the ever intrepid Beck assured Sanford that he would be fine with just "one native guide…and a boat." He also passed on Pendleton's Chief of Staff's request for verification from the AMNH that he was actually collecting for them.[89] Perhaps to placate Ida, Rollo told his wife that Dr. Sanford would pay for a "sufficient bodyguard of American soldiers to insure protection in the interior of the country." "But," as she wrote to her mother after receiving this assurance, "Rollo and others say the country people are friendly enough—it's the political people [that] start the riots…".[90]

It is unclear how Beck was finally able to gain General Pendleton's acquiescence to his new plan of collecting in the remote interior of the Dominican Republic, but on December 21, 1916, the General forwarded orders to ten provincial governors in the interior of the country to be on the lookout for and welcome Beck. In a form letter composed in Spanish, the General requested that these officials offer assistance to the ornithologist who was on a "misión especial, con el propósito de recojer ejemplares de aves para colección del citado Museu."[91]

It is only logical to assume that General Pendleton expected a *quid pro quo*; Beck would nominally be under Marine protection (although he did <u>not</u> travel with a bodyguard), in return for which, presumably, he would pass on reconnaissance about local agitators and/or pro-German sentiments or activities. Beck's penetration of the more remote areas would be especially useful to those Marine Corps administrators who were attempting to impose governance on what was an admittedly tenuous situation. At this point, less than a month had passed since United States formally declared control over the entire Dominican Republic.[92]

One of the principal tasks facing Marine Corps administrators was keeping an eye on German residents of the Dominican Republic.[93] For the most part, German interests in the country centered on the growing and shipping of tobacco (especially from the La Vega region) back to the Homeland. Their business dealings meant that the expatriate community was in constant communication with Germany, and it was equally apparent that their business activities could easily camouflage intelligence networks. To counter the growing threat of German penetration of the interior, the Marines frequently requisitioned land from German owners.[94] The situation was no different in neighboring Haiti, where Beck planned to go next. There the Marine Corps suspected that German actors were actively agitating against the ongoing U.S. occupation.[95]

Undoubtedly, Beck's keen powers of observation (as well as the pictures he took throughout the interior) would provide a welcome source of data for Marine Corps intelligence. Beck must have favorably impressed Pendleton because the General not only granted the collector permission to travel to the interior and facilitated his access, but promptly invited the collector and his wife to spend Christmas dinner with the troops, an event which Beck was allowed to photograph.[96] Beck also captured what appears to have been an informal Christmas Day basketball game organized by the troops on the grounds of Fortaleza Ozama. This colonial Spanish fortification is clearly visible in Beck's images, as are the tents in which the Marines were bivouacing.[97] One can only imagine that knowledge of the whereabouts and activities of the troops would have been classified, or at least off limits, except to the most trusted of civilians; Beck was apparently one of them.

Shortly before the new year began, Beck began his trek to the uplands of the interior, determined to climb the highest *cordilleras* (mountain ranges). Sometimes (but not usually) he was accompanied by a guide, one of whom he immortalized in a photograph (Fig. 10.11).[98] During the time of this trek, Beck endured Spartan living conditions, often forced to camp outside. He subsisted on equally meager meals (some of them consisting of nothing more than cold mashed bananas with imported sausages), while he suffered through endless bouts of rain and mud, all of which thwarted his collecting efforts.

Beck's reconnaissance began in the southern city of Azua, a colonial settlement about 100 km from Santo Domingo which only a few months earlier had been the scene of a rebellion quelled by the Marines.[99] From there he made his way to the Cordillera Central, the mountain chain of the central-west region. His goal was to reach Mount Tina (today known as Pico Duarte) which, at 10,100 feet, is the highest peak of the country. On January 3rd, 1917, he began his ascent of the mountain, encountering sparrows, flycatchers, trogans, parrots, pigeons, as well as an unfamiliar

Fig. 10.11 "Guide and pack horse in mountains," Dominican Republic, January, 1917; AMNH Asset ID 236872

type of woodpecker along the way.[100] He noted in his journal that he thought he was probably was the first stranger [i.e., American] to reach the top. For the rest of the month he pushed on toward Tubano and Constanza, maneuvering between valleys and ridges where he mostly encountered yellow breasted finches.[101] From this locale he returned to a sugar plantation owned by Ramoncito Velásquez, who had kindly not only lent him a horse for his explorations but also offered him some very welcome hospitality.

Thus immersed in his collecting, for the rest of January, 1917 and into February, Beck traipsed up and down the canyons, finally reaching the highest peak in the Mt. Rusilla range, where the elevation was high enough for there to be frost in the morning, by the end of the month.[102] Still Beck found nothing new of interest.[103] As he was exploring Mt. Rusilla he finally came upon some crossbills, which he then proceeded to shoot and prepare.[104] On March 4th, to be certain he had not missed any opportunity, he climbed Mt. Penole[105]; by mid-March, with little to show for his efforts, he returned to Azua. Beck had not found any unusual specimens, but neither had he encountered any revolutionaries. From Azua, Beck made his way back to Santo Domingo, arriving in the city on March 24th where Ida, once again, waited for him.[106]

Ida and Rollo's reunion must have been quite brief because Beck was soon on his way back to St. Thomas, where as noted before, he arrived on April 7, 1917, just in time to photograph the first United States governor to come ashore since the takeover from the Danish.[107] Preserving that image must have been the extent of his assignment because later that same day, Beck left the Virgin Islands (and the Greater Antilles) on the Schooner *Lillian*, bound for the island of Dominica. Passing by Tortola in the British Virgin Islands on his way to the Lesser Antilles, he observed Wilson petrels and a few Audubon shearwaters.[108] His instructions, as defined by Dr. Sanford,[109] were to bypass these ordinary specimens in favor of locating the "rare Oestralata" [Aestrelata], although it is unclear which one. He had earlier searched in vain for this very petrel in the highlands of Santo Domingo, where Sanford had thought he might have some success. Now Beck was enjoined by his mentor to pursue the rare petrel on the island of Dominica where, according to an early monograph owned by Sanford,[110] the bird had been spotted several decades earlier. Although Sanford admitted that he did not actually believe this account, as long as Beck was in the region, he thought there was no harm in having him check it out.[111]

Beck's ventures into the Diablotin Mountains of Dominica (which coincidentally bore the local island name for the same bird Beck and Sanford were seeking) were futile. Talking to local inhabitants, Beck surmised that the petrels had perhaps flourished there 50 or so years earlier but, by 1917, a combination of animals similar to the mongoose, as well as some very large snakes, probably destroyed the young birds and doomed the population.[112] By May 5th, an unsuccessful Beck was on his way back to St. Thomas, *sans* Aestrelata.[113]

During the few days in which Beck had briefly visited the Lesser Antilles, the United States had finally declared war on Germany. On his return voyage to the Virgin Islands, Beck had the unusual luxury of time in which he could consider this new state of affairs. Somewhat dispassionately, he wrote to his neighbor (and fellow bird enthusiast) back home in California, about "some of the aspirations driving me forward to possible death at the hand of bepistoled Dominican revolutionists," although he did not mention intelligence gathering, or his arrangement with the USMC. He also voiced concern about circulating news stories suggesting that thousands of Japanese soldiers were said to be ready to invade California, as well as the possibility of German submarines off the coast of Venezuela and Mexico.[114]

At the AMNH, as if oblivious to the possible consequences of the Declaration of War, Sanford, Osborn and Chapman disagreed over the direction of Beck's future fieldwork for the Museum. Sanford emphatically declared that he wanted Beck to sail to the South Pacific, but the Museum was unsure that the large vessel his benefactor had demanded for this venture was actually warranted. Moreover, not everyone was completely satisfied with Beck's Latin American collections: according to Osborn, Director Lucas "regrets that so little anatomical material was obtained during the recent cruises of Mr. Beck." On the other hand, President Osborn reported that "Doctor Chapman considers the discovery of the white grosbeak in Antigua as one of the great ornithological discoveries of recent times."[115] Strangely, there is nothing in Beck's notes to indicate that he actually shot birds in Antigua. Rather, according to his lists of birds, Beck encountered the grosbeak during his time in Sánchez, in the Dominican Republic. Perhaps Sanford was annoyed by this misunderstanding, or Chapman's opinion that "his [Sanford's] keen desire to add to his collections rather inclines him…to keep Beck continuously at work in the field."[116] At any rate, despite Chapman[117] and Osborn[118] pleading with him to have Beck return to New York, Sanford steadfastly clung to his decision to have Rollo proceed to Haiti, although he did allow his collector to enjoy a brief respite with his wife in Santo Domingo.

As long as Sanford could not be persuaded to end Beck's expedition, Chapman sent instructions to the collector to plan at least one more trip into the "high-interior." The Museum curator was still under the impression that the coveted Aestrelata might yet be found in the La Hotte Mountain range of southwestern Haiti or perhaps on the Island of Navassa, which Chapman knew "to be a bird island of some importance."[119] Soon after receiving these instructions, Beck set off on his new task. He booked a steamer which would take him from Santo Domingo to his new destination in Aux Cayes, a port in Southwestern Haiti. We can be sure that he procured permission from General Pendleton to continue his explorations in Haiti, because as soon as he landed at the port he proceeded to Marine headquarters, where he encountered Col. Kincaid, USMC.[120] The commander of a local unit of gendarmes, Kincaid was also the senior officer in charge of the Marine unit which had occupied Haiti since 1915, and whose mission was to determine the full extent of German influence on the local economy.[121] With Haiti not yet having formally declared war on Germany, and German commercial interest in the nation, there was some legitimate concern that the country might harbor German partisans.[122] Having Haiti firmly in the U.S. camp was a linchpin of Washington's war strategy, if for no other reason than that the southern portion of the country was situated on the Caribbean Sea route to the economically vital Panama Canal. If Beck was willing to take his chances against the insurgents in the interior, so much the better for the Marine officers who stood to benefit from any post-action intelligence he might provide.

After checking in with Kincaid, Beck started immediately for the Massif de la Hotte where he encountered Captain Ackroyd, yet another Marine, who commanded a unit of local gendarmes. Interestingly, the collector was given permission to stay with this unit for two days with access to the troops, allowing him ample time to photograph the men in formation and record their drills.[123] From there, accompanied by a porter to carry his equipment, Beck wandered into the highlands. The difficult terrain meant hacking trails by hand to clear the foliage ahead of them; an exhausted Beck often spent the night sharing quarters in the house of a local family.[124] According to his notes, he was able to find a woman to cook and wash for him, and his accommodations cost him 40 cents a day.[125]

For the better part of a month Beck wandered through the Haitian mountain range in search of the elusive Aestrelata. As unsuccessful as he had been in his earlier attempts to find it, he was at least rewarded by finding a new bird, "the yellow spot,"[126] as well as several (white throated) collared swifts.[127] Beck subsequently returned to the coast, where he boarded a schooner and left Haiti bound for Navassa Island, off Hispañola. On July 13, 1917, Beck reached the destination which Chapman had expected would yield many interesting birds, because of its history with guano production.[128] As luck would have it, this turned out to be not at all the case. By the following day, Beck had already surveyed a good part of the barren island,[129] finding little besides the common booby, noddy and sooty terns, pigeons and ground doves.[130] Determining that Navassa wasn't worth more of his time, he was lucky to be able to swiftly leave the island for Cuba on yet another schooner, a courtesy extended to him by a contractor who was in the process of building a lighthouse on the defunct guano island. Beck finally arrived in Santiago de Cuba in the eastern part of the island toward the end of the month, having first spent some time in Guantánamo, an area ceded to the United States by the Platt Amendment in 1903.[131]

In the meantime, Beck was growing increasingly frustrated with his mentors for not providing him with a clear picture of their plans. He cabled Chapman from Cuba, asking "Hotte, Navassa collected, where next?" and followed this cable with a letter which suggested that while rebels were rumored to be active in the mountains near Santiago de Cuba, there were indeed U.S. Marines in the interior to provide him protection; everything seemed calm enough.[132] Beck waited impatiently for almost two weeks for guidance. A letter to his mother reveals that he finally received instructions on July 26, and these new orders called for him to climb Pico Turquino, the highest point in Cuba. He also informed her that he did not expect to get to Havana (on the western side of the island) although, curiously, his final photographic portfolio contained more than a few images of the capital city, including his usual views of harbors, transportation means, and other revealing aspects of the infrastructure.[133]

From August 6th to the 24th, 1917, Beck dutifully explored the high peaks of the Sierra Maestra, finding little that interested him in the way of unusual birds, but managing to snap an image of what he ironically referred to as "a lonely wireless station" (Fig. 10.12).[134] Did Beck suspect that this was a German transmitter? The answer is unclear from his notes but his prior interest in wireless stations was noted

Fig. 10.12 "Lonely wireless station," Cuban Sierra Maestra mountains, August, 1917; AMNH Asset ID 236614

earlier concerning the Falklands (and of course Spinden, and others, were perpetually on the lookout for innocent looking stations that disguised communication links to either Mexico or Germany). The constant rain in the mountain range, and the difficult terrain notwithstanding, while he explored Pico Turquino, Beck managed to shoot and capture a good number of by now familiar Caribbean highland birds.[135] Toward the end of the month he returned to Santiago de Cuba where he found Ida (who had come over from Santo Domingo) and a letter from Chapman directing him to return at once to New York.[136] The couple sailed from Santiago on August 28th and arrived in Key West on September 1.[137] If, as is likely, they visited Havana on this last leg of their voyage, that meant that they traveled westward to Havana before returning to Santiago to sail to the United States. Given the type of photographs which Beck took in the capital city, he clearly was documenting some type of information on factories, transportation, etc. For whom these images were meant remains an unanswered question.

By the time he returned to the AMNH, Beck and Ida had spent almost five years in search of pelagic birds, initially in the exclusive employ of Dr. Sanford and Mr. Brewster but ultimately responsible to the American Museum of Natural History and President Osborn. If nothing else, Beck had proven himself to be a consummate collector, capable of switching his focus from South American oceanic specimens to highland Caribbean birds apparently without any difficulty, and thus assuring the Museum that they had a more or less complete representation. As noted earlier, however, both director Lucas and, later, Robert Cushman Murphy had reservations about Beck's work and seemingly were not impressed with the results.

During his lengthy expedition, Beck was frequently in political hot spots, including countries which had declared themselves neutral in the European conflict, and it was clear that he understood that neutrality did not necessarily imply impartiality. Hyper aware of the potential dangers in which he found himself, Beck was well-positioned to report on and photograph any questionable activities he witnessed. The images he memorialized during his expedition testified to his sustained interests in military operations and fortifications, whether they were in Chile, Brazil, Cuba or at U.S. installations. For a man whose only supposed task was to make collections, and capture images, of oceanic (and later high-elevation) birds, his pictures of political meetings, wireless stations, harbor entrances and the like are suspiciously extensive, representing, in total, as much a part of the complete oeuvre of his trip as the agreed-upon biological images he had been tasked with taking. It is even more intriguing to note that Beck did not immediately turn over the negatives of his trips (including those of his future Pacific expedition which the AMNH co-sponsored in the 1920s) to the Museum archives but waited for years (to 1929) until he did so.[138] For whom then, besides Sanford, had these images, many of a seemingly classified nature, been intended? The only hint we have is Sanford's early warning to Beck to keep mum about the "Washington plans."

Seen in its broadest perspective, Beck's work did little to advance the role of the AMNH in Latin America with regard to "scientific Pan Americanism;" there is nothing to indicate that Beck made contact with any local collectors. In essence, nothing in his work helped further the objective of scientific diplomacy, so dear to the agenda of President Osborn, nor did it in any way facilitate future AMNH research in the region. As an independent contractor, Beck was the only representative of the AMNH operating in Latin America during the War to whom Osborn theoretically could neither dictate orders nor control his progress (although that hadn't stopped him from trying).

Nonetheless, if, as it appears likely, Beck passed on vital information to U.S. government officials and U.S. Marine Corps officers in the occupied zones of the Greater Antilles or to relevant personnel in the Southern Cone of South America, then, like Chapman, Spinden, Eastman and the other scientists who offered their services to the United States government in wartime, we must also include the peripatetic birdman in the rarified coterie of Museum-associated men who uniquely impacted the War effort.

Notes

1. Cf. Chap. 6, p. 104.

2. *Ibid.*, pp. 103–104.
3. See Chap. 3.
4. *Ibid.*
5. Cf. Chap. 6. As noted earlier, a personal communication from the archivist at NARA confirms that Brown worked briefly for the ONI. The fact that Brown was later recruited by the OSS during World War II is mentioned in Dingus and Norell, Barnum Brown…, pp. 285–287.
6. Cf. entry between April 8 and April 21, 1917 in Brown's "Memoirs," p. 4, Department of Vertebrate Paleontology Archives [hereafter DVPA], Brown file.
7. This occurred around mid-March, 1917 according to Brown's "Memoirs," p. 5, DVPA, Brown file.
8. *Ibid.*, p. 3.
9. Barnum Brown, "The Occurrence of Petroleum in Cuba," New York City, April 16, 1920.
10. Brown to W.D. Matthew, London, May 19, 1920, p. 1. DVPA, Brown file.
11. As noted in file page on Barnum Brown, compiled by Robin Smith, CA vertical files, AMNH Library.
12. Brown to W.D. Matthew, London, May 19, 1920, p. 1. DVPA, Brown file.
13. *Ibid.*, p. 2.
14. Cf. Delson, "The Bullet and…", p. 14.
15. NARA RG 38, E123. Roy Chapman Andrews file.
16. Frank M. Chapman to Chief of Military Intelligence. Washington, June 5, 1918. NARA RG 38, E123. Roy Chapman Andrews file.
17. See Chap. 6, note 54.
18. Cable via Signal Corps, United States Army, August 31, 1911. Nome, Alaska, Sanford to R. H. Beck. From files of Rollo H. Beck, California Academy of Sciences [hereafter, Beck CAS].
19. Letter of Dr. Sanford to Beck, New Haven, Connecticut, April 2, 1911. Beck, CAS.
20. *Ibid.*
21. Letter of Frank M. Chapman to Beck, New York, November 1, 1912. DOA, Beck file.
22. *Ibid.*, p. 1.
23. *Ibid.*, p. 5.
24. Brewster to Beck, New Haven, December 5, 1912. CAS, Beck file.
25. Sanford to Brown and Blanchard, Punta Arenas, Chile. March 29, 1916, CA, AMNH, file 551.
26. Beck field notes, February, 1913, CAS.
27. Letter of Frank M. Chapman to Beck, New York, November 1, 1912. DOA, Beck file.
28. Letter of Brewster to Beck, New Haven, February 26, 1913, CAS. Numerous exchanges between the administration at the AMNH and the State Department written to secure permits for Beck in Chile and Argentina between January and August of 1913 are housed at NARA, State Department Records, Box 291, folder 1, 031.11 Am 3/17. On at least two separate occasions, Beck wrote to his mother to complain that he had not yet received Peruvian permission to hunt birds: Beck to "Ma," Lima, February 19, 1913 and Beck to "Ma," Lima, April 23, 1913. (Beck CAS), Box 5/Folder 7, AR 8/C/4.
29. Distinguishing between display collections and study collections was part of the "New Museum Idea," a concept which took hold in the late nineteenth century and which very much influenced the AMNH. See Rieppel, Assembling the Dinosaur…, p. 62. It should be noted that although Sanford (an amateur ornithologist) knew Chapman personally, neither he nor Brewster were officially associated with the AMNH. Brewster became a trustee in 1916, and Sanford achieved that status in 1921.
30. Chapman to Sanford, Washington, D.C., September 20, 1917, CA, file 551. Interestingly, a series of articles supposedly written by Beck appeared in Natural History, Vol. XVIII, January, 1919, on his adventures in Cape Horn. Loosely based on his notes, these articles and photographic essays were probably heavily edited by the staff before publication. Cf. "Narrative of a Bird Quest in the Vicinity of Cape Horn," (pp. 5–16) and "Scenes from the Vicinity of Cape Horn," (pp. 17–24). In the February issue, Beck published the continuation of "Narrative of a Bird Quest in the Vicinity of Cape Horn," (pp. 111–120).

31. See John P. Dumbacher and Barbara West, "Collecting Galápagos and the Pacific: How Rollo Howard Beck Shaped Our Understanding of Evolution," Proceedings of the California Academy of Sciences, Series 4, Vol. 61, Supplement II, No. 13, 2010, p. 212.
32. Sanford to Osborn, New Haven, December 14, 1917. AMNH, CA, file 551.
33. Robert Cushman Murphy, Oceanic birds of South America: a study of the related coasts and seas, including the American quadrant of Antarctica, based on the Brewster-Sanford collection in the American Museum of Natural History. 2 vols. (New York, MacMillan Co., 1936).
34. Murphy to Osborn, Brooklyn, February 6, 1917. AMNH, CA, file 551.
35. See discussion in Chap. 4. The *Baltimore* Crisis of 1891 had resulted in the hardening of relationships between Chile and the United States.
36. See William F. Sater and Holger H. Herwig, The Grand Illusion: The Prussianization of the Chilean Army. (Lincoln: University of Nebraska Press, 2000).
37. Bisher, The Intelligence War…, p. 240.
38. *Ibid*., pp. 28–30.
39. *Ibid*., pp. 13–14.
40. *Ibid*., p. 16.
41. Beck to "Ma", Robinson Crusoe Island, Juan Fernandez, Chile, December 14, 1913. Box 5/folder 7, AR 8/C/4, CAS.
42. See Chap. 3. The images from the CAS are labeled "Roosevelt 1" and "Roosevelt 2," while slightly different images of Roosevelt aboard a small boat in the harbor of Valparaíso are Asset IDs 237788, 237789 and 237790 in the AMNH photo collection. Both sets of images were taken in November of 1913.
43. Bisher, The Intelligence War…, p. 19.
44. Beck's photos # 237545, 237546, 237547, 237548 of soldiers drilling in Punta Arenas. AMNH, photo collection.
45. Ida to Mother, Punta Arenas, Chile, August 31, 1915. CAS, Beck, Box 9, folder 1.
46. Bisher, The Intelligence War…, p. 17.
47. Beck to Ruth, North of Falkland Islands, Nov. 10, 1914. CAS, Beck, Box 5/Folder 7, AR8/C/4.
48. Bisher, The Intelligence War…, pp. 19–20.
49. *Ibid*., p. 37.
50. Ida to "Ina y familia," Punta Arenas, Chile, July 20, 1915. CAS, Beck, Box 9/Folder 1.
51. Ida to Mother, Punta Arenas, Chile, August 31, 1915. CAS, Beck, Box 9/Folder 1.
52. Ida to Ina, Port Stanley, Falkland Islands, October 28, 1915. CAS, Beck, Box9/Folder 1.
53. Ida to Mother, Port Stanley, Falkland Islands, December 19, 1915.
54. Ida to Mother, Port Stanley, Falkland Islands, January 29, 1916. CAS, Beck, Box 9/Folder 1.
55. *Ibid*.
56. Beck to "Ma," on board *Orissa*, 150 miles south of Montevideo, February 5, 1916. CAS, Beck, Box 5/Folder 7, AR 8/C/4.
57. Beck to "Ma," on board the SS *Vestris* nearing Rio de Janeiro, February 20, 1916. CAS, Beck, Box 5/Folder 7, AR 8/C/4.
58. Brewster to Assistant Secretary George Sherwood, AMNH, New Haven, May 10, 1916, and Sanford to Osborne [sic], New Haven, May 15, 1916. Both in CA, AMNH, File 551.
59. Osborn to the Brazilian Ambassador in Washington, D.C., New York, May 9, 1916. CA. AMNH, File 551.
60. Ambassador Da Gama to Osborn, Washington, D.C., May 18, 1916. CA, AMNH, File 551.
61. Robert Frazer, Jr., American Consul, Bahia, Brazil, May 26, 1916. CA, AMNH, File 551.
62. Bisher, The Intelligence War…, pp. 13–15.
63. Ida to Mother, Bahia, Brazil, May 8, 1916. CAS, Beck, Box 9/Folder 1.
64. Beck Journal [hereafter BJ] entry, June 27, 1916, p. 296. DOA, AMNH. Ida had also mentioned the possibility in her letter to her mother, May 8, 1916, *ibid*., as did Beck in a letter to his mother, February 20, 1916, *ibid*.

65. Beck photographic collection, AMNH special collections [hereafter BPC/AMNH]: Asset IDs 237265–237270, May, 1916.
66. BJ, August 10, 1916, p. 302. DOA, AMNH. Both the CAS and the AMNH hold copies of Beck's journal of the Brewster-Sanford expedition.
67. Luther K. Zabriskie, The Virgin Islands of the United States of America: Historical and Descriptive Commercial and Industrial Fact, Figures and Resources. (New York: G.P. Putnam's Sons, 1918), p. 252.
68. BPC/AMNH: Asset IDs 236992–236994, August, 1916.
69. BPC/AMNH: Asset IDs 101889–101890, August, 1916.
70. BPC/AMNH: images #'s 236995–236997, April, 1917.
71. In Zabriskie, The Virgin Islands…, Beck contributed images of hotels, "rooms to rent" and an old school house.
72. BJ entries August 14, 21 and 22, 1916. DOA, AMNH.
73. BJ entry September 26, 1916. DOA, AMNH.
74. Carl M.J. von Zielinski, American Vice Consul in charge, Santo Domingo to R.H. Beck, Santo Domingo, September 23, 1916. CAS, Box 9/Folder 1.
75. Entry of September 26, 1916. BJ, DOA, AMNH.
76. BJ, entries of September 27 through October 19, 1916.
77. Beck to Chapman, Santo Domingo, October 26, 1916. DOA, AMNH. In this letter Beck explains why he has left the capital.
78. Entry of October 27, 1916. BJ, DOA, AMNH.
79. BPC/AMNH: Asset IDs 236962–236967, November, 1916.
80. BPC/AMNH: Asset ID 236970.
81. BPC/AMNH: Asset IDs 236935, 236923. A paper by Mark R. Folse, "The Impact of the Great War on Marines in Hispañola, 1917–1919," presented at the 2014 Temple University Jame A. Barnes Graduate Student History Conference, comments on racism in the Marine Corps through the Haitian occupation.
82. Letter of Ida to Mother, San Juan, Puerto Rico, December 5, 1916. CAS, Beck, Box 5/Folder 1916.
83. For a review of the Marine Corps activity in the Dominican Republic, see Captain Robert J. Kane, USMC, A Brief History of the 2d Marines, revised. Washington, D.C., Historical Division Headquarters, U.S. Marine Corps, 1961, pp. 7–11.
84. BJ, DOA, AMNH, entry December 4, 1916, p. 335. Major Marix granted a shooting permit on November 27, the day Beck entered La Vega, p. 331.
85. BPC/AMNH: Asset IDs 236901–236903. La Vega, December, 1916. Images of "guard watching roadway," "U.S. Marine Barracks," and "U.S. Marines as Barrack doors."
86. Chapman to Beck, New York, December 16, 1916. DOA, AMNH. For Chapman's return to New York, see Chap. 9.
87. There is a document to this effect in CA, File 551. Dated December 17, 1916 it is entitled Sanford-Brewster Expedition, Conference with Dr. Leonard Sanford.
88. Ida to Mother, Santo Domingo, December 20, 1916. CAS, Beck, Box 5/Folder 1916.
89. Beck to Sanford, Santo Domingo, September 20, 1916. DOA, AMNH.
90. Ida to Mother, Santo Domingo, December 20, 1916. CAS, Beck, Box 5/Folder 1916.
91. Translation - re: Beck "who is on a special mission, with the purpose of collecting examples of birds for the collection of the above cited Museum [AMNH]." General Pendleton, Brigadier General U.S. Marines for the Military Government to the governors of the provinces of Barahona, Las Matas [Azua], San Juan, Bánica, Ecomcmdador, Espaillat, Santiago, La Vega, Sao Pedro de Macorís, Monte Cristo. CAS, Beck, Box 5/Folder 1916. General Pendleton seemed to think that the Dominican Republic was quieting down. See Martin L. Gordon, Joseph Henry Pendleton, 1860–1942: Register of His Personal Papers, History and Museums Division, Headquarters, U.S. Marine Corps, Washington, D.C., 1975 who notes in Pendleton's letter to Major General, Commandant George Barnett, HQ, December 18, 1916 (folder 17) that the "island was basically quiet" p. 69.

92. Beck briefly remarked on the December 1st takeover in his journal. BJ, La Vega, December 4, 1916, p. 335.
93. Ellen Davies Tillman, "Imperialism Revised: Military, Society, and U.S. Occupation in the Dominican Republic, 1880–1924," Dissertation in the Graduate College of the University of Illinois at Urbana-Champaign, 2010, p. 160.
94. *Ibid.*, p. 161. See also See Michiel Baud, "German Trade in the Caribbean: The Case of Dominican Tobacco, 1844–1940," Jahrbuch fur Geschichte von Staat, Wirtschaft und Gesellschaft, Vol. 25 (1988).
95. Tillman, "Imperialism Revised...", p. 162.
96. BPC/AMNH: # 101888 - "American Marines at Christmas dinner." Santo Domingo, December 25, 1916.
97. BPC/AMNH: Asset IDs 236869, 236870. Santo Domingo, December 25, 1916.
98. BPC/AMNH: Asset ID 236872, "Guide and pack horse in mountains." January, 1917.
99. The 15th company of the 2d regiment mounted an expedition to Azua, Dominican Republic in September, 1916 to quell a rebellion. Cf. Kane, A Brief History ..., p. 11. By January 11, an arms collecting trip in the province had quieted local conditions. See Gordon, Joseph Henry..., letter of William Fuerlein, Operator, HQ, 2nd Provisional Brigade, USMC, Santo Domingo City to Capt. F.A. Ramsey, Chief Intelligence Officer, 11 January, 1917 (folder 18), p. 71.
100. BJ, DOA, AMNH, entry January 3, 1917, p. 343.
101. BJ, DOA, AMNH, entry January 11, 12, 1917, p. 346.
102. BJ, DOA, AMNH, entry February 24, p. 365.
103. BJ, DOA, AMNH, entry February 26, p. 365.
104. BJ, DOA, AMNH, entry March 3, p. 366.
105. BJ, DOA, AMNH, entry March 14, p. 369.
106. BJ, DOA, AMNH, entry March 24, p. 370.
107. BPC/AMNH: Asset IDs 238995–238997, "First governor (American) arriving." April 1917 [incorrectly identified as April'16 in the Museum filing papers].
108. BJ, DOA, AMNH, entry April 15, 1917.
109. Sanford to Beck, New Haven, January 21, 1917, CAS, Beck, Box 5/folder 1916.
110. Sanford appears to be referring to the account in Frederick du Cane Godman, Monograph of the Petrels (order Tubinares) (London, Witherby and Company, 1907–1909).
111. Sanford to Beck, New Haven, January 21, 1917, CAS, Beck, Box 5/folder 1916.
112. BJ, DOA, AMNH, entry April 26, 1917, p. 374.
113. BJ, DOA, AMNH, entry May 5, 1917, p. 376.
114. Beck to Holmes, "Caribbean Sea," April 14, 1917, CAS, Beck, Box 5/folder 1916. Beck was already concerned about the Japanese while he traveled through the Falklands, cf. note 45.
115. Osborn to Sanford, New York, April 12, 1917, CA, AMNH, File 551.
116. Chapman to Osborn, New York, April 13, 1917, CA, AMNH, File 551.
117. *Ibid.*
118. Osborn to Sanford, New York, April 20, 1917, CA, AMNH, File 551.
119. Chapman to Beck, New York, May 21, 1917, DOA, AMNH.
120. BJ, DOA, AMNH, entry June 11, 1917, p. 380.
121. See Brenda Gayle Plummer, "The Metropolitan Connection: Foreign and Semiforeign Elite in Haiti, 1900–1915," Latin American Research Review, Vol. 19, No. 2 (1984), p. 122. See also Mary A. Renda, Taking Haiti: Military Occupation and the Culture of U.S. Imperialism, (Chapel Hill, University of North Carolina Press, 2001), p. 30.
122. They did so on July 12, 1918. Cf. Martin, Latin America and ..., p. 518.
123. BPC/AMNH: Asset IDs 236698 and 236699, "Captain Ackroy [sic] drilling rurales," June, 1917.
124. BJ, DOA, AMNH, entry June 11, 1917, p. 380.
125. BPC/AMNH: images #236696, "the mountain cook and dishwasher," June, 1917.
126. BJ, DOA, AMNH, entry June 22, 1917, p. 382.

127. BJ, DOA, AMNH, entry June 28, 1917, p. 384. Beck also informed Sanford of his find on July 9, 1917. DOA, AMNH.
128. See note 119.
129. BJ, DOA, AMNH, entry June 14, 1917, p. 386.
130. BJ, DOA, AMNH, entry June 19, 1917, p. 386.
131. BJ, DOA, AMNH, entry June 30, 1917, p. 386.
132. Beck to Chapman, Santiago de Cuba, July 28, 1817. DOA, AMNH.
133. Beck to "Ma," Santiago de Cuba, July 27, 1917 CAS, Beck, Box 5/1916. The images of Havana, clearly labeled, are in BPC/AMNH: images 236615–236620, August, 1917.
134. BPC/AMNH: Asset ID 236614, 1917 (no month indicated but probably August).
135. BJ, DOA, AMNH, entry August 12, 1917, p. 387–388.
136. Chapman to Beck, August 8, 1917 DOA, AMNH.
137. BJ, DOA, AMNH, entry August 27, 1917, p. 394.
138. Letter of Murphy to Sherwood, New York, October 5, 1929. CA, folder 1130.

Part III
The Aftermath of War

Chapter 11
The Post-war Years: Whither Latin America?

As everyone used to be required to learn in grade school, World War I was formally declared over on the eleventh hour, of the eleventh day, of the eleventh month of 1918. But the date of November 11, 1918, watershed that it is, only signaled the beginning of a world-wide process of analyzing and identifying the changes wrought by years of conflict. Coming to terms with the end of a conflagration of such magnitude would take months, if not years, to complete and, as Chapman so cogently observed while he served as a Red Cross Commissioner, hostilities did not necessarily end with the Armistice. One thing was blatantly clear to everyone who had been involved in the War: the world which had emerged from the struggle differed substantially from its pre-bellum status. With the clarity gained by the distance of time, it became increasingly obvious that the war had been experienced at levels ranging from micro to macro; moreover, the alterations which had occurred as a result of years of warfare affected everything from local institutions all the way to the governments of those nations involved in prosecuting the conflict. Seemingly, nothing was immune from change.

The same observation may be made of the post-War years at the American Museum of Natural History, where the conflict's impact on the institution was profound, widespread and felt across the spectrum of the Museum's day to day activities. By 1919, the AMNH had so significantly diverged from the way it had operated before the War began that President Osborn felt forced to lament in his mid-year report to the Trustees that "the Museum as a whole is now going backward, not forward."[1]

Objectively considered, it is not at all clear if Osborn's assessment was a realistic portrait of the institution, nor whether his pessimism about the future was shared with the broader Museum community. But with regard to the questions raised in this study, including the value of having staff from the American Museum of Natural History serve the interest of politics and politicians in Washington during the War, as well as determining what, if any, tangible benefits had occurred *ex post facto* to the institution from having participated in such secret transactional agreements,

R. M. Delson, *Scientists or Spies?*, The Latin American Studies Book Series,
https://doi.org/10.1007/978-3-032-18117-6_11

Osborn's post-War analysis is disturbing. It may have truly seemed to the Museum president that his dream of advancing his institution's status through collaborative wartime activities (both public and hidden) had somehow gone off the rails.

To determine whether Osborn was correct in his assessment, we can single out four separate arenas in which the Museum experienced profound change between the end of the War and 1922. The first and arguably the most important of these modifications involved the reorientation of the Museum's research program. Where once scientists from virtually every department of the AMNH flocked to Latin America to conduct research (after first jockeying for the express approval of the Museum's administration), that trend was halted at the War's conclusion. In its stead a new geographic emphasis was substituted: tracking the geology and collecting the fauna of Asia. As a result, the carefully planned expeditions to Latin America and the Caribbean, which had dominated the AMNH field agenda for more than three decades preceding the United States Declaration of War in 1917, were now replaced by ventures which sought to explore Asia.

A second departure from the pre-War posture of the AMNH after 1918 saw the ending of the Museum's brief, but consequential, flirtation with the Progressive Movement's agenda. In place of its once enthusiastic embrace of forward-looking exhibitions (which had provided the Museum with a public forum in which to advocate for health concerns such as control of communicable diseases and nutrition), discourses on these subjects (let alone exhibits) were now considered too provocative and, frankly, far too much of a detraction from the avowed "scientific mission" of the institution to be pursued in the future.

A third aspect of change at the post-War AMNH occurred in the mindset of those few members of the Museum hierarchy who had been privy to President Osborn's intentional blurring of the lines between science and politics. While the decision to allow the Museum's staff scientists to engage in secret intelligence gathering and espionage in Latin America had ultimately been his, a select inner circle of colleagues, including Director Lucas, Secretary Sherwood and Curator Dean, had not only sanctioned Osborn's plans, but had enabled him to proceed with the subterfuge. As the War drew to a conclusion, their after-action analysis of this practice turned overwhelmingly negative. First, it had become obvious that such arrangements had not been advantageous in regard to the Museum acquiring new specimens for its collections. Second, the administration's sanction of spying arrangements never provided an adequate rationale for the potential dangers faced by Museum scientists as they went about their government missions. Further, whatever self-justifications they consoled themselves with concerning their personal blessing of such secret schemes, these Museum administrators could not obliterate the ignominy of having knowingly, and perhaps even recklessly, eschewed the Museum's self-imposed and self-protective moratorium on overseas research.

A final dimension in which the Museum's post-War status suffered substantial alteration is represented by the faltering allegiance of its scientists toward the institution itself. Several of the staff members who had been involved in spying assignments, including some of the Museum's most distinguished scientists, had chafed under the Museum Administration's constant meddling; they had found it impossible to both satisfy President Osborn and his cohort and still proceed with their overseas intelligence work. As a consequence, several of the scientists, most of whom had worked in Latin America before the War, found themselves uncomfortable in their former positions upon returning to the institution. Still obliged by the Federal government to keep their wartime extracurricular experiences secret from their home-bound colleagues (who inevitably asked questions), such notable scientists as Crampton, Spinden and Winslow (Eastman, of course, was dead, and Beck was not a Museum staffer) charted their exodus from the institution just a few years after 1918.

Each of these changes to the AMNH deserves to be looked at in depth for what they reveal about the post-Armistice milieu at the Museum, and what they signal about the men who had willingly violated the Museum's own moratorium by working abroad during wartime.

The Push Toward the East

Arguably, the most fundamental change affecting the post-War scientific community of the AMNH was the decision to redirect their efforts from Latin America to Asia, a shift that likely reflected Osborn's personal disappointment in not having garnered sufficient evidence to advance his "grand Latin American evolutionary project." Although he had successfully managed to embed AMNH scientist/agents in the region, moratorium notwithstanding, his enthusiasm for research in the region was on the wane by the War's end. Where once Osborn had enthusiastically championed a young Barnum Brown and encouraged him to prospect for large mammal specimens in Patagonia while he simultaneously paid homage to the evolutionary theory of the Ameghino brothers that South America was the point of origin of large mammalian expansion throughout the hemisphere, his attention was now focused elsewhere. Indeed, Osborn's former eagerness for ferreting out the best biological and archaeological specimens of Latin America fizzled, almost in direct proportion to his new-found enthusiasm for exploration and collection on the Asian continent.

What had occasioned Osborn's about-face on pursuing evolutionary research in Latin America? The answer to that question lies in the Museum President's slow (but perhaps inevitable) acceptance of alternative hypotheses which challenged and ultimately undermined Hooker's theory, which had proposed an Antipodean source for mammalian evolution. Hooker's ideas (which had initially whetted Osborn's interest in South America exploration) gave way to other theories, most notable of which was the one advanced by Osborn's former student and now Museum colleague, William Diller Matthew. As discussed earlier, Matthew postulated, *contra* Hooker, that global migrations of large mammals had originated from the northern hemisphere, i.e., the Holarctic. Matthew further proposed that such migrations had moved in a North–South direction toward the southern hemisphere, rather than the reverse, opposing the premise which Hooker (and Osborn) had initially accepted. When the scholarly community acknowledged the plausibility of Matthew's hypothesis, an event which occurred in conjunction with the publication of his seminal paper, "Climate and Evolution",[2] in 1914–1915, it became likely that Latin America would lose its primacy as the locus for natural history research. By the time World War I had concluded, Osborn and his AMNH scientific staff were already planning a totally new geographic focus of research, one which would now concentrate on East Asia. They anticipated that exploration of this region would not only confirm mammalian migration in a southward direction, but would provide concrete evidence of the direction of mammalian evolution, in the form of large fossils.

It is worth noting that up to this point, the Museum's attention to Asia had been spotty. It was not until 1916–17 that the Museum sponsored its First Asiatic Zoological Expedition, during which staff scientist Roy Chapman Andrews had looked for specimens in two provinces of China (Yunnan and Fukien).[3] In addition to his new belief that collecting in Asia would lead to the discovery of the large mammalian fossils which had eluded him in Latin America, Osborn now also speculated that evidence of human origins would be uncovered in the region (which came to be known as the "Out of Asia" theory).[4] His enthusiasm for Far Eastern research continued to mount in the immediate post-War period when expeditions to great swaths of China (including Inner and Outer Mongolia) were put into motion by 1922. What had once seemed the immutable importance of Latin America had thus effectively been supplanted.

When Osborn looked to sponsor expeditions to the Far East, he was able to rely on the expertise of several of the same Museum scientists who had worked in intelligence gathering capacities during the War. Foremost was Roy Chapman Andrews, who arrived in Shanghai, China in 1918 courtesy of the ONI,[5] and was later transferred by that agency to Mongolia in 1919. Luckily for the AMNH and Osborn, Andrews was able to take advantage of his posting to sketch out plans for the Museum's future Central Asian expeditions to Mongolia, including a future lengthy junket in which William Diller Matthew would participate.

Other Museum research also looked to the Far East and Pacific regions. Although he was associated with the AMNH only in an emeritus capacity after 1920, Henry Crampton's research and explorations in the Mariana Islands still reflected Museum interests. Crampton's wanderings in search of snails and other invertebrates brought him back to the Pacific, where he had done his initial zoological research, post-1920. Like Andrews, while he was in the Pacific, Crampton was still working in his dual capacity as an ONI agent: his work of collecting snails proceeded in conjunction with his collecting of intelligence for the United States Government. With a portion of the Mariana Islands having come under the control of the Japanese at the conclusion of World War I, Crampton was able to use his cover as field zoologist to keep an eye on Japanese incursions in the area for the United States government from his expedition base in Saipan.

Lastly, while Rollo Beck remained a "contract" employee for the Museum, the AMNH sponsored yet another "official" mission for him in the 1920s, but this time to the Pacific. As seen earlier, the plans for this future expedition were already in motion in 1917 while Beck was still on assignment in the Caribbean. Once again he was funded primarily by private sources, as his previous explorations in Latin America had been, and as before, Rollo was accompanied by his wife Ida. Together the couple collected extensively as the principal investigators of the "Sanford-Whitney Pacific Expedition." Already wary of Japanese activities in Latin America for what he considered warlike tendencies (as revealed by several references in his diaries and personal letters), in the 1920s, while on this new expedition, Beck was conveniently positioned to report on their movement in the Pacific. It is more than likely that he passed on his observations to Washington.

The post-War decade at the Museum thus began with an emphasis on Asian exploration rather than Latin American. However, ultimately, the institution failed to uncover the evidence necessary to support Osborn's theories on the emergence of early humans in the region and, despite the underwriting of some "fantastically expensive expeditions to Mongolia,"[6] the Museum's best efforts in the region toward explicating the history of early humans were essentially disappointing. Characteristically, even as most researchers were considering that human origins could be found in Europe, or possibly Africa, rather than in Asia, Osborn stubbornly clung to his new agenda, emphasizing the importance of Asian fieldwork in the 1920s over other global regions, in much the same manner that he had previously championed exploration in Latin America.

Influenced by the new Asian focus of its research agenda, the AMNH consequently scheduled far fewer expeditions to Latin America in the 1920s than it had during the previous three decades. At the same time it seemingly dropped its interest in "scientific Pan Americanism". It would be a mistake, however, to suggest that the region was entirely forsaken by the Museum. In fact, one of the institution's more interesting expeditions involved two "old hands" of the Latin American scene, George Cherrie and Harold Anthony, who together traveled to Ecuador in the early 1920s to collect birds and mammals for the AMNH.

As noted earlier, unlike Cherrie, Anthony had not joined the ONI or other intelligence agency but had instead enlisted in the United States Army where he advised others on aviation. After his discharge from service and return to the AMNH, Anthony was once again ready to collect for the Museum; by late 1919, he had planned out a new venture to Ecuador with his Museum colleague, George Cherrie. At the same time, and apparently without much coaxing, Anthony simultaneously agreed to supply the Military Intelligence division of the U.S. War Department with any relevant intelligence he might uncover during this new expedition.

Anthony's decision to pass on data to the U.S. government had come about as a result of an apparently long-standing relationship between Colonel Sherman Miles of the War Department and Director Frederic Lucas of the Museum. Writing to compliment Lucas (and the Museum) for "your interest in our work," Colonel Miles wondered if it might be possible during Anthony's Ecuadorian trip (of which he had obviously been apprised) to persuade the scientist to stop at headquarters in Washington and share what he had observed about the region. In exchange, the

Colonel assured Lucas, Anthony would first be supplied with the War Department's "Strategic Estimate, which embraces the Political, Economic, Psychologic and Military situation up to date."[7] In what amounted to an informal, but once again indisputably transactional agreement, Anthony (apparently with the backing of Lucas) agreed to report on what he had observed in exchange for receiving the War Department's background information on local conditions in Ecuador. Moreover, his partner in this new expedition, George Cherrie, still a bona fide ONI agent, either volunteered, or was requested, to photograph Ecuadorian naval vessels. As a result Cherrie spent at least part of his time in country taking numerous images of Ecuadorian military ships in harbor; these, too, were undoubtedly passed on to Washington, and were probably taken at the U.S. government's request.[8] It is unclear whether the Museum knew of or sanctioned this aspect of Cherrie's trip, although Lucas certainly knew what Anthony had promised to the War Department.

Other than the time spent fulfilling government requests, during their expedition to Ecuador, Cherrie and Anthony collected specimens in the remote and sparsely populated region of Zamora, an area which previously had held little interest for the United States Government. However, beginning in the 1920s, and coincident with their expedition, Washington had unexpectedly become very interested in this backwater, largely because it was known to contain substantial gold deposits. With the U.S. mulling over a return to a modified gold standard, obtaining a reliable gold supply source from Latin America might prove critical to this effort.[9]

As had been previously agreed upon, when his work concluded, Anthony dutifully reported to Military Intelligence headquarters in Washington, D.C.[10] There is no available evidence to confirm exactly what he briefed the military on, but it is more than likely that his presentation contained his assessment of the availability of Zamoran gold.

While it is tempting to view this expedition as an extension of the kind of secretive activities which had taken place during the War, in fact it differed substantially from previously sponsored ONI (or other) intelligence gathering in Latin America. Despite the fact that Anthony and Cherrie relayed intelligence back to government agencies, it was the Museum which had approved of the Ecuadorian expedition in the first instance, rather than the U.S. government. Moreover, while the expedition certainly provided cover for their sleuthing, it was not the principal reason that Cherrie and Anthony were in the area. On the contrary, in this case, their work for the government was incidental to their primary goal of collecting.

Another South American expedition organized in the 1920s under the AMNH banner was led by Museum mammalogist George H.S. Tate. Tate's field party reached the top of Mount Roraima; Crampton and Lutz had earlier been to its base. In 1928, Tate's team went back to the region to collect specimens from the nearby locale of Mount Duida, an area populated with rare and unusual birds, plants and insects.[11] The collections made by his party in this locale greatly improved the quality of the Museum's holdings. Additionally, Frank Chapman continued his studies of South American birds, this time working primarily in Ecuador and Peru rather than Colombia, while Angelo Heilprin, another AMNH scientist and an active Latin Americanist, collected in the Dominican Republic. None of these expeditions, however, appear to have had any governmental connection.

In retrospect, notwithstanding the activity cited above, there were far fewer AMNH expeditions to Latin America overall in the 1920s than had been the case during the years preceding and during World War I. Broadly speaking, Osborn's former enthusiasm for uncovering the fossil evidence needed to illustrate the process of evolution in South America, and his quest for representative material of the complete flora and fauna of the Americas, had given way to his new found passion for potential evidence of mammalian and human evolution in Asia.

Retreat from Progressive Policies and Government Cooperation

Paralleling the shift in Museum research from Latin America to Asia, the institution simultaneously scaled back its advocacy of the progressive ideas which had been *au courant* in the years leading up to the War in favor of more "scientific" projects. In the early months of the U.S. entrance into the fight, it was the AMNH acceptance of the Progressive Movement's ideology which led to the institution's Public Health Department staging twin blockbuster exhibitions on nutrition and military hygiene in the Spring of 1917. Undertaken with Osborn's approval, it must have seemed to the administrator that if the Museum were to get involved in the important public matters of the day (including hygiene and food sustainability), such actions might constitute an argument in favor of the institution's continuing relevance during wartime. This same motivation presumably was also the impetus behind the decision to put the AMNH imprimatur on wartime volunteerism by providing printed guidelines to encourage citizens to engage in food conservation.[12]

But adoption of progressive themes had also meant purposely skewing away from the Museum's traditional biological, morphological, paleontological and geological mission. Inevitably, the Museum's focus on public issues exposed the institution to unsolicited scrutiny and, just as inescapably, to fierce public censure.[13] When the War concluded in November of 1918, the Administration belatedly recognized that the initiatives it had enthusiastically endorsed (allocating space at the institution to be used for a military recruitment center, assigning the engineering staff the task of creating plans for a moveable field hospital, and printing health pamphlets destined for city-wide distribution which recommended certain dietary choices over others) were all fraught with the likelihood of calling unwanted attention to the institution. As a result, in 1922, the administration deliberately reduced the Museum's footprint in the Progressive Movement by permanently closing the Department of Public Health. Its chair, Dr. Winslow, quickly left the Museum for a more promising career at Yale University.[14]

In hindsight, it is abundantly clear that the patriotic mien in which all of these non-traditional activities had been couched had simultaneously provided camouflage for the institution's, and Osborn's, self-serving ambition. What is unclear from this reflection, however, is how much mileage the Museum had actually garnered from these varied wartime efforts. The abandonment of the Progressive ideology at the Museum post-1918 would seem to suggest that President Osborn understood the need for caution when venturing beyond the traditional mission of the Museum. But if this were the case, how can we make sense of Osborn's fierce attachment to the highly controversial, and racist, Eugenics movement of the 1920s?[15] Was he naïve enough to imagine that his own attraction to Eugenics might be regarded as apolitical, or that it would not reflect on the institution that he served? Or had his embrace of this new movement simply been reckless?

The post-War period also witnessed the re-evaluation of the Administration's part (Osborn at first, Lucas and Sherwood later) in allowing Museum scientists to work in tandem with United States government agencies. On the one hand, involvement of scientific institutions with Washington was not new: the AMNH had begun interacting with politicians on the state and Federal level well before World War I began and prior to Washington's search for scientist/intelligence agents. As noted earlier, from the early 1910s onward, for example, President Osborn had deliberately solicited the help of the State Department in facilitating the research of both Beck and Chapman in South America.

Moreover, the dependence of the Museum on the United States government for facilitating research in Latin America had been characterized by transactional aims right from the outset, a phenomenon which had strengthened during the War in the face of recalcitrance toward foreign scientists from countries professing to be neutral in the War. To fulfill its part of the *quid pro quo,* the Museum accommodated the successive Secretaries of State by not only advocating for "scientific Pan Americanism," but by making this mantra a critical part of the institution's mission to Latin America. The hosting of Latin American delegates to the Second Pan American Scientific Congress at the Museum in early 1916 at Secretary Lansing's specific request, for example, should be interpreted as part and parcel of this mutual self-serving arrangement.

As the 1910s progressed, the State Department's aid would become a critical factor in expediting AMNH expeditions to Latin American nations which continued to harbor suspicions of the United States. As a result of this growing relationship between the AMNH and the State Department, Washington-appointed consular officials serving in politically sensitive Latin American nations might unexpectedly find themselves assigned the duty of negotiating permits for Museum scientists to carry guns and collect specimens in those countries. The obvious goal of the AMNH was not just to curry favor with Washington officialdom, but to gain greater access into Latin America than had been achieved by rival North American natural history institutions. Ironically, such interactions at the same time reinforced the centrality of the role which the AMNH had assumed for itself early on in promoting both Secretary's Root's and Leo Rowe's vision of an idealized "scientific Pan Americanism." That this posture might ultimately provide Museum scientists with cover to access extensive collections of Latin American specimens at the expense of other institutions was the unacknowledged, but clearly accepted, *sub rosa* reality of their actions. Of course, such bi-lateral arrangements between the State Department and the Museum were warmly welcomed by President Osborn, and it is likely that most of the AMNH scientists were aware of the broader parameters of these agreements by the time the War ended, even if they had not personally engaged in any interactions with Washington.

But all of these arrangements with the State Department were in the open, or at least there was no attempt to hide (as in the case of Rollo Beck) the Museum's growing dependence on government intervention to conduct research in Latin America. By contrast, Osborn's decision in the Spring of 1917 to permit some of the Museum's best scientists to become U.S. government intelligence agents in Latin America was very much a private judgment call. Approving the loan of scientists to the government for the express purpose of conducting government (i.e., intelligence) work was a decision of a very different order of magnitude from that of the previous arrangements with the State Department. Further, the number of agencies with which the AMNH had made "arrangements" expanded beyond the ONI to include such entities as the Council of National Defense, the War Trade Board, the War Department and the Red Cross. The fact that the inner circle of administrators including President Osborn, Director Lucas and Secretary Sherwood (and later Curator Dean) maintained absolute silence about the participation of these AMNH scientists in government intelligence programs during the course of the War and afterwards was an indication that they appreciated the wider-ranging implications of their actions. They also likely understood that their decisions to authorize such ventures would have proven controversial within the broader Museum community had they been made public, given the existing Museum moratorium on overseas research to which both the Science Faculty and the Trustees had subscribed. Osborn, Lucas and Sherwood especially would have faced widespread condemnation or recriminations, as well as exposure to charges of hypocrisy, had they been exposed.

But ironically, their actions were shielded from exposure by the government's insistence upon complete secrecy; the Museum administration had nothing to fear as long as they kept quiet. As a result, most of the staff at the Museum (including the scientists who remained in the United States) were not privy to the covert political assignments of Crampton, Eastman, Cherrie, Spinden, Andrews and Brown, nor the work done for the U.S. government by Harold Anthony. Likewise, the full nature of Chapman's political observations made as a Red Cross Commissioner for Latin America, as well as Beck's unusual relationship with the United States Marine Corps, were similarly kept concealed from the Museum staff during the War and for decades afterwards.

It is thus somewhat curious that Osborn felt comfortable enough about these transactional *quid pro quos* to openly acknowledge in his 1917 Report to the Trustees that some assignments (which he conveniently did not identify) given to AMNH scientists were potentially "hazardous."[16] In reality, Osborn's admission (even if it was only to the Trustees) only magnified his need to keep quiet about his staff's espionage dealings to maintain their covers, and their very lives while in the field, as well as his personal cooperation in the scheme. Considering the crude training and rudimentary spycraft tools with which these volunteers were dispatched, which typically consisted of nothing more than a simple cipher and code book, it was indeed remarkable that all of them returned to New York unscathed.

We can speculate that the problematic nature of his decision probably provoked commentary among the higher-ups at the Museum, although there is nothing in the files to confirm this. What had led Osborn in the first instance to ignore the obvious dangers of allowing his staff scientists to participate in such schemes?

Did Osborn justify the authorization of his staff's cloak and dagger undertakings as payback for the assistance which had been rendered to the Museum in Latin America by the State Department and /or the Council of National Defense and other agencies? If so, there was nothing that the ONI could have offered the Museum except allowing them to collect specimens in the field while engaged in spying. Or might he have consoled himself by rationalizing that the participation of his scientists in these schemes was completely voluntary, even though such behavior amounted to sabotaging the Museum's moratorium on overseas research? Notwithstanding these possible defenses which he may have offered, neither a heightened sense of obligation to the government, nor an overblown spirit of patriotism, including volunteering for espionage, provides a convincing explanation for why the president of the AMNH agreed to let his ill-trained scientists pursue potentially dangerous assignments in the first instance. One can only conclude that Osborn's approval of the participation of his scientists in potentially dangerous ONI and other U.S. government intelligence projects, was as self-serving as it was duplicitous, in that it allowed for expansion of the Museum's collections of specimens and artifacts against all odds.There is no question that the dangers had been real enough. Even a cursory look at the "General Instructions" provided to ONI agents prior to reaching their field assignment implies that they were more than likely to face real as well as existential crises while alone on assignment. To wit: each agent was instructed to maintain constant vigilance to avoid being detected and warned against confiding in others. All agents were advised to "always have a good reason for being anywhere," and "avoid any action which might bring you [them] under suspicion." If they suffered the misfortune of being arrested, each agent was given a name to call.[17]

Tellingly, none of the above instructions even remotely considered the ethical challenges faced by AMNH scientist/agents while they were in the field, especially as each man understood a priori he had willingly violated the Museum's overseas research moratorium. Unlike professional spies who worked unencumbered by other obligations than securing intelligence data, Museum scientists were simultaneously at the beck and call of Osborn and the U.S. government. The ONI, and other agencies, did not attempt to address the friction that such divided loyalties might produce, nor the uneasiness such agents would experience when they returned to the Museum and were still prohibited from discussing their wartime assignments. The subsequent departure of some of the best of these scientist/agents from the AMNH serves to confirm the high level of anxiety to which they were exposed. Finally, the ONI failed to consider how the scientists/agents might personally feel about spying against the colleagues and countries to which they had been posted or where they had previously conducted research. In essence, these ONI assignments were in diametric opposition to the goodwill supposedly inherent in "Scientific Pan Americanism."

While it might be argued that specific references to "danger," or personal security concerns, did not appear with frequency in the official correspondence of the scientists/agents while in the field, there were numerous, albeit subtle, indications from their writings of just how exposed they felt while on assignment. For example, when Spinden feared his cover had been blown in Honduras, he quickly withdrew from that country to the safer borders of Nicaragua, where the presence of occupying U.S. Marines had provided some stability since 1912. Beck and Cherrie also commented warily on the activities of the revolutionaries operating in the Dominican Republic and the *pantanales*, respectively, in areas in which they were working and where rebels represented clear and present threats. And, if his assignment had not been perilous, how else can we understand why Red Cross Commissioner Frank Chapman felt compelled to travel under the cover of night, or in private train cars, where he could remain anonymous and unseen?

Clearly when Osborn approved of his scientists' spying activities he likely was not thinking of the dangers which might befall them. Perhaps his individual conversations with the men chosen for these spying tasks concluded with heartfelt wishes for success and a casual "take care of yourself" warning. But Osborn seemingly was not interested in dissuading these scientists from proceeding with their secretive assignments. On the contrary, he was instead fixated on the ongoing Museum-imposed restrictions which effectively meant there could be no more overseas expeditions to pursue either the evolutionary history of the Western Hemisphere or collect its unusual fauna. His determination to find a way to circumvent onerous prohibitions and to allow his Latin American project to proceed on schedule was foremost in his mind, and in turn, it depended on his ability to maintain his scientists in Latin America during the War. The ONI and other agencies conveniently provided him with the mechanism which would allow him to do so.

One other point needs to be considered: had Osborn needed to justify the participation of his staff scientists in the ONI or War Department (or any other federal agency for that matter) on the basis of their individual scientific expertise, he would have found it nearly impossible to do so. Despite the fact that each of the men selected for these postings (with the exception of Eastman) had previously done extensive fieldwork in the region and consequently possessed a familiarity with the territory, cultures and languages of Latin America, these were natural scientists; <u>none</u> of them had been recruited to spy because their area of expertise was needed to help the war effort. Nor was the government interested in obtaining from them what would come to be known as "scientific intelligence." There was little expectation that AMNH scientists would ferret out knowledge about the technical side of Latin American science.

Instead, recruits from the Museum were instructed to find evidence of collaboration with the enemy in nations to which they were assigned, as well as confirmation of existing supplies of commodities vital to the war effort. When he received his assignment to track the outflow of manganese from Brazil, to cite one instance, Eastman (a specialist in Devonian fishes) likely knew little about its actual use in armament. But he was more than capable of tracking down those business firms which dealt in the material as well as relaying his observations on the wavering Brazilian mindset toward the War to Washington.

Likewise, Chapman's firsthand knowledge of the ornithology of South America did not prepare him to analyze labor unrest when the Red Cross requested that he personally assess the state of worker/owner relations in the Chilean mines, nor did it equip him to address the Versailles Peace Conference on what he had observed. On the other hand, Chapman was not asked to make observations on the armament or technological innovations of the Peruvian, Chilean or Argentinean military, but merely to provide descriptions of the political climate which he had witnessed during his travels. By the same token, Mayan specialist Spinden had no previous experience in scoping out German wireless stations, nor in securing the stores of mahogany needed for airplane propellers or platinum for the War effort, when he was assigned those tasks by the ONI. Similarly, he also was not asked to report on technology or the armament capacities of the countries to which he had been assigned, but to similarly outline the potential for enemy activity.

Perhaps the only exception to this pattern was Crampton's assignment during his tenure at the Council of National Defense. But while it is the case that Henry Crampton's opinion was solicited by the Council on such timely matters as materiel needed to equip an army of a million men and the various aircraft in use during the war, this was hardly his area of expertise, nor was he an arms expert. The same observation may be made about Barnum Brown, who undoubtedly had some training in stratigraphy during the course of his paleontological studies, but still had to be tutored in the rudiments of oil exploration before the Treasury Department considered him competent to prospect for petroleum in Cuba. Although by the time he was assigned to Cuba he was qualified to analyze oil production technology, even his remarks could not be construed as "scientific intelligence;" Brown was not spying on new methodologies of oil extraction but merely assessing the possibility and difficulty of extracting that commodity from Cuba.

Ironically, over the entire course of the AMNH involvement in wartime sleuthing, there was only one project which might conceivably come under the rubric of "science/defense oriented." This proposal, actually put forth by Director Lucas, called for the scientific staff to develop a method to hunt German submarines. According to Dr. Lucas, seagulls could be trained to act as submarine spotters since, in his words, "as various sea gulls follow schools of fish, and other whales, to pick up such food as may be left in their wake, there is no reason why gulls might not be led to associate food with submarines and thereby indicate their presence, because a gull flying high above the water and having specially sharp eyesight would detect any such craft at considerable distances below the surface."[18] Lucas' proposal was met with skepticism at the Museum (a similar idea had been summarily rejected by the British Admiralty), but he continued to argue for his scheme as late as January, 1918.[19] Strangely, Rollo Beck, certainly the most qualified individual associated with the Museum with regard to oceanic birds, was never consulted on this extraordinary proposal; there is no mention of it in his extensive journal nor in his frequent exchanges with Chapman or Dr. Sanford. Had he known about it, no doubt Beck would have had a good laugh.

In short, the AMNH scientists who were recruited as spies for government agencies during World War I were neither arms experts, nor trained in warfare or military technology. While in the field, the only defense available to them, other than keeping a low profile, was their ability to shoot, with considerable accuracy, the hunting rifles or revolvers which they carried with them for bringing down specimens. Thankfully, they were not assigned the task of obtaining "scientific intelligence" in the countries to which they were assigned except for what they might have fortuitously observed on the ground (for example, Beck's images of Chilean soldiers in Punta Arena).

Paradoxically (although this observation in no way exonerates Osborn), their lack of military training might have inadvertently served to shield the AMNH recruits. If their performances as agents in Latin America had a kind of helter-skelter quality, as judged by the locals observing them in situ, their awkwardness only served to reinforce the idea that these were merely curious scientists, interested solely in Latin American research. Obviously, their observations of wireless stations, or military installations (in the case of Beck or Cherrie) or whatever else they were assigned to keep track of (manganese, nitrates, platinum and copper) was only to be passed on to trusted contacts. There were, consequently, few outward signs forewarning their Latin American colleagues of their dual status as scientists and spies, nor reason to suspect they were simultaneously secret government agents. To their credit, to a man, each of the AMNH recruits adapted brilliantly to their espionage assignments, responsibly procuring and forwarding to headquarters the intelligence requested of them exactly as they had been instructed, without being detected.

From the vantage point of the United States government bureaus which recruited them, therefore, the benefits of utilizing AMNH scientists in wartime Latin America accrued precisely as had been anticipated. These intelligence "newbies" were seen as appealingly malleable, educated and intelligent men, capable of learning and adapting quickly on the job. Conversely, the scope of the "scientific work" done by the AMNH recruits was so far removed from the day to day activities of the government bureaucracies which employed them that their in-field supervisors never thought to question them on that score; as long as the agents performed credibly and delivered their reports in a timely fashion, they were left alone. To have done otherwise would have been counterproductive; indeed, such meddling might even have placed the agents' cover stories in jeopardy. Overall, seen from the perspective of the agencies of the United States government which recruited them, the scientists from the American Museum of Natural History who served as intelligence assets were an unparalleled success.

But the reverse was true when we consider the impact of these arrangements on the Museum. The consensus of the very few AMNH administrators who knew of the transactional arrangements negotiated with the various Washington agencies was that the net return from "loaning" AMNH scientists to the spying programs had been decidedly limited. Osborn, Lucas, and Sherwood (and later, Dean) were often left disappointed, if not downright exasperated, by what they considered the mediocre results of the time spent in the field by their men. With the possible exception of George Cherrie's superb 1916–17 assemblage from the *pantanales*, which he pluckily managed to collect as he made the rounds of Southern Cone cattle stations, the Museum administration was largely dissatisfied with the specimens procured by their scientists during their long interludes in wartime Latin America. This realization left Osborn, Sherwood and Lucas not only feeling betrayed, but led them to conclude that these scientists had largely failed to fulfill their transactional contracts with the Museum.

That it all ended badly, with disillusionment experienced on both sides, should not have been unexpected. Nor was it surprising that several of the scientists who had served as intelligence assets during the War departed from the AMNH community soon after they returned to New York. After spending months fretting about losing his position at the American Museum as he traipsed through Central America, a fear heightened by Dr. Wissler's incessant hectoring to produce something of "value" while absent from New York, "Joe" Spinden no longer felt comfortable in his old position when he returned to the AMNH in the Spring of 1919. His sense of uneasiness increased until, in 1921, he decided to leave the AMNH to accept a position at a rival institution, the Peabody Museum of Harvard University.

Henry Crampton had also been perturbed by the acrimonious squabbles he had endured with Osborn during his tenure at the Council of National Defense. He viewed these spats not only as demonstrative of a lack of confidence in his abilities, but unworthy of the esteem in which he was held by the Council. The elevated importance he had enjoyed as a policy maker during his 1916–1917 tenure at the Council continued to haunt him; as seen in his micro-history, the zoologist relentlessly petitioned various intelligence agencies based in New York for additional tasks. Perhaps no one who really understood the degree of Crampton's disillusionment should have been surprised when the zoologist chose to formally "retire" from his curatorial duties at the AMNH in 1920, although he apparently had no reluctance in holding onto his professorship at Columbia.

Always the consummate negotiator, Crampton managed to finagle an "honorary curatorship" from the Administration when he left his formal position. This meant that although he had officially parted company with the Museum, he would now be able to maintain his professional association with the institution without experiencing the pressure to "do the Museum's work." Free to pursue his own research, Crampton returned to the Pacific during the year in which his curatorship ended. As noted before, he managed to maintain his affiliation with the ONI, but the difference now was that he could participate in government service and at the AMNH simultaneously on his own terms. It was fortuitous for him that the Museum's focus shifted to the East just as Crampton renewed his interest in collecting in the region.

In the case of Frank Chapman, while he was "on loan" to the War Department-run Red Cross in Washington, D.C., he found little time to pursue his own scientific interests. When the ornithologist was unexpectedly posted back to Latin America as a Red Cross Commissioner in the Fall of 1918, Osborn seized upon his new assignment as a way of obligating Chapman to devote at least some of his time to collecting for the Museum. But, given the broad demands of his government portfolio, this would clearly be an impossible ask, and as was observed earlier, the ornithologist did not do any Museum work during this trip. Undoubtedly dissatisfied with this outcome,

Osborn was likely even more frustrated by Chapman's outright refusal to write the administrator's much desired essay for the American Museum Journal on the state of Latin American natural history museums.

What Osborn had not realized was that Chapman was not only reluctant to do Osborn's bidding, but as his letters to the Museum president confirm, Chapman had become increasingly disillusioned with the state of science in the Americas, as well as the failure of the nations to which he was posted to support the Allies in the War. The post-War atmosphere did little to mollify his bitterness. After his brief appearance at the Versailles Peace Conference in 1919, Chapman returned to his secure curatorship in Ornithology at the Museum, and continued collecting South American birds, but with a new emphasis on Ecuador and Peru, rather than Colombia and the Southern Cone countries as in the past. Despite the fact that Chapman was much lauded, and remained an important member of the Museum staff, there is no doubt that Osborn had been put off by (and was perhaps jealous of) the ornithologist for prioritizing his political agenda over that of the Museum.

Surprisingly, not even a contract collector was immune from the scrutiny and censure of the Museum's top administrators. Considering the breadth of Rollo H. Beck's multi-year collections from South America and the Caribbean, it is surprising that Dr. Lucas, for one, openly questioned the "anatomical" value of Beck's work. Within the Museum community it was Chapman who rose to Beck's defense, saving his reputation by proclaiming the importance of Beck's five years of fieldwork and perhaps influencing the decision to send him to the Pacific in the following decade, once again under the aegis of the Museum.

However, it is fairly safe to say that no one's fieldwork caused the Museum administrators greater consternation than that of Charles Eastman, whose hastily put together Brazilian fish assemblage was judged far less than stellar. Director Lucas not only privately believed that Eastman's piscine material had little scientific value, he did not hesitate to openly characterize Eastman's promise of future exchanges of AMNH casts with Brazilian museums as foolhardy. Finally, when Eastman returned to New York, Bashford Dean displayed his contempt for his protégé to the Museum administration, an assessment which ultimately resulted in the ichthyologist's removal from the staff. Given the disdain in which his work had been held, the outcome was hardly surprising.

But, in the end, Eastman parted ways with the AMNH on his own terms, preferring to devote his time to intelligence work at the War Trade Board, rather than finish the promised bibliography of fishes. Given the bitter feelings his absences had engendered, it was unlikely that Eastman could have successfully resumed his duties at the Museum at the conclusion of the War, but his untimely death in September of 1918 of course rendered a future reunion moot.

Viewed cumulatively, the unique arrangements engineered between the Museum and U.S. government agencies, which saw staff scientists recruited as intelligence assets while they ostensibly collected for the AMNH in Latin America, fell far short of Osborn's expectations. For those in the upper echelons of the Museum, the hazards endured by AMNH scientists/agents while spying for the U.S. government were compensated for neither by outstanding assemblages of specimens, nor by solid inroads into the Latin American scientific community. On the contrary, "scientific Pan Americanism" advanced quite slowly during the War years or in some instances stalled outright during the period while the AMNH scientist/agents were in the field. In the case of Chapman and Cherrie, moreover, one could further argue that their opinions of their colleagues in Latin America seriously deteriorated during this period; neither of these men attempted to hide their contempt for what they encountered in Latin American natural history museums. And that contempt often extended to Latin Americans of the working classes, either in the mines in Chile (where Chapman remarked they were mollified by music), or on the streets of Rio de Janeiro, where Cherrie wrote critically of the masses. Is there a hint of racism in these observations? Perhaps, but there is little to outwardly confirm these points of view as based in racialist diatribe in either the Museum records or personal diaries of both men. Curiously, in contrast to his subordinates, Osborn was seemingly more bent on the necessity of seeing Latin American scientists as equals than some of the men he sent into the field.

In retrospect, therefore, considering the impact on the institution as a whole, one can conclude that the secret government missions of AMNH scientists to Latin America during World War I justified neither the risks to their lives, nor the blatant policy breach of the Museum-imposed research moratorium. In the end, what was accomplished by this precocious, but evidently intentional, blurring of the boundaries between politics and science?

What Benefit from Linking Science and Politics?

What conclusions should be drawn from this study of the AMNH and its role in the intersection of the tropes of War, Science and "Scientific Pan Americanism"? President Osborn's decision to loan scientists to Washington agencies for intelligence work failed to deliver the institutional benefits he had prioritized and anticipated, although it might be argued in the abstract that his actions resulted in momentarily placing the AMNH ahead of rival U.S. natural history museums in the pursuit of Latin American research. Nonetheless, the secrecy mandated by the U.S. government in association with this activity confounded Osborn's broader aim of publicly advancing the Museum's scientific reputation while it simultaneously vitiated any claim of the Museum to super-patriotism. Finally, it was fairly obvious that having scientists spy on their host countries was hardly a basis for the cooperation implied by the concept of "scientific Pan Americanism." As noted earlier, a new generation of scholars from Latin America has recently begun to examine the activities of the AMNH scientists through the lens of imperialism and colonialism.

The paucity of faunal materials collected in Latin America during the War years as well as the lack of advancement on the question of large-mammal evolution suggests that the overall benefits to the Museum of having scientists work in tandem with the government were minimal. Yet, oddly enough, in a single piece of commentary about the appropriateness of agreements between scientific institutions and the U.S. government, the November, 1918 issue of the American Museum Journal boldly asserted that "The war has centered attention on what can be gained by cooperation between scientific men and the government of a country."[20] Ironically, this was not a reference to the covert activities of the Museum's scientists involved in Latin American espionage, but rather a nod to an obscure State University of Iowa expedition to Barbados and Antigua in the summer of 1918 which had been conducted under official government auspices.

In the final analysis, Osborn, Lucas, Sherwood and Dean could ill afford to reveal their own roles in advancing the elision between science and government at the Museum and, of course, under the government-negotiated terms of these arrangements, these administrators were expressly prohibited from doing so. Consequently, the participation of Museum scientists in government intelligence gathering in World War I Latin America has remained a closely guarded secret up to the present day. Over one hundred years later, there are still those within the Museum community who would vehemently deny that such a program ever existed. Their stubbornness presumably stems from the success that the U.S. government had in cloistering records concerning the activities of these wartime scientist/agents, only choosing to declassify them in the 1990s, when, of course, no one was left who could authoritatively discuss the era.

Even so, despite the hyper-secrecy, some inferences about AMNH scientists who had acted as agents during the War evidently leaked, as witnessed by Franz Boas' scorching condemnation of Herbert Spinden in his 1919 article in The Nation referenced at the beginning of this book. One imagines that had Osborn's complicity in this scheme been known, then he too would have been singled out for Boas' vehement scorn.

Did Osborn envision the activities of the War years, and the melding of scientific and political aims, to be a template for the future? If so, then he must surely have been disappointed, a conclusion supported by his own assessment of the period referenced at the beginning of this chapter. But pragmatist that he was, Osborn continued to view the "volunteerism" of his staff scientists as a windfall, an opportunity to parlay their patriotism into gains for the Museum and his own ambition.

With the advantage of hindsight, however, it would be unfair to hold Osborn singularly responsible for Museum-cum-government intrigue: each of the scientist/agents discussed in this study not only volunteered for their spying assignments, but each had personally solicited Osborn's blessing or, at least, his acquiescence. Earlier the question was raised concerning the belief held by the AMNH scientist/agents in the importance of their work for the government. Surely these men were convinced of the value of the cause they served, apparently to the point of being willing to prevaricate when it came to explaining their absences to others at the Museum; as recruits, they surely construed their missions as patriotic service, a characterization which perhaps allowed them to rationalize what each man knew to be a callous violation of Museum policy. In addition, many of these agents exhibited a strong anti-German bias, notwithstanding the fact that some had received scientific training in Germany.

Yet it is hard to reach any conclusion other than that Henry Fairfield Osborn ultimately bore the primary responsibility for permitting AMNH scientists to work hand in glove with the U.S. government in Latin America; his actions may have been reckless and/or naïve, but they were certainly not innocent. By knowingly placing into jeopardy the Museum's reputation, as well as the lives of its highly respected scientists, Osborn's decisions nullify any possible attempt to justify his behavior.

What has this discussion revealed about the conjunction of scientific impartiality with political imperative? As discussed earlier, today the interconnection between political actors (and/or businesses) and scientists is no longer secret, but instead is understood as normative. This very sentiment was recently expressed in the prestigious science journal Nature, which asserted that in the twenty-first century "science and politics are inseparable."[21]

At the same time, the idea of scientific neutrality has not altogether disappeared; an opposing perspective continues to argue for the separation between science & scientists and politics & politicians.[22] Absent a clear consensus, the question of whether science and scientists should be kept insulated from politics and politicians, versus the acceptance of the overlap of their purposes as inevitable in the modern world, remains an important if not critical conundrum.

If this question is pertinent today, then it was most assuredly relevant in the historical past when these elisions first occurred. Throughout antiquity and the Renaissance as well as in the colonial period of Latin America, and into the hegemonic, informal empires which defined the nineteenth and early twentieth centuries in that region (and globally), there are numerous examples of science working in the service of, and in tandem with, the state. But it was in the World War I era that the collaboration of scientists with state actors (especially with regard to Latin America) was not only overtly encouraged, but became institutionalized.

This book has aimed to define this period of collaboration within the history of the American Museum of Natural History as its staff wrestled internally with the exigencies of war. Admittedly, the picture which emerges from this book regarding the purposeful obfuscation of what several AMNH scientists were actually doing in Latin America during wartime (other than their own scientific research) reveals what will undoubtedly be for many readers new and uncomfortable "truths". Yet Osborn's willingness to obscure the distinctions between governmental policies and scientific objectives at the Museum must be understood as having overshadowed and shaped a pivotal era in the history of the institution. His actions must not be construed solely as an example of his personal fallibility or ambition, nor dismissed on the grounds that such decisions were merely representative of the thinking of the time. Rather, the interactions of AMNH scientist/agents with politicians during World War I must be analyzed to understand how this history can help formulate a set of guidelines for a natural history museum in the twenty-first century. We need to interpolate the actions of these scientists against the background of a century-old historical past, to understand how they might foreshadow the future of the institution. Given that the relationship of science and politics has never been closer than at the present, the question must be asked if there should now be stipulated policies for any future collaboration between the institution and politicians? With over thirty years having passed since the declassification of materials pertaining to World War I intelligence gathering, it is more than timely to open an honest conversation about the meaning and the outcome of Osborn's "experiment" for the AMNH.

The engagement of natural history scientists with the political machine of war and espionage was unlikely but clearly successful. Yet did it have an enduring impact on the AMNH? Was that impact anything more than a resolve not to repeat the practice? The transactional behavior of the Museum through the duration of hostilities remains a very cautionary tale. A thoughtful evaluation of the long term repercussions of such policies within the AMNH scientific community is not only long overdue, but urgently needed, in the hope that engaging in such soul-searching might result in important new directions for the future of the AMNH (and other natural history museums), and, perhaps, even for science itself.

Notes

1. "Report of the President," Fifty-First Annual Report of the Trustees of the American Museum of Natural History for the Year 1919 (New York, May 1, 1920), p. 18.
2. William Diller Matthew, "Climate and Evolution," Annals of the New York Academy of Sciences, (New York, 1915), reprinted as Climate and Evolution (New York, New York Academy of Sciences, 1939).
3. "Central Asiatic Expeditions: 1921–1930" in https://data.library.amnh.org/archives-authorities/id/amnhc_2000167.
4. Speculation about Asia as the source for human evolution gave rise to the "Out of Asia" theory, as opposed to what has more recently been termed the "Out of Africa" theory. See discussion of Osborn's ideas in Rexer and Klein, American Museum..., pp. 54–55. Osborn received support for his ideas on the origin of early humans from Museum paleontologists William D. Matthew and William K. Gregory. See *Ibid*., p. 55.
5. NARA RG 38, E123, Roy Chapman Andrews file. See also Chap. 10 for discussion of Andrews and his ONI assignment (as "Mr. Reynolds") in China. In addition to keeping one eye on the fighting between Russian factions in Mongolia, there was also menacing Japanese interest in the region which concerned the ONI.
6. Clark, God–or Gorilla...,p. 117. Although partial skulls of Peking Man ("*Sinanthropus*") were found in China in 1929 and 1930, Osborn was disappointed that his men had not found the anticipated examples of much earlier human ancestors. Even the discovery by Raymond Dart of *Australopithecus* in South Africa in 1924 did not dissuade Osborn from his Asian quest. See also Rieppel, Assembling the Dinosaur..., pp. 221–222.
7. Sherman Miles, Colonel, General Staff, Military Intelligence Division of the War Department to Director F.A. Lucas, AMNH, Washington, May 10, 1920. Department of Mammalogy Archives [hereafter DMA], Anthony file.
8. See images 212944–212949 in AMNH photo collection.
9. For a discussion of the gold standard see Lawrence H. Officer, "Gold Standard," EH.Net Encyclopedia, ed. Robert Whaples, March 26, 2008, https://eh.net/encyclopedia/gold-standard/ (accessed on October 23, 2020).
10. Harold Anthony, Associate Curator of Mammals, Western Hemisphere to Colonel Sherman Miles. New York, May 13, 1920, DMA, Anthony file. Anthony wrote that he would either visit before proceeding to Ecuador or after, and "in any event, I shall be able to report to you upon my return from Ecuador." Interestingly, Harris and Sadler, The Archaeologist...,pp. 308–9, suggest that an arrangement similar to Anthony's with the military was made by Roy Chapman Andrews and Director Lucas. The War Office (MID) sent an engineer and a topographer

to accompany Andrews and produce maps during his 1925 and 1930 expeditions to Outer Mongolia.

11. Rexer and Klein, American Museum…, p. 172.
12. See Chap. 4 and Delson, "'The Bullet and…".
13. The Museum was in fact highly criticized by the Socialist newspaper, The Call, for representing an idealized daily menu for New Yorkers which, in that newspaper's opinion, was far more costly than the public could afford. See Delson, "The Bullet and…", p. 13.
14. See *Ibid.*, p. 18.
15. Eugenics advocates argued for, among other ideas, the "selective breeding" of humans. See Clark, God–or Gorilla…, p. 30 and pp. 116–118, for a discussion of Osborn's role in the Eugenics Movement.
16. HFO, Report to Trustees of AMNH, May 7, 1917. See Introduction.
17. "General Instructions" given to ONI agents. Cf. Appendix 5, Harris and Sadler, The Archaeologist…, pp. 381–382.
18. Director Lucas to Dr. A.D. Pentz, Jr., Santiseptic Manufacturing Company, Staten Island, July 16, 1917. CA, file 1069. Sonar only became operational as a method for detecting submarine activity after the War. See Katzir, "In War and in Peace…', p. 226.
19. Director Lucas to Dr. A.D. Pentz, Jr., January 22, 1918. CA, file 1069. It is also somewhat ironic that in 2020, the idea of using sea birds for surveillance once again surfaced. By implanting sensors in albatrosses, which then fly over illegal fishing boats in the southern Indian Ocean, there has been some success in controlling piracy. See <<www.audubon.org>> , "Albatross Wearing Data Trackers Are Exposing Illegal Fishing Boats," January 27, 2020 (accessed on October 23, 2020).
20. See American Museum Journal, Vol. XVIII, November 1918, p. 730, "Notes…".
21. "Why *Nature* needs to cover politics now more than ever" (Editorial). Nature, Vol. 586, October 8, 2020, pp. 169–170.
22. William Roper, "Science, politics, and public health," Science, October 23, 2020, vol. 370, p. 385.

Correction to: Convergence: The Declaration of War, the ONI in Latin America and Osborn's Change of Heart

Roberta Marx Delson

Correction to:
Chapter 4 in: R. M. Delson, *Scientists or Spies?*, The Latin American Studies Book Series, https://doi.org/10.1007/978-3-032-18117-6_4

The original version of the chapter "Convergence: The Declaration of War, the ONI in Latin America and Osborn's Change of Heart" was inadvertently published with incorrect figure caption for Figure 4.3. The correct caption has been updated in the chapter.

The updated version of this chapter can be found at
https://doi.org/10.1007/978-3-032-18117-6_4

R. M. Delson, *Scientists or Spies?*, The Latin American Studies Book Series,
https://doi.org/10.1007/978-3-032-18117-6_12

Epilogue

As I was completing this study, two mentions of the AMNH in the press caught my attention. In the first, Dr. John Flynn, Frick Curator of Fossil Mammals in the Division of Paleontology and Dean of the Richard Gilder Graduate School, was quoted in a discussion of very early mammals from Madagascar. Flynn now agrees with those scientists who view the ancestors of modern mammals as likely originating in the southern continents rather than in the north. Intriguingly, this represents somewhat of a return to Osborn's (and Hooker's) thinking before Matthew rejected these views.[1]

Even more directly related to the thesis of this book, a recent interview conducted by the Editor of Science with the new President of the AMNH, Dr. Sean Decatur, revealed the latter's concern over the modern day role of museums. "There's more to returning items as a repatriation process than just putting them in the mail," in his view. How things are collected, why and by whom are stories Decatur thinks need to be told.[2] This is exactly my reason for urging a conversation about the participation of AMNH scientists in the political arena of World War I Latin America. There is still much to be told and much more to be learned.

Notes

1. Anthony Ham, "Why Scientists Are Debating Mammal Origins," New York Times, July 18, 2023, D3.
2. H. Holden Thorp, "Museum ethics–more than lip service," Science, Vol. 381, July 7 2023, p. 7.

R. M. Delson, *Scientists or Spies?*, The Latin American Studies Book Series,
https://doi.org/10.1007/978-3-032-18117-6

References

(I) Archives

American Museum of Natural History

Gottesman Research Library: AMNH vertical biography files, Eds. Robin Smith and John C. Pallister; Central Archives (CA): Files 1000, 1069, 906, 632; special collection Osborn files, 0837, Box 2, Fls. 12 and 13; Iconographic/Digital Archives
Division of Anthropology Archives (DAA): Herbert Spinden files
Division of Mammalogy Archives: Harold Anthony Files
Division of Ornithology Archives (DOA): Bibliography and correspondence of Frank M. Chapman; George Cherrie correspondence; Cherrie Diary, 1916, Peru-Ecuador, Matto Grosso
Division of Vertebrate Paleontology Archives (DVPA): Barnum Brown Memoirs and Files

Other Institutions

American Geographical Society Archives, New York (AGSA)
American Philosophical Society, Philadelphia, Pennsylvania: Boas Papers
California Academy of Sciences, San Francisco (Beck CAS): Files of Rollo H. Beck
Carnegie Institution, Washington, D.C.: Crampton Files
Columbia University Rare Book Collection, New York (CUSMC): Special Manuscripts Collections - Henry E. Crampton Files
National Archives (NARA): Washington office, Record Group 59, Scientific expeditions; Record Group 127, Entry 38, C 298 Cuba; United States Marine Corps Intelligence Reports; Maryland office; New York office
New York Historical [Society]: Diaries of Henry Fairfield Osborn, Box IX, folder 1917, folder 1918

R. M. Delson, *Scientists or Spies?*, The Latin American Studies Book Series,
https://doi.org/10.1007/978-3-032-18117-6

(II) American Museum of Natural History Official Documents

American Museum Journal, 1902.Vol. II, – 1918. Vol. XVIII, New York.
American Museum Journal, Vol. XVIII, November 1918, p. 730, "Notes…".
American Museum of Natural History. 1900. Annual Report of the President for the year 1899. New York.
American Museum of Natural History. 1911. Annual Report of the President for the year 1910. New York.
American Museum of Natural History. 1913. Annual Report of the President for the year 1912. New York.
American Museum of Natural History. 1917. Annual Report of the President for the year 1916. New York.
American Museum of Natural History, Faculty Minutes.
American Museum of Natural History, Minutes of meetings, Executive Committee
American Museum of Natural History. "Report on the Mitla Restaurant." 1908. Executive Committee Report. arch. 20, 106. New York.
Collecting and Collaborating (1857–2017). New York: American Museum of Natural History, Special Publication.
Forty-third Annual Report of the Trustees of the American Museum of Natural History for the year 1911. 1912. New York.
Forty-eighth Annual Report of the Trustees of the American Museum of Natural History for the year 1916, 1917. New York. "Report of the President."
Fiftieth Annual Report of the Trustees of the American Museum of Natural History for the year 1918, 1919. New York.
Fifty-first Annual Report of the Trustees, of the American Museum of Natural History for the Year 1919, 1920. New York. "Report of the President," p. 18.
"Central Asiatic Expeditions: 1921 – 1930 in https://data.library.amnh.org/archives-authorities/id/amnhc_2000167.
Minutes of the Regular Meeting of the Board of Trustees, American Museum of Natural History.
Nichols, J.T. 1915. Memorandum on the Possibilities of Field-work in Porto Rico, CA, AMNH, File 1000.

(III) Other Publications and Documents

(A) Institutional and Newspaper Reports

"Albatross Wearing Data Trackers Are Exposing Illegal Fishing Boats," <<www.audubon.org>>, January 27, 2020 (accessed on October 23, 2020).
Annual Report 2006. The American Anthropological Association.
Annual Report of the Board of Regents of the Smithsonian Institution showing the Operations, Expenditures and condition of the Institution for the year ending June 30, 1914. 1915. Washington.
Boletín de la Unión Panamericana. 1915. Vol. 40, No. 4.
"C. R. Eastman Acquitted." May 12, 1901, New York Times.
"Council of National Defense created by order of President Wilson on October 11, 1916." October 12, 1916, New York Times, 10.
"Dr. Crampton Faced Death in the Jungle." September 20, 1911, New York Times.
"Eastman Case From Shooting to Verdict." 12 May, 1901. Boston Sunday Post,
Harvard College Class of 1906. Fifteenth Anniversary Report. Printed for the Class at the University Press, 1921.

"Inquiry of the American Geographical Society for the Information of the Peace Commissioners." 1918. Science. Vol. 48, Issue 1250, 590–592.
"Obituary of Charles R. Eastman." 5 October, 1918. Cambridge Tribune. Vol. XLI, No. 32.
"Obituary of Charles R. Eastman." September 30, 1918. New York Times.
Proceedings of the Second Pan American Scientific Congress, Washington, U.S.A., Section III, Conservation of Natural Resources, 1917. Washington.
Report of the Delegates of the United States to the Pan-American Scientific Congress (Santiago, Chile). December 25, 1908, to January 4, 1909. 1909. Washington.
Report of the War Trade Board. 1920. Washington: Government Printing Office.
Scientific Survey of Porto Rico and the Virgin Islands, 1919. Vols. 1–10. New York: New York Academy of Sciences.
"The American Museum of Natural History," May 29, 1903. Science. N.S. Vol. VII. No. 439.
"The Following Assistants in the American Museum of Natural History have enlisted in the army…Harold E. Anthony…and Leo. E. Miller." 1917. Science, No. 8, Vol. XLVI, No. 1177, 58.
"The victim with his dying breath is said to have accused Eastman of murdering him." 23 April, 1901. The Press Democrat. Vol. XXVII, No. 139.
"Trial of C.R. Eastman." 27 April, 1901. The New York Times.
"Why *Nature* needs to cover politics now more than ever" (Editorial). 2020. Nature, Vol. 586, pp. 169–170.

(B) Authored Works

Aguirre, Robert D. 2005. Informal Empire: Mexico and Central America in Victorian Culture. Minneapolis: University of Minnesota Press.
Alberti, Samuel J.M.M. 2011. "The Status of Museums: Authority, Identity, and Material Culture." In David N. Livingstone and Charles W. J. Withers, eds. Geographies of Nineteenth-Century Science. Chicago: University of Chicago Press.
Allen, Joel A. 1916. List of Mammals Collected in Colombia by the American Museum of Natural History Expeditions, 1910–1915. Bulletin of the American Museum of Natural History, Vol. XXXV, 191–238.
Allison, R. Bruce, Anderson, James R., Androff, Amy L., Ross, Robert J., Senalik, C. Adam, Hagan, Jeffrey, Oehrlein, Mary L., Kamprath, Nathan and York, Keith, "Lumber from World War I Research at the Forest Products Laboratory Helps Repair the U.S. Capitol Building," General Technical Report FPL-GTR-296 (Forest Products Laboratory, U.S. Department of Agriculture Forest Service), (December 2022), pp. 1–12. https://www.fpl.fs.usda.gov/documnts/fplgtr/fpl_gtr296.pdf
Andrews, Roy Chapman. 1943. Under a Lucky Star: A Lifetime of Adventure. New York: Viking Press. Reprinted 2007, Speath Press.
Ayala, César J. 1999. American Sugar Kingdom: The Plantation Economy of the Spanish Caribbean, 1898-1934. Chapel Hill: University of North Carolina Press.
Baecker, Thomas. 1973. "The Arms of Ypiranga: The German Side," The Americas, vol. 30, no. 1, pp. 1–17.
Barrero-Osorio, Antonio. 2006. Experiencing Nature: The Spanish American Empire and the Early Scientific Revolution. Austin: University of Texas Press.
Barrett, John. 1915. "Practical Pan-Americanism," North American Review, September, 1–11.
Bashkow, Ira. 2023. "There's more to anthropology's past than most of us know." American Anthropologist, Vol. 125, pp. 177–180.
Baud, Michiel. 1988. "German Trade in the Caribbean: The Case of Dominican Tobacco, 1844–1940." Jahrbuch fur Geschichte von Staat, Wirtschaft und Gesellschaft, Vol. 25.
Beck, Rollo H. 1919. "Narrative of a Bird Quest in the Vicinity of Cape Horn." Natural History, Vol. XVIII, no. 1 (January), pp. 5–16; continued Vol. XVIII, no. 2 (February), pp. 111–120.

Beck, Rollo H. 1919. "Scenes from the Vicinity of Cape Horn." Natural History, Vol. XVIII, no. 1 (January), pp. 17–24.

Bello, José Maria. 1966. A History of Modern Brazil, 1889–1964. trans. James L. Taylor. Stanford: Stanford University Press.

Bernstein, Harry. 1961. "Some Interamerican Aspects of the Enlightenment." In Arthur P. Whitaker, ed., Latin America and the Enlightenment. 2nd ed. Ithaca, New York: Cornell University Press.

Bill, Albert. 1988. South America and the First World War: The Impact of the War on Brazil, Argentina, Peru and Chile. Cambridge: Cambridge University Press.

Bisher, Jamie. 2016. The Intelligence War in Latin America, 1914–1922. Jefferson, North Carolina: McFarland and Company.

Boas, Franz. 1919. "Scientists as Spies." The Nation, 109, 797.

Bowler, Peter J. and Morus, Iwan Rhys. 2008. Making Modern Science: A Historical Survey. Chicago: University of Chicago Press.

Breem, William J., 1984. Uncle Sam at Home: Civilian Mobilization, Wartime Federalism, and the Council of National Defense, 1917-1919. Westport, Connecticut: Greenwood Press.

Brockmann, Sophie. 2020. The Science of Useful Nature in Central America: Landscapes, Networks and Practical Enlightenment, 1784-1838. Cambridge: Cambridge University Press.

Browman, David L. 2011. "Spying by American Archaeologists in World War 1 (with a minor linkage to the development of the Society for American Archaeology." Bulletin of the History of Archaeology, Vol. 21, No, 2, 10–17.

Brown, Stephen R. 2005. A Most Damnable Invention: Dynamite, Nitrates and the Making of the Modern World. New York:Thomas Dunne Books.

Burleigh, Nina. 2007. Mirage: Napoleon's Scientists and the Unveiling of Egypt. New York: Harper Collins.

Burns, E. Bradford. 1970. A History of Brazil. New York: Columbia University Press.

Castañeda, Quetzil E. 2005. "The Carnegie Mission and the Vision of Science: Institutional Contexts of Maya Archaeology and Espionage." In Regna Darnell and Frederic W. Gleach (Eds.) Histories of Anthropology Annual. Vol. 1. Lincoln: University of Nebraska Press, 27–60.

Chapman, Frank M. 1918. One Year with the American Red Cross: June, 1917-June,1918. A Record of How the American People Came to the Support of Their Country at War. Washington, D.C., I.S. No. 2.

Chapman, Frank M. 1918. "Notes from a Traveler in the Tropics." Bird-Lore. Vol. XX, No. 6, 393–397.

Chapman, Frank M. 1933. Autobiography of a Bird-Lover. New York: Appleton-Century.

Cherrie, George. 1917. "To South America for Bird Study" The American Museum Journal. April, 269–273.

Cherrie, George. 1930. Dark Trails: Adventures of a Naturalist. New York: G.P. Putnam's Sons.

Clark, Constance Areson. 2008. God--or Gorilla: Images of Evolution in the Jazz Age. Baltimore: Johns Hopkins University Press.

Colón, Julio C. Figueroa (ed.) 1996. The Scientific Survey of Puerto Rico and the Virgin Islands: An Eighty-Year Reassessment of the Island's Natural History. New York: New York Academy of Sciences.

Crampton, Henry E. 1916. Studies on the Variation, Distribution and Evolution of the Genus *Partula*: The Species Inhabiting Tahiti. Carnegie Institution of Washington, Monograph 228.

Crampton, Henry E. 1925. Studies on the Variation, Distribution, and Evolution of the Genus *Partula*. Vol. II, The Species of the Mariana Islands, Guam and Saipan, Carnegie Institution of Washington, Monograph 228b.

Crampton, Henry E. 1942. Department of Zoology of Columbia University, 1892-1942. New York: Columbia University.

Cuddy, Thomas W. 2007. Political Identity and Archaeology in Northeast Honduras. Denver: University Press of Colorado.

Dabbene, Roberto. 1916. "Recepción del Dr. Frank M. Chapman dada por la sociedades Argentina de Ciencias Naturales y Ornitológica del Plata, el 11 de septiembre de 1916." Physis (Revista de la Sociedad Argentina de Ciencias Naturales). tomo II, 439–448.

Dean, Bashford, 1896. "Notes on the Ventral Armoring of Dinicthys." New York: New York Academy of Sciences.

Dean, Bashford. 1919. "Charles Rochester Eastman." Science, N.S., Vol. XLIX., No. 1258, 139–141.

Dehne, Phillip. 2014. "How Important was Latin America to the First World War?," Iberoamericana, vol. 14, pp.151–164.

Delson, Roberta Marx. 2018. "'The Bullet and the Crust': A World War I Exhibition on nutrition and food conservation at the American Museum of Natural History," Museum History Journal, vol. 11, No. 1, 1–23.

Dingus, Lowell and Norell, Mark. 2010. Barnum Brown: The Man Who Discovered *Tyrannosaurus rex.* Berkeley: University of California Press.

Dorwart, Jeffery M. 1979. The Office of Naval Intelligence: The Birth of America's First Intelligence Agency, 1865-1918. Annapolis: U.S. Naval Institute Press.

Duarte, Regina Horta. 2006. "Pássaros e Cientistas no Brasil: Em busca de proteção, 1894-1938." Latin American Research Review. Vol. 41, No. 1, 3-26.

Duarte, Regina Horta. 2013. "Birds and Scientists in Brazil: In Search of Protection, 1894-1938," In Martha Few and Zeb Tortorici, Eds. Centering Animals in Latin American History. Durham: Duke University Press, 270–301.

Dumbacher, John P. and West, Barbara. 2010. "Collecting Galápagos and the Pacific: How Rollo Howard Beck Shaped Our Understanding of Evolution." Proceedings of the California Academy of Sciences, Series 4, Vol. 61, Supplement II, No. 13, p. 212.

Dyck, Ian. 2016. "Scientific Archaeology in Central North America: The Life and Work of W.B. Nickerson (1825–1926)." In Archaeology paper, 177, Mercury Series. Ottawa: Canadian Museum of History and the University of Ottawa Press.

Ely, Richard T. 1917. "Conservation and Economic Theory," in Proceedings of the Second Pan American Scientific Congress, Washington, USA. Section III, Conservation of Natural Resources. Washington. 10–21.

Fernos, Rodrigo. 2003. Science Still Born: The Rise and Impact of the Pan American Scientific Congresses, 1898–191. New York: iUniverse.

Ficker, Sandra Kuntz. 2014. "El impacto de la Primera Guerra Mundial sobre el comercio exterior de México," Iberoamericana, vol. 14, pp. 117–137.

Folse, Mark R. 2014. "The Impact of the Great War on Marines in Hispañola, 1917–1919." Paper presented at the Temple University Jame A. Barnes Graduate Student History Conference, Philadelphia, Pennsylvania.

Freed, Stanley A. 2012. Anthropology Unmasked: Museums, Science, and Politics in New York City. 2 vols. Wilmington, Ohio: Orange Frazer Press.

Galchen, Rivka. 2022. "Why Scientists Become Spies: Access to information only goes so far to explain the curious link between secrets and those who tell them." The New Yorker, January 5.

Gelfand, Lawrence E. 1963. The Inquiry: American Preparations for Peace, 1917-1918. New Haven, Connecticut: Yale University Press.

Gil¸ Frederico G. 1971. Latin American-United States Relations. New York: Harcourt, Brace, Jovanovich.

Godman, Frederick du Cane. 1907–1909. Monograph of the Petrels (order Tubinares), (London: Witherby and Company).

Goldberg, Suzann L., Novacek, Michael J., Alayón, Giraldo. 2017. The History of Scientific Relations between Cuba and the American Museum of Natural History (AMNH), 160 Years of Collecting and Collaborating (1857–2017) (New York: American Museum of Natural History Special Publication, 2017)

Gonzales, Robert Alexander. 2011, Designing Pan-America: U.S. Architectural Visions for the Western Hemisphere. Austin: University of Texas.

Gordon, Martin L. 1975. Joseph Henry Pendleton, 1860-1942: Register of His Personal Papers. Washington, D.C.: History and Museums Division, Headquarters, U.S. Marine Corps.

Gould, Stephen Jay. 2018. "Crampton, Henry Edward." Complete Dictionary of Scientific Biography. https://enclyclopedia.com.

Graham, Elyse. 2024. Book and Dagger: How Scholars and Librarians Became the Unlikely Heroes of World War II. New York: Ecco Press.

Gray, Christopher. 2012. "The 'Feather King': A NoHo Prequel." New York Times, May 27.

Gregory, William King. 1948. Biographical Memoir of Frank Michler Chapman, 1864-1945. Washington: National Academy of Sciences.

Ham, Anthony. 2023. "Why Scientists Are Debating Mammal Origins," New York Times, July 18, D3.

Harris, Charles H. III and Sadler, Louis R. 1988. The Border and the Revolution: Clandestine Activities of the Mexican Revolution, 1910-1920 (Silver City: High-Lonesome Books).

Harris, Charles H. III and Sadler, Louis R. 2003. The Archaeologist was a Spy: Sylvanus G. Morley and the Office of Naval Intelligence. Albuquerque: University of New Mexico Press.

Hodge, F.W. 1917. "What the United States Government has done for the Science of Anthropology" in Proceedings of the Second Pan American Scientific Congress, Section I, Anthropology. Washington. 168–174.

Huggins, Martha K. 1998. Political Policing: The United States and Latin America (Durham: Duke University Press).

Jaffe, Steven. 2012. New York At War: Four Centuries of Combat, Fear, and Intrigue in Gotham. New York: Basic Books.

Kane, Robert J., Captain USMC. 1961. A Brief History of the 2d Marines, revised. Washington, D.C.: Historical Division Headquarters, U.S. Marine Corps.

Katz, Friedrich. 1981. The Secret War in Mexico: Europe, the United States and the Mexican Revolution (Chicago: The University of Chicago Press).

Katzir, Shaul. 2017. "In War and Peace: The technological promise of science following the First World War." In Jaume Navarro, ed., special issue "The Promises of Science. Historical Perspectives," Centaurus, Vol. 59. Issue 3, 223–237.

Keegan, John. 2003. Intelligence in War: Knowledge of the Enemy from Napoleon to Al-Qaeda (New York: Alfred A. Knopf).

Kember, Joe, Plunkett, Joseph, Sullivan, and Jill A. Sullivan. 2012. Popular Exhibitions, Science and Showmanship, 1840–1910. London: Pickering and Chatto.

Kennedy, David M. 1980. Over Here: The First World War and American Society. Oxford: Oxford University Press.

Kleinman, Samuel D. 2016. "State's Spies: The Bureau of Secret Intelligence and the Development of State Department Bureaucracy in the First World War." Honors Thesis, History Department, Georgetown University.

Kumar, Prakash. 2007. "Plantation science: improving natural indigo in colonial India, 1860-1913." British Journal of the History of Science. Vol. 40, 537–565.

Lael, Richard L and Killen, Linda. 1982. "The Pressure of Shortage: Platinum Policy and the Wilson Administration During World War I." Business History Review, Vol. LVI, No. 4, 545–558.

Lalli, Roberto and Navarro, Juame. 2024. "How physicists strove for peace." Nature, Vol. 633, 515–517.

Langfur, Hal. 2023. Adrift on an Inland Sea: Misinformation and the Limits of Empire in the Brazilian Backlands . Stanford: Stanford University Press.

La Rocca, Donald J. 2017. "Bashford Dean and the Development of Helmets and Body Armor during World War I." In Heilbrunn Timeline of Art History. New York, The Metropolitan Museum of Art.. http://www.metmuseum.org/toah/hd/bash/hd_bash.htm (March 2017).

Lansing, Robert. 1931. "Drama of the Virgin Islands Purchase," New York Times Magazine, July 19, 4–5.

Lansing, Robert. 1935. War Memoirs of Robert Lansing, Secretary of State, Indianapolis: Bobbs-Merrill Company.

LaPorte, Nicole. 2003, "Da Vinci Was a Double Major," Town and Country Magazine, May, 2003, 93.

Levine, Robert M. 1989. Nineteenth and Early Twentieth Century Latin American Photographs as Documents. Durham: Duke University Press.

Lewis, Herbert S. 2001. "The Passion of Franz Boas," American Anthropologist, Vol. 103, No. 2, 447–467.

Lopes, Maria Margaret and Podgorny, Irina. 2001, "The Shaping of Latin American Museums of Natural History, 1850-1990." Osiris. Vol. 15, No. 1, 108–118.

Luebke, Frederick C. 1987. Germans in Brazil: A Comparative History of Cultural Conflict During World War I. Baton Rouge: Louisiana State University Press.

Lyons, Michael J. 2000. World War I: A Short History (Upper Saddle River, New Jersey: Prentice Hall, Inc.).

McCalman, Iain. 2009. Darwin's Armada: Four Voyages and the Battle for the Theory of Evolution. New York: W.W. Norton and Company.

McCann, Frank D. 1981. "Brazilian Foreign Relations in the Twentieth Century." In Wayne A. Selcher, ed. Brazil in the International System: The Rise of a Middle Power, Boulder, Colorado: Westview Press, 1–24.

McCreery, David. 1993. "Wireless Empire: The United States and Radio Communication in Central America and the Caribbean, 1904–1926." South Eastern Latin Americanist. Summer, 23–41.

McPherson, Alan and Wehrli, Yannick. eds. 2015. Beyond Geopolitics: New Histories of Latin America at the League of Nations (Albuquerque: University of New Mexico Press).

Martin, Percy Alvin. 1925. Latin America and the War: The Albert Shaw Lecture on Diplomatic History, 1921. Baltimore: Johns Hopkins University Press.

Matthew, William Diller. 1915."Climate and Evolution." Annals of the New York Academy of Sciences. Vol.24, No.1, 171–318; reprinted as Climate and Evolution (New York, New York Academy of Sciences, 1939).

Mayer, Gregory C., Coyne, Jerry A., Losos, Jonathan B., Foufopoulos, Johannes, Shubin, Neil, Futuyma, Douglas J., Campbell, Benjamin C., and Edwards, Scott V. 2013, "Museums' Role: Increasing Knowledge." Science, Vol. 339, 1148–1149.

Meyer, Michael C., Sherman, William L., and Deeds, Susan M. 1999. The Course of Mexican History, 6th ed. (New York: Oxford University Press).

Millard, Candice. 2005. The River Of Doubt: Theodore Roosevelt's Darkest Journey. New York: Doubleday.

Miller, Clark A. 2006. "An Effective Instrument of Peace: Scientific Cooperation as an Instrument of U.S. Foreign Policy, 1928-1950," OSIRIS. Vol. 21, 133–160.

Miller, Leo E. 1918. In the Wilds of South America: Six Years of Exploration in Colombia, Venezuela, British Guiana, Peru, Bolivia, Argentina, Paraguay, and Brazil. New York: Charles Scribner's Sons.

Mills, Bill. 2017. Treacherous Passage: Germany's Secret Plot against the United States in Mexico during World War I (Lincoln: University of Nebraska Press).

Mitchell, Nancy. 1999. The Danger of Dreams: German and American Imperialism in Latin America. Chapel Hill: University of North Carolina Press.

Murphy, Robert Cushman. 1936. Oceanic birds of South America: a study of the related coasts and seas, including the American quadrant of Antarctica, based on the Brewster-Sanford collection in the American Museum of Natural History. 2 Vols., New York: MacMillan Co.

Neagle, Michael E. 2016. America's Forgotten Colony: Cuba's Isle of Pines. Cambridge: Cambridge University Press.

Nugent, David. 2010. "Knowledge and Empire: The Social Sciences and the United States Imperial Expansion." Identities: Global Studies in Culture and Power, Vol. 17, No. 1, 2–44.

Nunn, Jack. 1979. "MIT: A University's Contribution to National Defense." Military Affairs, vol. 43, Issue 3, 120–125.

O'Brien, Thomas. 1999. The Century of U.S. Capitalism in Latin America. Albuquerque: University of New Mexico Press.

O'Toole, Patricia. 2005. When Trumpets Call: Theodore Roosevelt After the White House, New York: Simon and Schuster.

Ocasio, Rafael. 2022. "Franz Boas's Trip to Puerto Rico (1915) and the Contemporary Boriken Nation." The Latin Americanist. Vol. 66, No. 1, 81–94.

Officer, Lawrence H. 2008. "Gold Standard," EH.Net Encyclopedia, ed. Robert Whaples, March 26, 2008, https://eh.net/encyclopedia/gold-standard/, (accessed on October 23, 2020).

Packard, Wyman H. 1996. A Century of U.S. Naval Intelligence. Washington, Department of the Navy.

Paget, Karen M. 2015. Patriotic Betrayal: The Inside Story of the CIA's Secret Campaign to Enroll American Students in the Crusade Against Communism. New Haven, Connecticut: Yale University Press.

Parikh, Sudip S., McNutt, Marcia K., and Gil, Darío. 2024. "A new vision for American science," Science, Vol 386, Issue 6725, 947.

Pearcy, Thomas L. 2006. The History of Central America (New York: Palgrave Macmillan).

Plummer, Brenda Gayle. 1984. "The Metropolitan Connection: Foreign and Semiforeign Elite in Haiti, 1900-1915." Latin American Research Review, Vol. 19, No. 2, p. 122.

Podgorny, Irina. 2005. "Bones and Devices in the Constitution of Paleontology in Argentina at the end of the Nineteenth Century." Science in Context. Vol. 18, No. 2.

Price, David. 2000. "Anthropologists as Spies," The Nation, November 20, 24–27.

Rainger, Ronald. 1986. "Just Before Simpson: William Diller Matthew's Understanding of Evolution." Proceedings of the American Philosophical Society. Vol. 130, No. 4, 453–474.

Rainger, Ronald. 1991 An Agenda for Antiquity: Henry Fairfield Osborn and Vertebrate Paleontology at the American Museum of Natural History, 1890-1935. Tuscaloosa: University of Alabama Press.

Rausch, Jane M. 2014. Colombia and World War I: The Experience of a Neutral Latin American Nation during the Great War and Its Aftermath, 1914–1921 (Lanham: Lexington Books).

Regal, Brian. 2002. Henry Fairfield Osborn: Race, and the Search for the Origins of Man. Aldershot, England: Ashgate (Routledge).

Renda, Mary A. 2001. Taking Haiti: Military Occupation and the Culture of U.S. Imperialism. Chapel Hill: University of North Carolina Press.

Rexer, Lyle and Klein, Rachel. 1995. American Museum of Natural History: 125 Years of Expedition and Discovery. New York: Harry N. Abrams, Inc.

Rice, Mark. 2012. "Transnational Business and U.S. Diplomacy in Late Nineteenth Century South America: W. R. Grace & Co. and the Chilean Crisis of 1891." Journal of Latin American Studies, Vol. 44, No. 4, 765–792.

Rieppel, Lukas. 2019. Assembling the Dinosaur: Fossil Hunters, Tycoons and the Making of a Spectacle. Cambridge: Harvard University Press.

Rink, Stefan. 2017. Latin America and the First World War, trans. Christopher W. Reid (New York: Cambridge University Press).

Roosevelt, Theodore. 1914. Through the Brazilian Wilderness. New York: Charles Scribner's Sons.

Roper, William. 2020. "Science, politics, and public health." Science, Vol. 370, p. 385.

Rowe, Leo Stanton. 1909."Our Interest in a United America." North American Review, April, 587–593.

Saldaña, Juan José. 2006. "Science and Public Happiness during the Latin American Enlightenment." In Juan José Saldaña, ed. Science in Latin America: A History. Austin: University of Texas Press, 51–92.

Salvatore, Ricardo D. 2016. Disciplinary Conquest: U.S. Scholars in South America, 1900–1945. Durham: Duke University Press.

Salzman, Neil V. 1991. Reform and revolution: the life and times of Raymond Robbins. Kent, Ohio: Kent State University Press.

Sater, William F. Holger and Herwig, H. 2000. The Grand Illusion: The Prussianization of the Chilean Army. Lincoln, Nebraska: University of Nebraska Press.

Savala, Joshua. 2019. "Ports of Transnational Labor Organizing: Anarchism along the Peruvian-Chilean Littoral, 1916–1928." Hispanic American Historical Review. Vol. 99, No. 3.
Schuler, Friedrich. 2010. Secret Wars and Secret Policies in the Americas, 1842-1929 (Albuquerque: University of New Mexico Press).
Schulze, Frederik. 2014. "German Immigrants (Brazil)." In Ute Daniel, Peter Gatrell, Oliver Janz, Heather Jones, Alan Kramer, and Bill Nasson, eds. 1914–1918-online. International Encyclopedia of the First World War. Berlin: Freie Universitat Berlin.
Scott, James Brown. 1916. Second Pan American Scientific Congress: Final Act and Interpretive Commentary Thereon. Washington.
Secord, Anne. 1994. "Corresponding Interests: Artisans and Gentlemen in Nineteenth Century Natural History." The British Journal for the History of Science. Vol. 27, No. 4, 383–408.
Setzekorn, Eric. 2017. "Intelligence in Defense of the Homeland: The Office of Naval Intelligence in World War I: Diverse Threats, Divergent Responses." Studies in Intelligence. Vol. 61, No. 2 (Extracts, June 2017), 43–54.
Skidmore, Thomas E. 1999. Brazil: Five Centuries of Change, New York: Oxford University Press.
Sloan, Douglas. 1980. "Science in New York City, 1867-1907." Isis. vol. 71, No. 1, 35–76.
Slotkin, Richard. 2005. Lost Battalions: The Great War and the Crisis of American Nationality. New York: Holt.
Smith, Joseph. 2000. "The First Conference of American States (1889-1890) and the Early Pan American Policy of the United States." In David Seining, ed. Beyond the Ideal: Pan Americanism in Inter-American Affairs. Westport, Connecticut: Praeger, 24–26.
Smith, Neil. 2003, American Empire: Roosevelt's Geographer and the Prelude to Globalization. Berkeley: University of California Press.
Snow, C.P. 1959. The Two Cultures and the Scientific Revolution (London: Cambridge University Press)
Spinden, Herbert J. 1917. Ancient Civilizations of Mexico and Central America. American Museum of Natural History Handbook Series. Number 3. New York.
Spinden, Herbert J. 1917. "Travel Notes in Western Venezuela." American Museum Journal, Vol. XVII, No. 1, 15–23.
Stein, Glenn M. 2009. "General David L. Brainard, U.S. Army: Last Survivor of the United States' Lady Franklin Bay Expedition (1881–84)." downloaded from Academia.edu, 1–26. Originally published in The Polar Times, July 2008/January 2009.
Stocking, Jr., George W. (ed.) 1974. The Shaping of American Anthropology, 1883-1911: A Franz Boas Reader. New York: Basic Books Inc.
Sullivan, Paul. 1989. Unfinished Conversations: Mayas and Foreigners Between Two Wars. New York: Alfred A. Knopf.
Taylor, Tom and Taylor, Michael. 2011. Aves: A Survey of the Literature of Neotropical Ornithology. Baton Rouge: Louisiana State University Libraries.
Teixeira Vinhosa, Francisco Luiz. 1990. O Brasil e Primeira Guerra Mundial: A Diplomacia Brasileira e as Grandes Potências (Rio de Janeiro: Instituto Histórico e Geográfico Brasileiro).
Thorp, H. Holden. 17 July 2020. "Science has always been political," Science, Vol. 369, 227.
Thorp, H. Holden. 7 July 2023, "Museum ethics-more than lip service," Science, Vol. 381, 7.
Tillman, Ellen Davies. 2010. Imperialism Revised: Military, Society, and U.S. Occupation in the Dominican Republic, 1880–1924. Dissertation in the Graduate College of the University of Illinois at Urbana-Champaign.
Tuchman, Barbara W. 1958. The Zimmerman Telegram (New York: Ballantine).
Veit, Helen Zoe. 2013. Modern Food, Moral Food: Self-Control, Science, and the Rise of Modern American Eating in the Early Twentieth Century. Chapel Hill: University of North Carolina Press.
Watriss, Wendy and Zamora, Lois Parkinson. 1998. Images and Memory, Photography from Latin America. Austin; University of Texas Press.
Weidensaul, Scott. 2007. Of a Feather: A Brief History of American Birding. New York: Mariner Books.

Zabriskie, Luther K. 1916. The Virgin Islands of the United States of America: Historical and Descriptive Commercial and Industrial Fact, Figures and Resources. New York: G.P. Putnam's Sons.

Index

R. M. Delson, *Scientists or Spies?*, The Latin American Studies Book Series,
https://doi.org/10.1007/978-3-032-18117-6

B

C

MIX
Papier aus verantwortungsvollen Quellen
Paper from responsible sources
FSC® C105338

If you have any concerns about our products,
you can contact us on
ProductSafety@springernature.com

In case Publisher is established outside the EU,
the EU authorized representative is:
Springer Nature Customer Service Center GmbH
Europaplatz 3, 69115 Heidelberg, Germany

Printed by Libri Plureos GmbH
in Hamburg, Germany